S0-BDP-385

INTERNATIONAL PROJECT MANAGEMENT: LEADERSHIP IN COMPLEX ENVIRONMENTS

INTERNATIONAL PROJECT MANAGEMENT: LEADERSHIP IN COMPLEX ENVIRONMENTS

Thomas W. Grisham

John Wiley & Sons, Inc.

This book is printed on acid-free paper. ♾

Published by John Wiley & Sons, Inc., Hoboken, New Jersey

Published simultaneously in Canada

Library of Congress Cataloging-in-Publication Data:

Grisham, Thomas W.
International project management : leadership in complex environments / Thomas W. Grisham.
p. cm.
Includes bibliographical references and index.
ISBN 978-0-470-57882-7 (cloth)
1. Project management. 2. Leadership. I. Title.
HD69.P75G755 2010
658.4′04—dc22

2009041803

Printed in the United States of America

10 9 8 7 6 5 4 3 2

Contents

Chapter 1

Introduction

International project management is the leadership of projects that are conducted in multiple countries and cultures. Projects may include:

- An international Swiss-based nonprofit organization providing humanitarian assistance to people in Somalia; with a local political partner; a value chain in Kenya, Egypt, and India; and funding from multiple donor countries.
- A Singaporean for-profit organization building a new manufacturing facility in Cambodia; with a Chinese partner; value chain organizations in Germany, Morocco, Vietnam, and Brazil; and a government agency in Cambodia.
- A U.K. multinational organization with partners in South Korea, the United States, and Japan that is performing a public private partnership project in Botswana. Funding is being provided by a consortium of organizations, including private equity from the United Kingdom, bonds issued in the United States, and a government guarantee. The project will be operated by a joint venture of organizations from Botswana, South Africa, Spain, and Germany.
- A Russian charitable organization providing emergency medical and food services for people in various countries in the Balkans and the Persian Gulf. Pharmaceuticals and food sourced and delivered from value chain organizations in 10 different countries, physicians from all over the European Union and Russia, and donations from Central Asia and the subcontinent.

- A global organization that will launch a new information technology platform. The platform will be used in the Mexican division, and the architecture and code is being provided from Finland, Estonia, and India. The concept was developed from lessons learned, or best practices, in Russia.
- An Italian organization constructing a new prison facility for the U.S. government.
- A transnational organization that markets through divisions in 82 countries. The products are designed in Italy, Sweden, and France, with customization done in each country. The product components are manufactured in Ukraine, Sri Lanka, Laos, and Pakistan; are shipped to Malaysia for subcomponent assembly; then shipped on for final assembly in China. The organization is preparing to develop a new product line as a project.

International projects differ from domestic projects by their complexity of culture, politics, law, local practice, language, time zones, holidays, processes, resources, and more. As you can see from the list, they can be incredibly complex even if internal to a single organization. This is one reason many global organizations have embraced project management as a way to increase quality and decrease price. Projects that are conducted with multiple organizations are truly fascinating undertakings. International projects are similar to domestic projects in that they share basic project management processes, but it does not follow that a project manager with only domestic experience can lead an international project. For example, stakeholder management is required on both domestic and international projects.

On international projects, attitudes are influenced by culture. Think of changes in scope in a culture that values long-term relationships (the Chinese) compared to one that places more importance on the contract itself (the United States). Project success can be jeopardized by failing to account for such cultural differences. This is part of leadership, and thus the title of the book. Project management is a profession, so we use that term to refer to the profession. However, we must be clear from the beginning that *managing* an international project only will greatly diminish the probability of success; *leading* an international project requires a different set of skills and a different attitude, and is the focus of this book. We discuss the differences between leadership and management later.

International markets are places of rapid change and severe economic pressure; the pressure today is to drive quality up and prices down. To manage change effectively requires a foundation of leadership first, then processes, and metrics. Taken together, they provide a platform for imbuing a culture of change within organizations. When firms and organizations become flatter (less distance between the chief executive and a worker), there is a greater need for people to enter the workplace fully trained, especially for international project managers. The project management profession has responded with professional certifications that include practical experience, project management training, and a code of ethics. For these reasons, many firms and organizations are seeking project management professionals (PMPs), including governments, nongovernmental organizations (NGO's), United Nations groups, development banks, and of course private industry. Many firms and agencies are requiring that suppliers of project management services name a PMP in their tender documents as a prerequisite for bidding.

International business and project management practice have converged in the last 10 years. Organizations are tending toward hiring multitalented people who are self-motivated, intelligent, and willing to take responsibility. Some of the reasons are:

- The need for leaner and flatter organizations to reduce costs
- The need for leadership skills throughout the organizational food chain from top to bottom—lead one day, follow the next, and be comfortable personally in either role
- The need for knowledge workers throughout the organization
- Globalization and the need to improve quality while reducing cost
- Kaizen to keep quality high while reducing cost
- Diversity

This book is designed to describe the confluence of theory and practice for international projects and to highlight the need for leadership skills in such environments. It is written for these groups:

- International project and program management practitioners who wish to enhance their skills and broaden their understanding of international project management.
- Domestic project and program management practitioners who need to learn how to lead international projects.

- Academics who want to extend the research into topics that impinge on international project management and techniques.
- Students of project or program management. From our experience, few projects now have only a domestic component, and so the broader view of leading international projects will serve students well in domestic settings as well.

To meet the needs of these groups, we have incorporated a wide range of material taken from the literature on project management, leadership, business management, conflict management, psychology, sociology, anthropology, the arts, religion, history, geography, political science, and more. Successful international project managers need to be inquisitive and need to have a broad range of interests and capabilities. These attributes are inherent in cross-cultural leadership intelligence (XLQ), and key to the success of international project managers. XLQ combines the concepts of intelligence quotient (IQ), emotional intelligence (EQ), and cultural intelligence (CQ). It is a measure of the leadership attributes possessed by a project manager. We describe it XLQ detail in Chapter 4, but for now think of it as the ability of an international project manager to inspire the desire of others to follow her/him regardless of their societal, organizational, or group cultures.

The structure of the book mirrors the *Project Management Body of Knowledge* (*PMBOK* 2004), so that project and program managers with that credential can relate back to this standard. The fourth edition of the *PMBOK* was available in early 2009, but the differences between the third edition that we refer to in this book and the fourth edition are not important for our purposes. We selected the *PMBOK* knowledge areas as the standard because far more people belong to the Project Management Institute (PMI) than the other international project management organizations combined. The book is therefore organized around the nine knowledge areas—integration, scope, time, cost, quality, human resources, communication, risk, and procurement—and includes discussion of the process groups—initiating, planning, execution, monitoring and control, and close-out.

The *PMBOK* provides a generic process, written from an internal organizational view, and, presumably, for both domestic and international projects. To validate this, in 2006 the PMI offered a grant to determine how "internationalized" the *PMBOK* truly is. Our answer to this question lies in the pages of this book.

We should make perfectly clear one more item about the *PMBOK*. This book, as you will see, is not intended to expand on or explain the *PMBOK* or the International Competence Baseline (ICB) of the International Project Management Association (IPMA). We do not favor one over the other, and we refer to them to enable our readers to connect their knowledge back to these established standards. That is the *only* reason these works are mentioned here. In fact, we see both as works in progress and in need of significant adjustment to meet the needs of international projects.

International projects managers have the same requirement, in our view, as do physicians, and that is to stay current with the knowledge base. International project management requires people to be curious and adventuresome. To that end, and others, this book is approximately 80% focused on professional best practices and 20% academic. In some sections, the reader will find more references, or citations, to other sources. The reason is to provide those of you who are curious about the topic and want more information a place to begin your search and to provide those who need to hone their skills references that will provide you with more detail. Although the book is a blend of practice and theory, it is also our hope that it exposes the reader to more than just project management thinking, for that is what international project managers must possess.

What exactly is an international project? The *PMBOK* defines a project as "a temporary endeavor undertaken to create a unique product, service or result" (p. 368). This distinguishes a project from the ongoing operations of a business. Starting with this definition as a basis, we need to add that an international project is one that utilizes resources from or provides services in more than one country, physically or virtually. At one extreme, a project performed in a single country utilizing local people that have international backgrounds is an international project. At the other extreme, a project that utilizes resources from, and provides services in, multiple countries is also an international project. One common denominator is the need for XLQ. This book provides processes and techniques for managing international projects across this spectrum of possibilities; we recognize that many will not be able to reach the transparency goal we suggest.

What you will find in these pages is a description of a goal. Leaders of international projects have different levels of expertise, and their organizations have different cultural norms. Countries have different cultural, political, and legal systems. Some countries have a higher quality of life, some have corruption issues, some have civil strife, and some face refugee challenges. Some government agencies have strict regulations regarding procurement of goods and services, and in some countries, the rule of law is not yet fully established. It is a wonderfully challenging and complex environment. For this reason, each project will be unique and will have its own set of challenges. The processes that we describe in this book are a goal that must be modified to fit the conditions encountered. The processes are not *the* way but rather *a* way to increase the probability of success. The measure of an expert international project leader in our view is the ability to adjust the processes we describe to fit the conditions.

The book uses "we" rather than "I," simply because a team of people are required to complete an international project successfully. So in our examples—some of which are true, some a composite of our experience, and some hypothetical—we use "we" to recognize the many people who may have participated in the projects we were associated with. "We" here means the author's view of the composite. Actually, I recommend that project managers consciously limit their use of the word "I" and replace it with "we." In this book, the teams on the projects in my experience are speaking through me. The use of "we" is my way to recognize them.

The book uses the *PMBOK* knowledge areas as chapter headings. The chapters are, however, in a different order from those in the *PMBOK* for a reason. The chapters in this book follow the sequence we suggest in practice for planning a project or program. A brief description of each is presented next.

Chapter 1, "Introduction," provides basic considerations for international projects, such as sustainability, ethics, laws, compensation, culture, knowledge management, and of course project structures. The sustainability section includes more references than in the remainder of the book. We feel this is a critical topic for international project managers, and currently the literature for project management is thin on this topic.

Chapter 2, "Framework," provides some basics for international projects, such as the stakeholders, participants, project life cycle and contracting methods. These first two chapters lay the groundwork for the remainder of the book.

Chapter 3, "Project Basics," discusses the first of the *PMBOK* knowledge areas: integration management. Beginning with this chapter, the book discusses the international project requirements of each *PMBOK* knowledge area. We have not internationalized the *PMBOK* but rather have organized our ideas on international project management using the *PMBOK* format. Integration management spans all of the *PMBOK* process groups starting with initiation, through planning, execution, monitoring and controlling, and ending with close-out of the project. It also connects these process groups to each of the knowledge areas—the thread binds the quilt of project management. In this chapter, we discuss the foundation concepts for successful international project management. We cover cross-cultural issue from an enterprise point of view and introduce societal culture.

We consider Chapter 4, "Leading Diversity," to be the most important aspect of international project management. Leading people is the first job of a project manager, followed closely by communication skills. Managing the project takes third place in our experience. In this chapter, we address cross-cultural leadership in detail and provide our model for XLQ. We also discuss teambuilding, motivation, values and ethics, knowledge sharing, and education and discuss communication management. For international projects, there are the obvious language issues that must be addressed, but perhaps more important are the cultural, sociological, and psychological aspects of communications. As with sustainability and risk, the project management literature is not deep in the area of communications, and we need to look to other disciplines for help in honing these skills. We present ways that international project managers can improve their effectiveness when communicating across cultures, business or cultural.

In Chapter 5, "Integration Management," covers the purchase order structure, charter, planning, and ethics.

In Chapter 6, "Scope Management," we discuss the considerations for scope management, including development of a scope split document, a change management plan, and a dispute resolution process.

In Chapter 7, "Cost and Progress Management," we discuss the issues of cost and progress. For international projects, this knowledge area can be quite a challenge. We provide, as in all of the chapters, the challenges and our suggestions on how to meet them more effectively.

Chapter 8 is called "Risk Management." On international projects, the key is to select the most critical project risks and actively manage them, for in today's world there are many risks indeed. This chapter presents techniques to assist in focusing on the critical project risks.

Chapter 9, "Time Management," addresses the issues relating to time management, float, and transparency. On international projects, addressing this issue requires a basic understanding of time zones, religions, cultures, and customs. It also enables a 24-hour view of project planning.

Chapter 10 is titled "Quality Management—Customer Satisfaction." On all projects, quality has both a service and product component. On international projects, establishing standards for these components and developing quality plans and dashboards is complex. As we show, there are at least four major international standards for project management: those promulgated by the Project Management Institute (U.S.), the International Project Management Association (Swiss), Association for Project Management (U.K.), and Projects In Controlled Environments, the second version (PRINCE2) (U.K.). In addition, multiple technical standards exist. In this chapter, we provide guidance on how to manage this diversity.

Chapter 11 is titled "Procurement Management." In a global economy, most international projects take advantage of international value chains. Suggestions on how to manage these systems better and how to integrate them into the project are provided.

Chapter 12 provides a summary review of the concept of the collaborative project enterprise and some final thoughts about leading international projects.

The book offers a blend of theory, practice, and example projects from our personal experiences. As noted, where the project management literature is thin, we have provided more references to point to theory that can be used to improve the knowledge of international project managers. On the practice side, we have translated our experience into the format of the *PMBOK*, but with a close eye toward the other international standards. Where we find non-*PMBOK* standards to be more effective, or where we find no information in the *PMBOK*, we have used the other international standards. We have used short stories of our experience, throughout the book, to provide a frame of reference for the reader. People have always used stories to communicate with one another, and they are a powerful medium, along with metaphors, to convey complex ideas in a short time frame. Perhaps most important, stories are simply enjoyable.

In conclusion, the book takes the view that international project managers need to be leaders with high cross-cultural intelligence, creative communication skills, the ability to establish and maintain dependable project management processes, and compelling curiosity.

Chapter 2

Framework

> The chameleon changes color to match the earth, the earth doesn't change color to match the chameleon.
>
> *Senegalese proverb*

> The best leaders of all, the people know not they exist. They turn to each other and say we did it ourselves.
>
> *Zen proverb*

In this chapter, we begin by discussing the palette upon which international projects are painted. We introduce our view of an international project manager and the professional standards. Then we turn our attention to the international environment and discuss sustainability, ethics, laws, value chains, e-business, culture, and human resources. We then discuss organizational structures, change management, knowledge management, and end with project structures or projects as collaborative project enterprises (or, as some call them, temporary organizations).

It is critical that an international project manager have an understanding of the context within which she or he works. The reason is that the international markets have not yet agreed on accepted standards. The diversity in everything from culture to technical standards is astounding, and it is what makes international project management both challenging and rewarding. If you celebrate change and diversity, this is the field for you.

2.1 INTERNATIONAL PROJECT MANAGER

The standards for project management, discussed in the next section, provide a structured systems approach to facilitate the development of repeatable processes. The advantage of those standards is that the

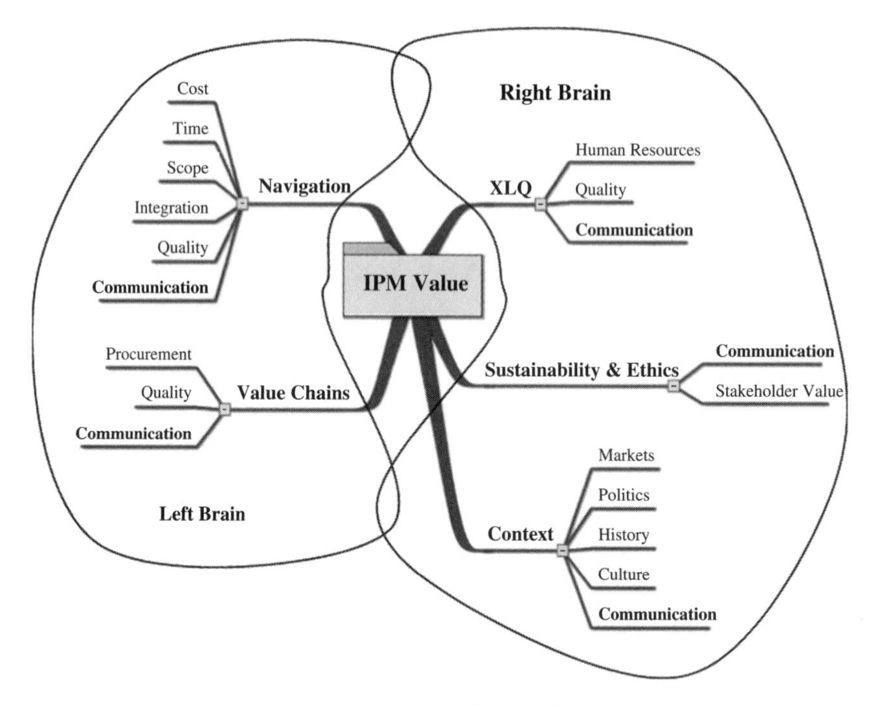

Figure 2.1 International Project Manager

discussion about each of the processes, or *PMBOK* knowledge areas, is made more direct and definable. The disadvantage is that the interrelationship between the processes is difficult to explain and to correlate. This book is organized in line with the *PMBOK* to make correlation to this recognized standard more direct.

Figure 2.1 provides a graphic look at the balance necessary between brain functions for international project managers. The view is from the top of the head, with the face being at the top of the figure. The figure is based on psychological, sociological, and medical research, which has found that found people use the right hemisphere of their brain to deal with more conceptual types of issues, holistic thinking, images, intuition, artistic creativity, and language intonation—the so-called soft skills. People use the left hemisphere to process more logical activities, analytical thinking, and language (grammar and words)—the so-called hard skills. An international project manager must find ways to fuse the leadership side—soft skills—with the managerial process side—hard skills.

This book emphasizes the importance of the right-brain leadership skills, for they are *vital*. The left-brain managerial aspects are necessary to the successful completion of projects as well, and need to be in balance, but they have a significantly smaller impact on the realization of successful international projects. Leadership is the key.

Figure 2.1 was created using mindmapping (Buzan 1993) software to show fuzzy relationships. The extreme branches shown in the left-brain represent the nine *PMBOK* knowledge areas. On the right-brain side, cross-cultural leadership intelligence (XLQ), sustainability and ethics, and context are the three major areas. Under context, it is especially important for an international project manager to understand the peripheral conditions surrounding the project, the teams, and the international marketplace.

On that point, imagine you are a lead project manager for a communications organization that is providing services in Iran to enlarge the telecommunications network. The government chooses to employ a "deep packet" (information technology [IT] system that offers filtering of data), an option that your organization offers, as part of that project. There is a disputed election, and the government uses the system in an attempt to disable Internet communications. Your project is a great success with the customer—assume it is the government—but the rest of the world sees your organization as playing a key role in the suppression of information. Your project could result in questions about the ethics of your organization. Our point is that the person who leads international projects must be curious enough to become informed about global issues, trends, and attitudes. Given enough pressure on your organization, you may be forced to disable the option and, in doing so, alienate the customer. The customer could then sanction your actions and affect your partners and value chains.

To use a more straightforward example, if the project is an IT endeavor that plans to connect financial markets in three countries, including the United States, the politics and history of the Sarbanes-Oxley Act may well be critical.[1] Again, the point is an international project manager

[1]Generally both U.S. and non-U.S. organizations are subject to the disclosure requirements of the Sarbanes-Oxley Act, with some exceptions. The quality and timeliness of organization information has been enhanced, including these provisions:

- Management and auditors must annually assess their organization's internal controls and related disclosures.
- Additional disclosure of off–balance sheet financing and financial contingencies is now required.
- The presentation of VFM information is now required.
- Disclosure under Section 16 of the Exchange Act of insider stock transactions has been accelerated to two business days. (This does not apply to non-U.S. issuers.)
- Disclosure of certain information will now be required in "real time."

must understand the international business environment and be current with international politics, economics, and societal trends. One excellent source we recommend is a British periodical, the *Economist* (www.economist.com).

On the left-brain side are the more direct managerial functions that an international project manager must lead. Value chains are discussed in Chapter 11 as part of the procurement knowledge area. Likewise, the navigation considerations are the subject of their own individual chapters. We use sailing as a metaphor to describe these processes, thus the word "navigation." These left-brain considerations are more process related, technical, and logical. The right-brain considerations provide a foundation or palette on which the left-brain activities are painted.

The left- and right-brain example is an attempt to draw a clear distinction between management and leadership. If you manage an international project, you will reduce your probability of success—radically. If you lead an international project, you will increase the probability of success radically. Many people ask what the balance should be between the two. Unfortunately, the answer depends on the project conditions and requires a leader to adapt skills to the environment. In general, however, we would suggest something in the range of 80% leadership, 20% management. We were fortunate enough to have had the opportunity to work for the extreme versions of manager and leader on the same project at different times. That and other experiences have led us to this heuristic.

Sustainability and ethics are central challenges on international projects, due to the variability between organizational and societal cultures. Consider the issue of corruption. Transparency International provides an annual corruption perception index (CPI) that ranks more than 150 countries by their perceived levels of corruption, as determined by expert assessments and opinion surveys of businesses (www.transparency.org). In 2006, Finland had a CPI of 9.6 (low perception of corruption), whereas Haiti had an index of 1.8. If you have a project to provide telecommunications in Haiti, and your organization considers its corporate social responsibility (CSR) as a core value, there will be significant tension between sustainability and ethics, and context.

International project management is complex and ever changing. In this book, we describe a perspective that is optimum. We also argue that the lead project manager must adapt to his or her environment, seek to understand all perspectives, and empathize. Leading international projects is not about fixed processes, black and white, yes and no, male or female, plus and minus, up and down. It is about balance. In Taoism, this notion is represented by the symbol, shown above, for yin and yang, or female and male. There is male in female (white dot in black area) and female in male. One would not exist without the other, and they are balanced and intertwined. When you see this symbol in the book, remember to consider what we have just said in this context. We cannot provide a fixed process that will work in every circumstance, but we can provide a vision that will increase your chances of success.

The next section provides a brief look at the Project Management Institute's *PMBOK*, the International Project Management Association (IPMA), the Association of Project Managers (APM), and Projects in Controlled Environments (PRINCE2). These are the primary international standards for project management, and throughout the book, we relate our approach to international project management back to the standards so that those who are credentialed by any of these groups can see the connections back to the standards. Remember, however, that the standards are only guidelines, and the standards do not address how to apply them on international projects. The goal of this book is to illustrate how to apply best practices.

2.2 STANDARDS

There are four major standards for project management. Each standard provides a knowledge base that promulgates a common language and methodology for the profession. Although they are similar to one another in content, they differ in organization, emphasis, and terminology. Each of the standards has its strengths and weaknesses, but the *PMBOK* is widely used and there must be some common benchmark; otherwise both the terminology and methodology can easily be misunderstood. This book uses the *PMBOK* as the benchmark for structure and for terminology, since it has the largest current membership. There are areas where the *PMBOK* is silent or where it places too little emphasis on particular skills; one of those areas is leadership. In such cases, this book looks to the other standards.

The Project Management Institute (PMI) has published the *PMBOK* (2004) as its standard of practice. PMI (www.pmi.org) was founded in 1969 in the United States and has over 420,000 members from 70 countries. The PMI has a project management professional (PMP) credential that requires experience, course work, and an exam. It also offers a credential for program managers. PMI also publishes the *Project Management Journal*, a scholarly magazine for project management professionals.

The International Project Management Association (IPMA) has published the *ICB-IPMA Competence Baseline* (2006) as its standard of practice. The IPMA (www.ipma.org) was founded in Vienna in 1965 and had over 40,000 members in 2007. It is composed of an alliance of member organizations in 45 countries. The IPMA has four levels of credentials: level A-certified project director; level B certified senior project manager; level C project manager; and level D certified project management associate. The IPMA also publishes a scholarly journal, the *International Journal of Project Management.*

The Association of Project Managers (APM) has published the *APM Body of Knowledge* (2006) as its standard of practice. The APM (www.apm.org.uk) was conceived of in 1972, adopted its name formally in 1975, and was a founding member of IPMA. The APM is based in the United Kingdom and as of 1998 had over 10,000 members in 40 countries. It utilizes the four-level IPMA certification approach. The APM publishes books, but no regular publications in magazine format.

Projects in Controlled Environments (PRINCE2®) was first developed as a British government standard for IT project management in 1989 by the Central Computer and Telecommunications Agency (CCTA), now part of the Office of Government Commerce (OGC). The PRINCE2 process model (www.prince2.com) provides a graphical process view of the project management process. As with the other standards, PRINCE2 offers a credential examination for "foundation" and "practitioner" levels.

The Australian Institute of Project Management (AIPM; www.aipm.com.au/html/default.cfm) was founded in 1976 and had about 9,000 members in 2008. It also has a registration program called "RegPM," or registered project manager.

We provide some of the definition used by PMI and IPMA for selected terms, again to refer people back to these standards. We also provide a glossary for easy reference to words and acronyms used in this book, including those from PMI and IPMA.

These are the primary standards for project management and for international project management. In the next section, we discuss the environment in which these standards are applied.

2.3 INTERNATIONAL ENVIRONMENT

Life is change, and life has been very good over the last 20 years as the rate of change has increased rapidly. The pace of change in international markets continues to increase along with Internet accessibility and the speed of global business. As connectivity increases, the ability to locate and exploit lower-cost resources provides significant opportunities for organizations, large and small. Organizations need people, and human resources are now available globally.

The demographics of the world are changing. Today developed countries, such as Japan, have aging populations that require young people to support their economic systems and to provide services. Advertising and satellite television are bringing images of development to the remotest corners of our planet while the economies in developing or undeveloped countries continue, in most cases, to lag significantly behind the developed countries. The disparities are now so great, and the pressures to reduce costs so intense, that it is possible for foreign nationals to increase income levels by factors of 10 or more by immigrating. These trends have also led to changes in social norms and a new dialogue about globalization.

Globalization places pressure on governments, firms, societies, and people to adjust to different values, economic conditions, norms, tempos, and foreigners. Societies have established words to describe foreigners, such as *farang* in Thai or *gaijin* in Japanese. Such words serve to differentiate and stereotype groups of people, and in doing so create the potential for conflict. The pressures created by globalization have in turn inspired international attention toward the concept of sustainability. Globalization brings the benefits of enhanced economic prosperity and the disadvantage of adverse impacts on society and the environment. However, perhaps most important, it underscores the differences between cultural values and norms.

In an article on globalization, the *Economist* ("Survey: World Economy", 2006) took the position indicates that globalization has provided a number of benefits to organizations: the reduction of labor costs gained from offshoring to low-wage countries; the flexibility of moving production curbs the bargaining power of workers in rich countries; and depressed wages in certain sectors, such as construction, due to increased immigration. The article points out that these labor issues are not just limited to low-wage workers. For example, the number of students graduating in India and China with science and engineering degrees in 2004 was over 1.2 million, compared with 0.9 million in the United States and the European Union (EU). On the economic scale, the article indicates that the emerging economies will produce over 60% of the world's gross domestic product (GDP) by the year 2025. The *Economist* also notes that since 2000, the GDP per capita has increased at 3.2% per year, offering more people better living conditions.

However, the article warns that the sharing of economic benefits is seriously disproportionate. Corporate profits and their percentage of developed economies are increasing while the share of wage earners is decreasing. For example, in 1980, wages in the G10 countries represented almost 63% of national income, and corporate profits in the G7 countries approximately 11%. In 2006, the wages were at about 59%, and the profits at about 15%.[2] The article suggests that governments may have to intervene to force sharing through taxes and benefits. This is an important piece of knowledge for an international project manager, for it warns of the potential for economic inequality on your team.

Organizations are on the leading edge of these challenges and must find creative ways to cope with them as part of their ongoing business operations. Many governmental systems in the twenty-first century are still relatively insular; often organizations must navigate through uncharted waters when working in multiple countries. Multinational organizations, such as the United Nations, the World Health Organization, and the World Bank, provide a forum in which the incredible diversity of our planet comes together to find a common way. Despite the efforts of such forums, ethical inconsistencies abound in the international markets. These uncertainties force organizations to define their own standards for the environment, the societies within which they work, and their employees. International ethical standards exist, but they are inconsistent and not in wide use by a plurality of organizations.

Thus, an offering of *baksheesh,* a gratuity for a service, in one country could be a bribe in a second country and a tax deduction in a third. The establishment of international cultural values has been an

[2]G10 (Group of 10) countries are eight International Monetary Fund (IMF) members—Belgium, Canada, France, Italy, Japan, the Netherlands, the United Kingdom, and the United States—and the central banks of Germany and Sweden. G7 countries are the United States, France, Germany, Italy, Japan, United Kingdom, and Canada. The G8 includes Russia.

ongoing human endeavor for the past 5,000 years, beginning with the recorded worship of deities (Grun 1982), and it continues to be a work in progress. As an international project manager, you must lead by setting an example for the team. Doing so requires balancing local custom against international laws and customs of other cultures.

Religion is one tool that societies use to define ethics, so commonalities run through most religions: People should not kill, people should not steal, and people should not lie. However, the application of these ethics varies and is manifested through different values and norms. The values and norms vary internationally for certain, but they also can vary widely within a country and within societal sects or tribes and organizations. To offer one example, Jainism is a religion that reveres life in all forms; most religions revere human life but not necessarily other life-forms. A devout Jain may refuse to consume food after dark to avoid the possibility of accidentally harming insects. Imagine that you are a project manager who is a Christian working on a project in India and wear leather shoes, leather belt, and carry a leather portfolio. What does your colleague who is Jain feel about this? How about your Hindu colleagues? A high level of XLQ would enable an international project manager to be aware of such considerations.

International legal systems often have evolved from religious beliefs and establish a formal definition of the ethics for a culture. As such, they carry great weight and importance for the people of a particular culture. In some legal systems, the connection between religion and laws is explicit; in others, it is more implicit. Norms in legal systems remain open for interpretation. Navigating the legal complexities of the international marketplace can be a daunting task, one that plays a critical role in the planning and execution of international projects. We discuss this issue in more detail in Section 2.3.4. General Electric has a published view of ethics in a document called "The Spirit and the Letter." The ethical values just described are the spirit, and the laws are the letter. The essence of this idea is at the heart of international ethics and sustainability: You can follow the law by the letter and still do unethical acts by the spirit.

The international marketplace is an environment of rapid change, high economic pressure, dispersed resources, and ethical diversity. To be successful, an international project manager must be cognizant of this diversity and must have a high level of XLQ. For an international project to be successful means that it satisfies the customer (quality), provides the scope agreed on, is within budget, and is within the time specified.

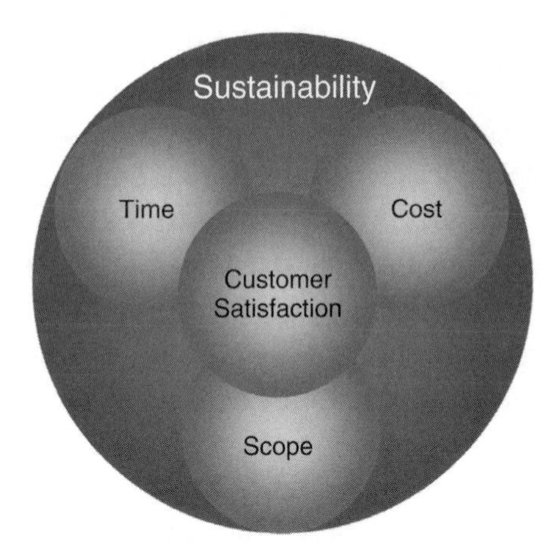

Figure 2.2 International Project Success

International projects also should be conducted with a view to long-term sustainability. Figure 2.2 combines the conventional "triple constraint" (time, cost, scope) into a view of the success criteria for an international project. If the project is not conducted in a sustainable way, or the product project is not sustainable, then society and history will judge it a failure. Equally important is customer satisfaction. Seldom, if ever, have we seen a project be successful if the customer was dissatisfied even if the scope, time, and cost were as planned. The International Standards Organization (ISO) definition of quality means that the project meets the stated and implied needs of a customer. Then there is the so-called triple constraint that a project must be completed to the scope defined, and within the time frame and budget established. What then is sustainability?

2.3.1 Globalization

Globalization is increasing global connectivity, integration, and interdependence in the economic, social, technological, cultural, political, and ecological spheres. For international project managers, this means greater access to resources at reduced prices and greater potential for conflict. Joseph Stiglitz (2003), a Nobel laureate, presidential advisor, and chief economist at the International Monetary Fund (IMF) and World Bank, understands globalization from a unique perspective: "globalization today is not working for many of the world's poor. It is not working for much of the environment. It is not working for the stability of the global economy" (p. 214). Stiglitz emphasizes that globalization undermines traditional values and that the pace of globalization matters because people need a more gradual opportunity to adjust. Many international project managers must confront the economic disparities in the global markets and must lead their teams in demonstrating that it is important to protect and cherish values in a time of rapid change.

People and materials for international projects are now sourced globally. We have seen a steel control module manufactured in Spain, shipped to Brazil where the transformers are installed, then shipped to Mexico where the motors are installed. Then they are shipped to Japan where the pumps are installed, and then they are shipped to the Philippines where the piping is installed. Then they are shipped to Malaysia for the computer and control equipment installation, then to Singapore, where the accessories from China, India, and Bulgaria are installed and the unit is tested. At each step, the competitive advantage of the country en route is leveraged, whether the advantage is low wages, high technical skill, or tax benefits. As Stiglitz points out, these value chain opportunities do not necessarily reach the people in the countries themselves, because global firms that repatriate the profits move the benefits.

Gladwell (2000) describes tipping points and uses epidemics as a metaphor to describe their effects. The three rules of tipping points are: (1) the law of the few (charisma and infectious behavior), (2) the stickiness factor (making information irresistible), and (3) the power of context (people are sensitive to their environment). The concept of tipping points applies to globalization and its potential for rapid radical changes rather than slow adjustments. In a world filled with both poverty and economic inequality, the potential for explosive change is increasing. This is of great concern for an international project manager who has resources in multiple countries. Consider a strike scheduled by the French labor unions that shut down transportation. If critical equipment needed in Vietnam is manufactured in Paris, this tipping point will impact the project. If avian flu breaks out in Hong Kong, it could well affect the team members from Albania who were traveling through Hong Kong on the way home. Our point is that an international project manager must stay informed.

As a direct result of the Internet, international projects also must confront the issue of equity in pay. Many organizations utilize purchasing power parity formulas to ensure that people of the same experience and skills from different countries are paid fairly. In this way, IT technicians working on the same project who live and work in Bangalore, Sophia, and San Jose are paid a salary that enables them to have equivalent purchasing power in their individual countries. More and more organizations are adopting this approach to provide equitable, not equal, reimbursement so that the people working on the project can enjoy a "living wage" in their home country. One of the dimensions of XLQ is empathy, or the ability to see the world through another's eyes.

Then there is the pressure to make money for the shareholders of the organizations. At best, this viewpoint yields the short-term goal of meeting quarterly metrics if the organization is publicly traded. At worst, it represents greed and a lack of concern for others. Competition reduces price, and reduced prices encourage growth and spending. It is natural then for organizations to seek the most effective location for producing their products and services. An international project manager should understand the effects of project decisions on project stakeholders.

One excellent example of globalization and the fundamental changes occurring in the 21st century is summarized in *Three cups of Tea: one man's mission to fight terrorism and build nations . . . one school at a time* by Greg Mortenson and David Oliver Relin (2006). Early in the book, Greg Mortenson (who was raised by missionary parents in Tanzania) is separated from a group evacuating a climber with high altitude sickness from a failed attempt on K2, one of the most dangerous climbs in the Karakoram. Pushed to the limits of endurance, he sings a childhood melody. In Taoism the term *wu-wei* means to act without acting, or said another way to be spontaneous without intention. In this wu-wei moment, Mortenson says to himself:

> Yesu ni refiki Yangu Ah kayee Mbingunie (what a friend we have in Jesus he lives in heaven) he sang in Swahili, the language they had used in the plain church building, with its distant view of Kilimanjaro, at services every Sunday. The tune was too ingrained for Mortenson to consider the novelty of this moment—an American, lost in Pakistan, singing a German hymn in Swahili (p. 18).

Globalization will be touched on throughout this book, because international project managers must manage it. As globalization evolves, international project managers must confront the issues raised and lead their teams. We therefore strongly encourage our readers to develop an in-depth understanding of this topic, and Stiglitz is a great place to start. The short-term aspect of globalization leads into our next discussion of sustainability.

2.3.2 Sustainability

The idea of sustainability can be traced back to 1983, when the United Nations formed the Brundtland Commission to address "the accelerating deterioration of the human environment and natural resources and the consequences of that deterioration for economic and social development." Driven in part by increasing globalization and its effects upon people, the report defined sustainable development as development that "meets the needs of the present without compromising the ability for future generations to meet their own needs." This definition then led to the concept of the triple bottom line (TBL) (Elkington 1998), and of corporate social responsibility.

The United Nations Global Compact (www.unglobalcompact.org) asks companies to embrace, support, and enact, within their sphere of influence, a set of core values in the areas of human rights, labor standards, the environment, and anti-corruption. More than 5,000 global companies in 130 countries take part in the initiative, which is made up of 10 principles:

Human Rights

1. Businesses should support and respect the protection of internationally proclaimed human rights; and
2. Make sure that they are not complicit in human rights abuses.

Labor

3. Businesses should uphold the freedom of association and the effective recognition of the right to collective bargaining;
4. The elimination of all forms of forced and compulsory labor;
5. The effective abolition of child labor; and
6. The elimination of discrimination in respect of employment and occupation.

Environment

7. Businesses should support a precautionary approach to environmental challenges;
8. Undertake initiatives to promote greater environmental responsibility; and

9. Encourage the development and diffusion of environmentally friendly technologies.

Anti-Corruption

10. Businesses should work against corruption in all its forms, including extortion and bribery.

CSR is, at this point, a self-regulated business policy that commits the organization to follows laws and international standards of ethics and norms. Normally it is concerned with stakeholders (e.g., profitability), society (e.g., social consciousness), employees (e.g., equal opportunity), and the environment. TBL, like CSR, commits the organization to focus on stakeholders, the environment, and society. Organizations that strive to meet a TBL or CSR goal recognize a responsibility to make money, to protect the environment, and to be good citizens. At RMIT University in Melbourne, Australia, the need for corporate governance, the care and nurturing of the workforce and transparency, was added to produce the TBL+1.

Elkington devotes an entire chapter in his book to the subject of time, which he believes is one of the most critical dimensions of sustainability, as do we. He was interested in knowing how long a view an organization can take using its life span as a benchmark. Interestingly, he found only 9 of the original 30 organizations in London's *Financial Times* Ordinary Share Index, launched in 1935, was still in business in 1984. In the United States, General Electric was the only survivor of the 12 organizations that made up the original Dow Jones Industrial Average, launched in 1900. The only firm with a longer history was Sweden's Stora Kopparberget, which could be traced back to a mine started in 850.

Elkington found that an average corporate life expectancy is about 50 years. For people, life expectancy is normally under 80, depending on the where one lives. For institutions, the oldest monarchy is likely that of the Japanese emperor, which dates back 2,600 years. Civilizations also have life expectancies, with the Chinese dating back at least 5,000 years. Likewise, projects have life expectancies that can range from a 3-day outage to a 500-year cathedral construction, the case with the Duomo Milano. What should, or could, be a sustainability time horizon, and does it stretch backward and forward? As Santayana (1953) said: "Those who cannot remember the past are condemned to repeat it." Project management is founded on the principle that the process of managing a project needs to be repeatable, plan-do-check-act, and should improve through the use of lessons learned.

Many project managers function on a short-term time horizon, a project reporting period. Such a focus will increase the probability of failure. Compare the attitudes that this engenders to a lifetime 80-year scale. The long-term view considers actions and short-term needs within the context of ethics, values, and ways to improve the human condition. Imagine driving a Formula One racer at 300 kilometers per hour (186.41 mph) and being able to see only 10 meters ahead of the car. You react second by second to the immediate and try to avoid crashing. If, however, you can see the entire track, you can place your actions within a larger context by anticipating what is coming up, using knowledge of the past, and planning for the future. If the driver is thought of as a project manager, imagine how the stakeholders (passengers) feel in each case. The international project manager, we argue, must take the long-term view of sustainability even if the firms involved in the project do not.

An international project manager must build trust by setting an example and by leading sustainability. That does not mean changing the world; it does, however, mean striving for something more than just short-term needs. We will recommend, time and again, that the lead project manager must set the standards that others may aspire to. There are no absolutes internationally, but that does not bar a lead project manager from encouraging behavior that inspires others to follow.

Another view of sustainability comes from psychological research. Eric Fromm (1976) argued that we have moved from a "being mode" to a "having mode" when we associate identity and self-worth with satisfaction gained through material objects. Pointing to globalization, he discusses consumption as one form of having (though transitory) and what he calls the marketing character of a person. This means that people view their value in terms of their marketability rather than their moral values. Marketing characters have no deep attachment to either themselves or others, and they respond to the world through manipulative intelligence. In the having mode, happiness lies in superiority over others, power, and, in the last analysis, one's capacity to conquer, steal, or worse, whereas in the being mode, happiness lies in loving, sharing, giving. By acting in the present and adopting a long-term view, an international project manager should lead, in the being mode, by example.

What are the global values that firms and organizations should strive to meet? John Hoyt (1998) said: "If we are to succeed in our quest for a humane future, we must begin by embracing and nurturing values that place far less emphasis on consumption and much greater emphasis on conservation" (p. 63). James Gustave Speth (1977), administrator of the United Nations Development Program (UNDP), reminds us that:

> Desperation and despair are still the lot of many people. Some 1.3 billion people live in absolute poverty, with incomes of less than a dollar a day, and poverty is growing as fast as populations. Over a hundred countries are worse off today than fifteen years ago. Each year 13 to 18 million people—mostly children—die from hunger and poverty-related causes. That computes to 1,700 human beings an hour—with only about 10 percent to 15 percent caused by emergencies (p. 1).

It is common to confront such issues when managing international projects, and it is our experience that feeling and showing empathy is one specific way to show long-term leadership.

Perhaps our global society has recognized the importance of values with the advent of globalization. For with the benefits of globalization and people being lifted out of poverty come the disadvantages, such as placing more emphasis on the "having" mode. This recognition may be placing pressure on people, societies, and organizations to return to a healthier balance between "being" and "having." Since economics are the driving force, and firms are the embodiment of our economic system, it makes sense that the focus should be on CSR, TBL+1, and other such aspirations. There are numerous standards for governance and corporate social responsibility, and the search continues for an international one. A few references that we can suggest to assist in getting started are: Global Reporting Initiative (2002). Millstein (2001), Social Accountability International (2001), and South Asia Human Development Report (2004).

Some organizations have adopted sustainability or CSR guidelines, but, unfortunately, many have not. On international projects, it has been our experience that there is a mixture of organizational policies, attitudes, and values on this subject. The international project manager of each organization that participates is responsible for knowing the standards that apply for her or his organization. The challenge then is for the lead project manager to negotiate, and establish, a set of standards that apply for the entire project team. Doing this may feel much like the dilemma of negotiating peace in the Middle East, where feelings run deep and a keen knowledge of the history is essential. The lead project manager cannot dictate the policies that each organization will follow but rather must set a standard that each organization can understand. Each organization, and person, can then choose to attempt to reach the project standard, partially or wholly. Of course, some will chose to eschew the standard. That is their choice, and while it is not encouraged, it should be a choice that the lead project manager respects.

We have seen organizations reject the notion of a project standard and insist on ignoring the issue of sustainability only to find later that taking a long-term view actually offers benefits for their organizations. Imagine two project managers. The first works for an organization that is entirely focused on meeting quarterly metrics, such as a return on investment of 12%. He is a young logical process-driven individual who has difficulty communicating with people, is aggressive, and does not appreciate it when people change plans or fail to meet deadlines. His firm promotes a "greenwashed" approach to CSR (advertises being environmentally friendly but shuns environmental change unless it is profitable).

The second project manager works for a firm that is founded on leading the way on CSR and the environment. The firm stresses ethics, focuses on a five-year plan, and engages all of its employees in knowing where the organization is headed. The organization is nimble—it must be—and ready to change at a moment's notice. This project manager is an older woman with strong opinions but who always listens to others actively. She understands processes but believes that they must be agile and is able to communicate effectively under conditions of uncertainty. She is strong in her beliefs but never aggressive with others.

Which project manager would you prefer to work with on a project that is chaotic, such as providing medical aid in Somalia, where the politics shift hourly, the value chain is wholly unpredictable, customers keep changing their needs, and the future is fuzzy?

Chaos, change, uncertainty, and a lack of vision can be avoided, or mitigated, through strong project leadership. Leaders thrive in circumstances like this, where there is uncertainty and no easy fix. As firms and organizations work in multicultural environments today, the need for XLQ can help by recognizing the attributes of strong leadership where uncertainty is rife. A closely linked consideration, which may be even more variable, is ethics.

2.3.3 International Ethics

Ferrell, Fraedrich, and Ferrell (2002) describe ethical responsibilities for business as "behaviors or activities that are expected by society, but are not codified in law" (p. 6). We would go further and include behaviors or activities that are intended by the law but not explicit—the spirit of the law as compared with the letter of the law. Unwritten social laws or customs also may carry the same weight as a written law in other cultures. Thus, ethics is intertwined with sustainability, CSR, the written law, unwritten social laws, and cultural values. The authors discuss the issue of cultural relativism, or the notion of adopting the ethical standards of the culture(s) in which the project teams function. This could compromise the international project manager's own ethics and leave her or him in a position of being unable to be trusted. It also has the disadvantage of attempting to balance multiple ethical standards on one project—one set of standards is difficult enough.

The ethics section of the International Competence Baseline (ICB) of the IPMA says that the "project manager should act according to accepted codes of professional conduct." The APM's *Body of Knowledge* is silent on the subject. According to the PMI's code of professional conduct, a project manager's responsibility is to "comply with the laws, regulations and ethical standards governing professional practice in the state/province and/or country." The Project Management Professional Exam Specification of 2005 says that a project manager should "ensure personal integrity and professionalism by adhering to legal requirements, ethical standards, and social norms, in order to protect the community and all stakeholders and to create a healthy working environment." Although we favor this last statement as a guide for international project managers, what does it mean in practice? Legal requirements and social norms are certainly obtainable, though arduous, but what of ethical standards?

In 2007, the PMI issued a revised code of Ethics and Professional Conduct to better define the ethical standards for global project managers (www.pmi.org/PDF/ap_pmicodeofethics.pdf). It separates the standards into mandatory and aspirational, strive to achieve rather than be required to achieve, categories. A few excerpts are as follows:

> 2.2 Responsibility: Aspirational Standards . . .
>
> 2.2.1 We make decisions and take actions based on the best interests of society, public safety, and the environment. . . .

2.3 Responsibility: Mandatory Standards…

2.3.1 We inform ourselves and uphold the policies, rules, regulations and laws that govern our work, professional, and volunteer activities….

3.2 Respect: Aspirational Standards…

3.2.1 We inform ourselves about the norms and customs of others and avoid engaging in behaviors they might consider disrespectful….

5.3 Honesty: Mandatory Standards…

5.3.1 We do not engage in or condone behavior that is designed to deceive others, including but not limited to making misleading or false statements, stating half-truths, providing information out of context or withholding information that, if known, would render our statements as misleading or incomplete….

Kholberg (1969) provides a model of cognitive development for individuals that places people at three levels of ethical behavior:

Personal Interests

- In Stage 1, punishment and obedience, a person responds to laws and norms. In Stage 2, instrumental purpose and exchange, a person behaves based on fairness to herself or himself.

Societies' Interests

- In Stage 3, mutual interpersonal expectations, a person emphasizes others rather than herself or himself. In Stage 4, social system and conscience maintenance, a person considers her or his duty to society. In Stage 5, social contract or utility, people are concerned with upholding basic rights and values of society, and of contracts.

Humanity's Interests

- In Stage 6, universal ethical principles, people strive to uphold principles and values that everyone should follow.

Kholberg's work provides benchmarks for international project managers. One can think about his or her own ethical level, that of the organization, that of the participants, and that of the societies in which the teams function. These benchmarks provide a means of thinking about the relative attitudes of the stakeholders as well. The Stage 6 level represents the TBL+1 attitude about how sustainable businesses should act. It also leads to the discussion to come regarding internationally recognized ethical standards.

Another way to look at the ethics issue is from the business perspective. Carroll and Meeks (1999) provided a simple structure for considering moral behavior in business:

- *Moral.* People who perceive the law (written and unwritten, social and professional) as a minimum standard and display ethical leadership. They seek to understand how stakeholders are negatively impacted by project activities and then to minimize those negative effects.
- *Amoral.* People who lack a moral barometer. The authors break this into two parts, unintentional and intentional. The unintentional amoral person is mindless and without consideration, ignorant of the law, or lacking ethical awareness. The intentional amoral person understands that there are moral and ethical issues, but everyone does it, so it is okay.
- *Immoral.* People who are devoid of ethical principles or values and who act in illegal ways. These are selfish people who put their needs, those of the project, or those of the organization above the needs of society.

Like Kholberg, Carroll and Meeks provide a way to think about the positions of the stakeholders on a project. Immoral behavior has been well represented in the press in recent years with numerous chief executives being convicted and imprisoned. Amoral behavior is also well represented in politics (spinning facts), in business (through advertising that is superficially true but intended to mislead), and in the management of projects (information is not complete or candid and lacks transparency).

Stanley Milgram(1969) said:

> Most men, as civilians will not hurt, maim, or kill others in the course of the day, but, place them into an organization with an authoritarian structure and a man may act with severity against another person. Because conscience, which regulates impulsive aggressive action, is perforce diminished at the point of entering the hierarchical structure (p. 42).

In other words, people of conscience can act without conscience if they believe they are only following orders. History is unfortunately replete with such thinking; two examples are Nazi Germany and Darfur. *The Corporation,* a recent film, addresses the issues of CRS and of how ethical people can be seduced into acting in unethical ways.

All international project managers should be required to undertake ethical studies—not to discover *the* ethical standard but to evaluate their own thinking and values and to become sensitive to the way different cultures view right and wrong. International project managers must first understand their own personal ethical standards. Then they must understand the ethical standards of the organization, the ethical standards of the other organizations in the value chain, and the ethical standards of the cultures and societies within which the project teams function. There will be ethical conflicts, but international project managers must select the standards to be used, articulate them to the entire team, and, most important, demonstrate their own in those standards through consistent practice. Ideally, international project managers should be moral persons, according to Carroll and Meeks, and should be working at Kholber's Stage 6 level of moral maturity.

There have been numerous attempts to devise a set of international ethical principles for business. One of them is the Caux Round Table (www.cauxroundtable.org/documents/Principles%20for%20Business.PDF), which was developed by business leaders in 1994. It is available in 12 languages and is reviewed each year. Another is the global Sullivan Principles (www.globalsullivanprinciples.org/principles.htm), which were developed from:

> [S]uggestions made to Reverend Dr. Leon H. Sullivan, by various world leaders, that he develop a successor code of conduct worldwide in orientation to the original Sullivan Principles (which were instrumental in helping to dismantle apartheid in South Africa). After consulting with leaders of business, government and human rights organizations in many different nations, and receiving strong encouragement to go forward, Reverend Sullivan accepted this challenge and worked with others to draft the Principles. Reverend Sullivan announced the inauguration of the Global Sullivan Principles during a special ceremony at the United Nations on November 2, 1999, with UN Secretary General Kofi Annan.

There are also the simple interfaith beliefs set forth by the Interaction Council (that every human being must be treated humanely and that one treats others as one wants others to do unto one).[3] This concept is part of every great religious tradition and is revisited in Chapter 4.

[3]Interaction Council, "In Search of Global Ethical Standards," meeting chaired by Helmut Schmidt, March 22–24, 1996, Vienna, Austria; www.interactioncouncil.org.

The Sullivan Principles are perhaps the most complete and straightforward beliefs at this time and can serve as a solid benchmark for international project managers. The following is quoted from the Sullivan Principles reference webpage:

> As a company which endorses the Global Sullivan Principles, we will respect the law, and as a responsible member of society we will apply these Principles with integrity consistent with the legitimate role of business. We will develop and implement company policies, procedures, training and internal reporting structures to ensure commitment to these Principles throughout our organization. We believe the application of these Principles will achieve greater tolerance and better understanding among peoples, and advance the culture of peace. Accordingly, we will:
>
> - Express our support for universal human rights and, particularly, those of our employees, the communities within which we operate and parties with whom we do business.
> - Promote equal opportunity for our employees at all levels of the company with respect to issues such as color, race, gender, age, ethnicity or religious beliefs, and operate without unacceptable worker treatment such as the exploitation of children, physical punishment, female abuse, involuntary servitude or other forms of abuse.
> - Respect our employees' voluntary freedom of association.
> - Compensate our employees to enable them to meet at least their basic needs and provide the opportunity to improve their skill and capability in order to raise their social and economic opportunities.
> - Provide a safe and healthy workplace; protect human health and the environment; and promote sustainable development.
> - Promote fair competition including respect for intellectual and other property rights, and not offer, pay or accept bribes.
> - Work with governments and communities in which we do business to improve the quality of life in those communities—their educational, cultural, economic and social well-being—and seek to provide training and opportunities for workers from disadvantaged backgrounds.
> - Promote the application of these Principles by those with whom we do business.
>
> We will be transparent in our implementation of these Principles and provide information which demonstrates publicly our commitment to them.

The second principle says that an organization or individual should operate without female abuse. In much of the western world, women are given (on paper at least) equal opportunities. In parts of the Middle East, Asia, and Africa, different attitudes exist. Therefore, if you are a project manager with people in these regions, you must establish a standard, or others will do it for you. If you permit the ad hoc approach, you will have failed the first leadership test: lead by example.

The discussion on ethics has been ongoing for thousands of years and no doubt will continue for thousands more. As with sustainability, there are no pat answers. Each project manager will have to confront this conundrum on each international project. Therefore, as we progress through the topics, we provide more discussion about ethics in practice. As we said at the beginning of this section, ethics is embodied in the spirit of the law, and it is the goal that international project managers should aspire to achieve.

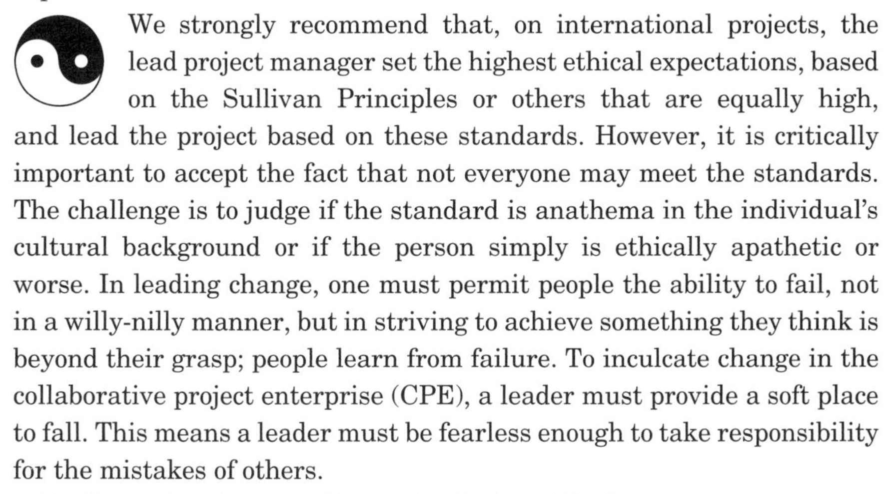

We strongly recommend that, on international projects, the lead project manager set the highest ethical expectations, based on the Sullivan Principles or others that are equally high, and lead the project based on these standards. However, it is critically important to accept the fact that not everyone may meet the standards. The challenge is to judge if the standard is anathema in the individual's cultural background or if the person simply is ethically apathetic or worse. In leading change, one must permit people the ability to fail, not in a willy-nilly manner, but in striving to achieve something they think is beyond their grasp; people learn from failure. To inculcate change in the collaborative project enterprise (CPE), a leader must provide a soft place to fall. This means a leader must be fearless enough to take responsibility for the mistakes of others.

In the next section, we discuss the letter of the law.

2.3.4 Laws and Regulations

On international projects, it is critical to understand the laws and regulations of all the countries where work is performed and for each link in the global value chain. International legal systems abound, and depending on the project's dispersion and complexity, full-time legal guidance may be required. For example, if the product of the project is to be provided in Saudi Arabia by firms in the United States, China, and Germany, there will be at least four different legal systems to consider and numerous government regulations. Some of the major legal systems that exist are shown in Table 2.1 (Ramlogan and Persadie 2004).

Table 2.1 International Legal Systems

Legal Systems	Basic Aspects
International	International laws or treaties must be ratified by a country. International laws may be adopted by businesses as well (e.g., Sarbanes-Oxley).
Civil	Statutory laws where judges apply and interpret the law, not make the law. Civil systems include the use of discovery. This system generally is used in western Europe, Quebec province, the Philippines, Puerto Rico, Asia, and Africa.
Common	Case law where laws are made by judges using precedents and current needs. This system generally is used in the United States, United Kingdom, former British Empire, English Caribbean, and West Africa.
Socialist	Similar to civil law but with a Marxist-Leninist twist. Property rights and gender rights are different. This system generally is used in Russia, China, and Cuba.
Shari'ah	The Qur'an provides the basic rules, the Sunnah provides Muhammad's practices and clarifies and interprets the Qur'an, the Ijma' provides consensus of the Islamic community of lawyers and laymen, and the Qiyas provide the intention of the laws. This system generally is used in the Middle East and East Africa.
Halakhah	Utilizes the Torah (or law of Moses), the Talmud (a record of oral law, and how to observe it), the Mishna (a collection of notions that forms the Talmud), and the Responsa (a record of how the common law was applied). This system generally is used in Israel.
Customary	Law consists of customs accepted by the members of a community, and it can be a blend of the other types of laws noted above. This system generally is in some use India, China, and Nigeria. Native title law in Australia and Canada has significant impact on cultural sensitivities, such as sacred sites, and can seriously affect some types of projects.

At a bare minimum, an international project manager must understand which legal systems are included in the project and the basic differences between them. He or she also must know the governing law of the contract(s) and be aware of any areas of conflict among laws.

Contracts for international projects themselves generally consist of special conditions, general conditions, and technical provisions. The special provisions deal with the particular requirements of a specific project, whereas the general conditions generally establish the basic legal and procedural rules of the contract. The general conditions normally address such issues as termination, insurance, payments, bonds, governing laws, indemnity, project management requirements, dispute resolution, and the like. Normally both special and general conditions are based on the prevailing laws in the country where the product is being provided or where the customer's procurement operation is based.

Organizations apply risk management techniques when reviewing potential project contract documents and emphasize the general conditions sections dealing with such issues as liability, payments, acceptance, and governing laws. Clauses that are receiving more and more attention are the requirements for resolving disputes. Many eschew resolution in local venues in favor of international mediation and arbitration. We review more about the laws and regulations relating to risk management and procurement management in Chapters 7 and 10.

On the regulatory side, the codes and standards on international projects vary as widely as do the legal requirements. The technical provisions of contracts provide most of the information on technical regulations and codes, but often additional requirements are provided on drawings and in specifications that are referenced in the contract. Most insidious though is the term "local practice," for it is assumed that responsible contracting parties know what that means. This is one of the many reasons that firms and organizations seek local partners. There is also the issue of having regulations and standards that potentially conflict with one another (European Union versus those of the United States, e.g.). International project managers need to understand the technical conditions of the contract and, as with the laws, where potential conflicts exist between codes and standards. Each international project manager should read the contract to look for these potential conflicts and attempt to resolve them long before the start of project implementation.

Every international project manager must develop an understanding of the legal systems and regulations that apply to the project. If the lead project manager participates in the initiation, planning, and the negotiations of the CPE, a thorough understanding of the legal issues can be fully developed, managed, and priced in to the project budget. In our experience, on international projects it is critical to project success for the proposed lead project manager to be involved in contract negotiations. Every international project manager also must get copies of *all* contract documents and regulations and codes, must understand the legal systems and the fundamental requirements for each, and must understand as much about local practice as possible. It is strongly recommended that this information be digitized and placed on a project server to enable any stakeholder access to the information from anywhere in the world (see Chapter 12). If there are 40 firms involved in a project imagine, the cost multiplier if everyone must do this. Also, imagine the risk multiplier if they do not.

This digitizing will save hundreds or thousands of hours of time and will help avoid needless confusion and conflicts. Too many projects have no basic contract documents available at remote locations. This manifests itself in the old adage that the only time one reads a contract is when there is a dispute. An understanding of the international legal systems and the requirements of the contract documents plays a direct role in the design and management of project value chains (see Chapter 11).

On this last point, we now look at the issue of value chains.

2.3.5 Competition and Value Chains

Economists contend that organizations benefit from the flexibility and global nature of competition and the economies of scale that globalization and the Internet offer. As we pointed out earlier, humanitarians contend that globalization harms societies and can harm organizations over the long term. International project managers need to appreciate both perspectives and the business and economic processes required to tap into global value chains. Most multinational organizations have well-established international value chains, but many firms must create them project by project. Today, small organizations can take advantage of lower costs for high-quality services internationally and forgo the need to add staff to fill short-term needs. However, to do so, they must construct the processes and nurture the relationships required to assemble such a team in a small organization.

We need to clarify a few terms that are bandied about frequently on projects, particularly international ones: value chain, value chain, critical chain, and INCO terms. The idea of a value chain was first popularized by Michael Porter (1998). His idea was that organizations add value to products through their processes. One example of this is taking raw steel ingots and milling them into pump components. He extended this idea to entire value chains for both manufacturers and services organizations. Using the same example, the organization that mills the pump components may then ship them to another firm, which assembles this component into a pump casing, and then ships the assembly to a project site for installation. In this example, each organization in the value chain adds value, thus becoming a value chain. We use the term "value chain" to capture this concept. If an international project manager sees the results of the value chain as being her or his responsibility to the customer, then you have it right.

Figure 2.3 provides a more detailed view of our example. In this case, the project manager relies on the pump manufacturer and the installation contractor to manage their value individual chains, but in the eyes of the customer, the project manager is responsible for managing the entire structure. At each step, the value added by the seller plays its own individual role in producing the quality (product and service) expected by the customer. In our experience, a value chain represents a hierarchy of buyers and sellers involved in transactions and self-interests that results in a quality project. Included in the value chain concept are issues of taxes, insurance, content (percentage content added in a specific country), ethics (bribery, societal respect, etc.), and transparency.

Look at the example of Nike, which became a case study for business schools for using child labor in the manufacture of its products. One such accusation had to do with soccer balls made in Pakistan. Nike attempted to refute the accusations, then to contend that it did not own the organizations that actually used the child labor. Company spokespersons argued that Nike was not responsible for the actions of its value chain. It did not work, and the criticisms were withering. Nike had its logo on the products and was considered responsible for what went into making them, regardless of the name of the organizations involved. In a properly functioning value chain process, value is *added*, not *diminished*. This case has nothing to do with the quality of the product but everything to do with the ethics, TBL+1, and transparency of the process.

The critical chain is a project's critical path that is task dependent and resource constrained (Wysocki 2007). The argument made is that the critical path alone does not take account of the statistical probability distribution of a task's duration; thus, buffers are built in to individual task durations. It also suggests that project critical paths are constructed from logic that does not take account of resource constraints. In our experience, the concept is correct but is seldom used in practice. We have seen few resource-loaded schedules in the past 35 years and

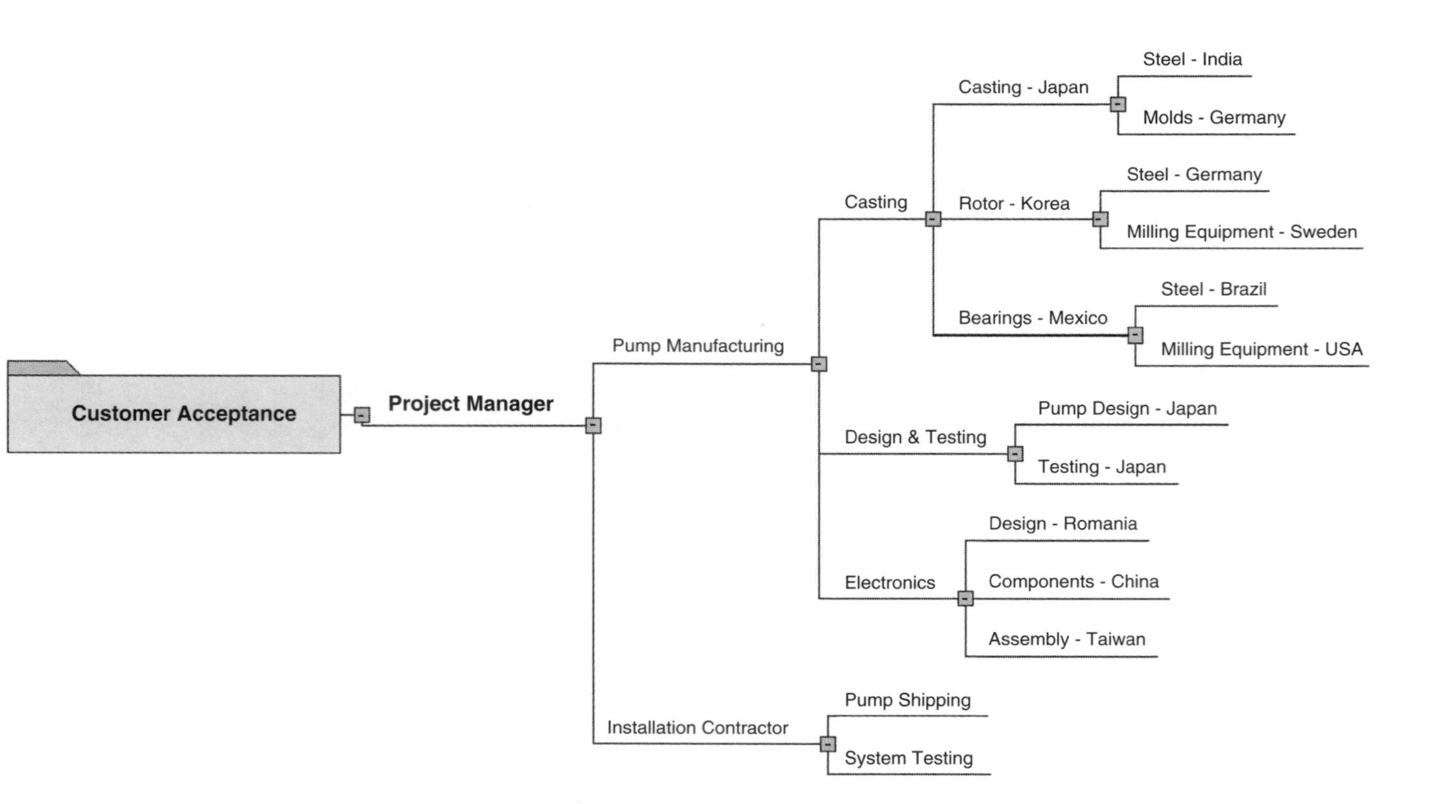

Figure 2.3 Value Chain Example

fewer yet with resource constraints built in to the logic. That is simply poor planning. We discuss more about the scheduling and resource issues in Chapters 5, 6, and 8. For now, recognize that the critical chain should not be confused with the supply chain or the value chain.

On international projects, it is particularly important to understand the terms of the transactions that occur throughout the value chain, especially if the project manager does not have a mature procurement department to rely upon or a value chain with which the organization has experience. Fortunately, there are recognized international standards promulgated by the International Chamber of Commerce (ICC) known as International Commercial Terms (INCO terms), which are also endorsed by the United Nations. A complete listing is available in book form, and on the Web page: www.iccwbo.org/incoterms/id3045/index.html.

Table 2.2 provides three benchmarks to illustrate the differences in obligations, with EXW (ex works) and DDP (delivered duty paid_ forming the extremes. If the project manager procures the pump from Japan (Figure 2.3) EXW, ex works, title to the pump transfers to her or him at the factory's loading dock. The project manager would be responsible for getting the pump to the port in Japan, shipping, insurance on the shipment, clearing customs in Japan and in the country of delivery, and transportation in the country of delivery. On large projects, the customer

Table 2.2 Sample INCO Terms

INCO Abbreviation	INCO Term	Description
EXW	Ex works	Sellers deliver when they place the goods at the disposal of the buyer, normally at the seller's premises, but also at a named location, not cleared for export, and not loaded on any collecting vehicle.
FOB	Free on board	The seller delivers the goods when the goods pass the ship's rail at the named port of shipment. This means that the buyer bears all costs and risks of loss or damage to the goods from that port. The seller is required to clear the goods for export.
DDP	Delivered duty paid	The seller delivers the goods to the buyer.

sometimes provides umbrella insurance to minimize the cost of insurance under a single policy. In such cases, the international project manager must adjust the terms of the purchase orders and contracts in the value chain accordingly. It is important to determine the responsibilities of the buyer and seller in the transactions and to write the general conditions accordingly. The risk of loss is defined by how the contract is written and by which organization is best equipped to manage the risk.

INCO terms address products and shipments, but what about services? Large organizations can take advantage of economies of scale and can cushion regional economic swings by utilizing people in other geographic areas as part of their value chain. Kerzner (2006) has suggested that firms seeking excellence in project management consider what he calls a project management maturity model (PMMM). The model includes five levels of maturity:

1. Common Language—broad understanding of international project management basics.
2. Common Processes—repeatability and recognition that international project management connects to the business.
3. Singular Methodology—combining all corporate methodologies with international project management at the center.
4. Benchmarking—process improvement requires a level of basic acceptability, or stretch goals.
5. Continuous Improvement—knowledge management, kaizen (Japanese for continuous improvement), lessons learned, centers of excellence.

The application of the PMMM for international projects offers some opportunities and challenges. Imagine that your organization wants to enter the market in Asia and needs local expertise on a long-term basis. The opportunity offered would be common language and an understanding of the market, values, norms, and processes considered "normal practice." It would benefit the organization to engage in a joint venture, partnership, or alliance with a local organization, as implied in Kerzner's model. The closer the relationship, the more sharing of processes, lessons learned, knowledge, customers, technology, resources, people, and the like. In addition, the ability to reduce the exposure to competition is a strategic business consideration for the type of relationship constructed.

The challenge side of the coin is, as always, the opposite of the benefits. Reconciling common processes and methodology will be extremely difficult, if not impossible, unless the relationship is long term and the transparency is extremely high. If the value chain includes more than two organizations, the PMMM concept becomes more complex and requires more attention on each of the five maturity levels. One must first establish and memorialize core internal competencies and processes (Krajewski and Ritzman 2005) of the organization. If one organization is at Level 4 and another is at Level 1, merely synchronizing the basic processes will be a challenge. This, of course, is another reason why organizations utilize value chains: to procure expertise. Krajewski and Ritzman propose that elements of a value chain for service providers include facilities (where the work is done), goods or materials, explicit services (producing the quality project product), and implicit services (PMMM). It is in these implicit services where disconnected or weak value chains display their fractures most often. More discussion about international project structures is provided in Section 2.7.

Producing a quality international project management service includes doing one's diligence as a trained project management professional and attending to the requirements of a project manager, according to the professional standards and ethics. In addition, it requires the project manager to be responsive, candid, focused on understanding the customer's definition of quality, and more. It also places upon the project manager the responsibility of articulating all of this throughout the value chain.

On most projects with external customers, value chains must be either constructed or customized for each project. An organization that has an existing value chain in place for a particular type of project should customize it, within reason, for each specific customer. This means that the implicit needs of a customer and its view of quality (product and service) should flow through the value chain, as noted in the example of the pump manufacturer. At the optimal, Level 5 in Kerzner's PMMM, the entire value chain would strive for customer satisfaction, with the project manager explaining exactly what that means.

If there are no existing value chains, the project manager will need to assess the risk of not creating one. On large, complex international projects, value chains take a lot of time to create. Supply chains can be created quickly, but to make them value chains (integrating them into a project, as a collaborative project enterprise) easily can require weeks or months of intense work. In initiating and planning international projects, a project manager must consider the options and evaluate the best sources of material and services at the lowest cost, highest quality, and least risk.

Chapters 4 and 11 review the value chain and considerations that need attention, beginning at the conceptual phase of a project.

One other aspect of international projects, competition for resources, needs to be considered on a macro scale during the conceptual phase. Foreign direct investment (FDI) inflows or market conditions can cause short-term aberrations in the local competition for resources. For example, in 2006, Trinidad and Tobago were enjoying a boom in revenues due to the price levels of oil and gas. FDI and governmental infrastructure investment increased concurrently, and a shortage of crushed stone on the islands ensued. Project managers then confronted a national dilemma of stone shortages, skyrocketing prices, and the need to import stone from other countries. The lesson here is to do one's homework regarding the economic conditions in a country during the conceptual phases of a project. As we often say, an international project manager must be part economist, part sociologist, part politician, and part clergy.

In addition to such physical considerations, there are also the virtual ones. That is the subject of the next section.

2.3.6 Virtual Environment

Two aspects of the virtual environment and e-business that a project manager must consider are communications and procurement. E-communications are critical for the virtual world of international project management. Thankfully, it is no longer necessary to rely upon fax and hard-copy documents for communications. E-mail, text messaging, project servers, and electronic transfers of financial, technical, and informational data can be utilized almost anywhere on the planet. However, connection speeds vary wildly and should be addressed during the initiating and planning phases of the project. One of the first things done in military campaigns is to establish the communication links, and international project managers should do likewise.

People rely on language, paralinguistics (tone, pitch, tempo, and volume), body language, and visual clues when communicating with other people. All of these can be utilized in a virtual environment as long as the bandwidth and technology are available. From our experience in synchronous education over the Internet, we have found that software which provides video, voice, file sharing, and whiteboards requires high-speed cable or fiber. By eliminating the video, the platforms will function on lower dial-up speed connections, but, depending on the software, systems can be unstable and slow. Today e-mail is available almost everywhere, so at worst an international virtual team would have to rely solely on language. Chapter 4 discusses communication management and the different considerations that international project managers need to address.

Although most all international projects make use of the Internet, it is not unusual for hard copies of contract documents to be used rather than electronic versions. Standards referenced in the technical specifications, such as codes and regulations, also often are not available online. Paper documents require each local team to acquire the correct versions for themselves, which adds costs to a project. Making the basic documents and information available must be part of a communication management plan. The plan also must take account of the availability and archival requirements of the project, both from a legal (e.g., Sarbanes-Oxley) and a productivity (retrieval of information) perspective. Communication is particularly important in the design and operation of value chains and on projects that have partnership, joint venture, or alliance structures. Communication is now easier on international projects that it was even 10 years ago. The downside is that this ease spawns more volume. Today's project manager must consider how to reduce and filter the information to get people what they need but not deluge them.

Another aspect of the virtual environment is the procurement of goods and services online. This industry is evolving, and businesses are striving to adopt e-business platforms to reduce costs. The dilemma is that moving to an e-business platform almost always requires a basic change in the way organizations do their business. If an organization on the project is making the transition, the lead project manager should be prepared to offer assistance. Organizations that offer e-business platforms can offer great advantages in value chains, but an experienced project manager will exercise some healthy skepticism when planning for the use of such systems. The lead international project manager should look at the maturity of a firm's e-business system—kick the tires, as the American saying goes (like when deciding to purchase an automobile to see if the basics are in place)—consider the benefits and disadvantages of utilizing an online procurement option and its ability to work in multiple currencies.

Being able to purchase just-in-time on-demand services can offer added flexibility and serve as a way to mitigate short-term procurement risk. Buying such services online means that the procurement system, electronic transfer systems, and authorization systems, in both buyer and seller organizations, must have been established in advance. Firms with an online option for ordering services require that they have internal processes for taking and confirming orders, reporting on progress, and processing payments. An international project manager must be familiar with the currencies used by the participants of a project and must plan for the likelihood of currency fluctuations.

The advent of the Internet has been a blessing for international project managers. Not being able to see other people is sometimes a blessing, for people misread paralinguistics and body language. The ability to communicate effectively also requires XLQ; more about that in Chapter 4.

2.3.7 Multiple Cultures

Embodied in international project management is the need to lead diverse teams from diverse cultural backgrounds. Anthropologist Margaret Mead (1955, p.12) described culture as "a body of learned behavior, a collection of beliefs, habits and traditions, shared by a group of people and successively learned by people who enter the society." If one substitutes "group," "team," "consortium," or the like for "society," the definition fits international projects perfectly. As we said earlier, culture can be societal, organizational, tribal, group, or, in our case, the CPE. Rosen, Digh et al. (2000) described twenty-first-century business cultures as an onion with each successive layer adding to, and possibly modifying, the core values of an individual. An adapted version of the onion is shown in Figure 2.4.

At the center is leadership culture (that of the individual)—born in Shinyang, moved to Calgary when 10 years old, educated there and

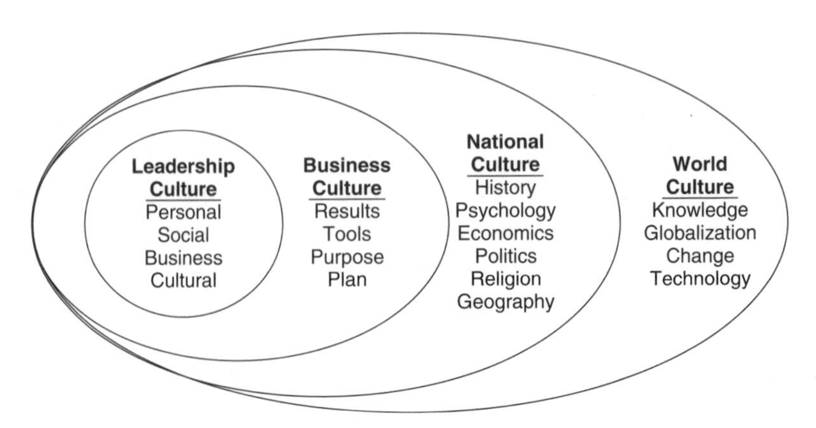

Figure 2.4 Cultural Layers

then in New York, first assignment in Singapore, second in Dubai, now located in London. The next layer is the business, the difference between General Electric and the International Red Cross, and that is followed by the world layer—the globalized person described and technology. The national culture is that most often thought of as a societal culture, and the world culture is a product of globalization. In this model, the leadership culture would be most important, since it is at the core. In our case, the CPE would be part of the business culture layer. The lead international project manager must create a culture for the CPE, for if she or he does not, the team will. That culture likely will require some adjustments for each personal, tribal, and organization culture, but it can, we argue, celebrate and leverage the benefits of the diversity. As the leader of the CPE, the lead project manager must be prepared to adjust, not change, her or his attitudes and cultural mores to set the standards for the team, and to do so by leading the way. Doing this requires emotional intelligence (EQ) (Lee and Templer 2003), cultural intelligence (CQ) (Earley and Ang 2003), and XLQ (Grisham 2006).

For example, on a project in the Netherlands, the customer was a Dutch multinational, and the primary players were contained in a consortium of two Dutch firms and one U.S. firm. The people with the Dutch multinational had worked around the world. In doing so, they had modified some of their societal cultural norms and had selectively absorbed some of those from the other cultures with which that they had worked. These people came from different global offices of the same organization and had acquired different organizational cultures along the way as well. The people with the local Dutch firms had limited experience outside of the Netherlands, and they displayed far fewer changes to their inherited societal and organizational cultural norms. The U.S. firm was also a multinational, but the team members had little experience outside the United States and no cultural training. In this case, the inherited societal and organizational culture was deep rooted, strong, and unyielding. The dilemma was that the consortium could not conclude an agreement for the project. What was missing was the cross-cultural leadership intelligence (organizational and societal) necessary to recognize the cultural differences and adjust them to a common CPE norm. Once that was done, the agreement moved forward in a matter of weeks, and the project was ultimately a success. The missing ingredient was trust, and it was missing in large measure because of a lack of understanding of culture and a lack of leadership. We cover XLQ in detail in Chapter 4.

Cultural diversity is a huge untapped reservoir of opportunity for international project managers. However, it demands effort and time to listen, and requires the lead project manager to assess the needs, aspirations, constraints, and abilities of stakeholders on the multiple cultural levels. It also requires a leader who has empathy for others, insatiable curiosity, the need to learn, compassion, and patience in abundance. In our experience, cross-cultural leadership offers the greatest return on investment of any other task that an international project manager undertakes. It takes time, patience, compassion, excellent communication skills, and consistency.

Unfortunately, we still see too many international project managers who have never had any training in leadership or culture, good people who simply have not been given the skills they need to be successful. One reason is that the project management standards do not emphasize these attributes as critical skills for a project manager. We believe this needs to change.

In the *PMBOK,* the word "leadership" appears 9 times; six of those times are thanking people for putting together the *PMBOK*, and the word "culture" appears 14 times, most often dealing with organizational culture. In the ICB, of the International Project Management Association, the word culture appears 16 times, and like the *PMBOK* has only one citation relating to societal culture. However, the ICB does have four references to team culture. Unless one is either very lucky or has an assistant who understands cross-cultural leadership, it will be difficult to succeed on an international project without the attributes that we have described. We have seen many international projects *managed* with great processes and procedures fail, because the lead project manager does not have these attributes. We call the cojoining of these attributes XLQ.

As we have indicated, international projects can take advantage of low-cost expertise wherever it is available. The globalization of human resources holds the possibility of a better life for many, but it requires policies and practices that are adjusted to the local circumstances.

2.3.8 Human Resources

Organizations strive to reduce costs as part of their efforts to increase profitability. As noted, globalization offers the promise of high-quality labor at lower prices. However, it also offers the promise of complexity in cultures, values, political pressures, immigration issues, and legal considerations. From the viewpoint of an organization, there are two basic types of labor: core and fluid resources. Core resources are full-time employees who likely receive some form of benefits and are engaged in a long-term relationship. However, many organizations no longer offer lifelong employment relationships, even in Japan. As each organization creates and nurtures its own culture, the relationship of a core resource to the organization is a more tribal sort of culture.

Core resources can more easily create and maintain their own communities of practice (CoPs), participate in the establishment of norms and values, create and use their own language, expect compliance with the unwritten rules of conduct, and more. These practices serve to differentiate "us" from "them." "Us" is a family or tribe, and the connection between the members is a strong one, especially in homogeneous national organizations. Transglobal organizations (see Figure 2.7) are more like different tribes in the same country. The organization provides a higher level of culture and norms, but the local culture is far more potent, understood, and practiced. "Them" are nontribe members and are different, dissimilar, and perhaps to be avoided. Figure 2.5 presents a typical structure of an international project conducted by a transglobal organization. Core resources exist in the London and Vilnius business centers, with the lead project manager (PM) located in London.

In addition, integral parts of international project teams are fluid resources, those that are hired temporarily by an organization. These resources do not receive benefits, and their remuneration is related to effort expended (hourly, monthly, and yearly). They are called when

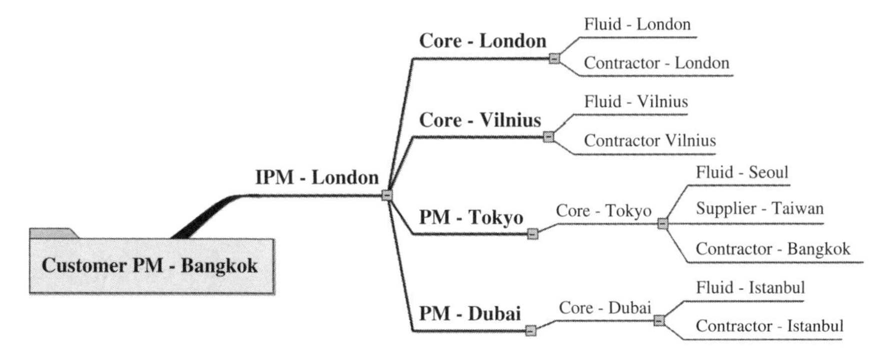

Figure 2.5 International Team

needed to provide expertise or to overcome temporary labor shortages, and are likely to be independent. They can be ex-employees of the organization who have taken early retirement. Certainly many organizations prefer this option, as the resources were once "us" and know the norms and customs. In Figure 2.5, these resources can be found in each location and can come from other geographical locations as well. For example, the Tokyo office might draw virtual fluid resources from Seoul. The core teams often view fluid resources as distant cousins, related but not closely, who are not current with internal policies and norms.

Contract services are a bit different from fluid resources; these are people who are not previous employees, so their relationship is even more distant. Those who are not ex-employees must acculturate themselves into the organization. Contracted services are those that have a defined scope, or an individual product, with no assured commitment to the values and norms of the core team members. To use a metaphor, fluid team members may be partially infected with the organizational culture, whereas contract employees are often immune. Purchasing design services from an engineering firm that results in system architecture for an information technology (IT) project, or a process and instrumentation diagram for a manufacturing project are two examples of contracted services. Contracted services may provide level-of-service support or a defined work product. Level-of service support is generally more tightly bound to the core team, for the relationship is more like that of a fluid team member. Defined services are generally less tightly bound as their work tends to be seen as a "black box" (inputs and outputs are known but not the processes) by the core team. Contracted services are part of most global value chains and require special leadership attention to blur the cultural boundaries.

There is another dimension to Figure 2.5, and that is the movement of people (immigration and emigration). People who are noncitizens can play core, fluid, or contracted team functions, and their needs are different at different cultural levels as shown in Figure 2.4. Leading and motivating collaborative project enterprises (CPEs) requires sensitivity to each of these subgroups and the demonstrated intention to pull them together into a project culture or a microculture (see Figure 2.4).

The impact of chasing lower cost and higher quality globally adds complexity in CPEs. An international project manager always must be mindful of this complexity and must plan projects to emphasize the benefits of this diversity, and minimize the disadvantages of serious

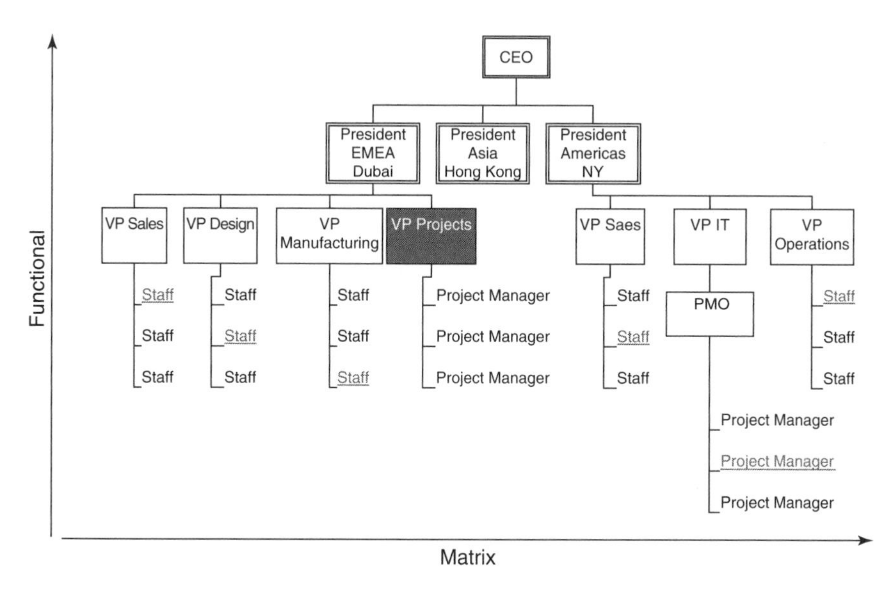

Figure 2.6 Organizational Structure

miscommunications, cultural disputes (corporate, social, etc.), and organizational goal disparity.

Another consideration for international project managers is the configuration of the organizations that are participants on the project. Look at Figure 2.6. All organizations are arranged in a functional hierarchy, and most, if not all, share resources horizontally. Organizations that share horizontally are called matrix organizations. Imagine that the VP Projects for the Dubai office is the customer and needs a new IT platform that will enable a project to be conducted on a single server. She requests services from the VP IT in New York, and the team is assembled as shown in italics and shaded.

The organizational structure shown is for what the *PMBOK* calls a "projectized" organization or "any organizational structure in which the project manager has full authority to assign priorities, apply resources, and direct the work of persons assigned to the project."(p. 437) We prefer the definition of any organization that has a Project Management Office (PMO) or similar function to systematize project management processes across the organizational boundaries. As you can see, the VP of Projects functions as a profit center in Dubai, and therefore the project managers would have direct access to upper-level management and thus more power in obtaining staff. Consider the difference between the VP Projects

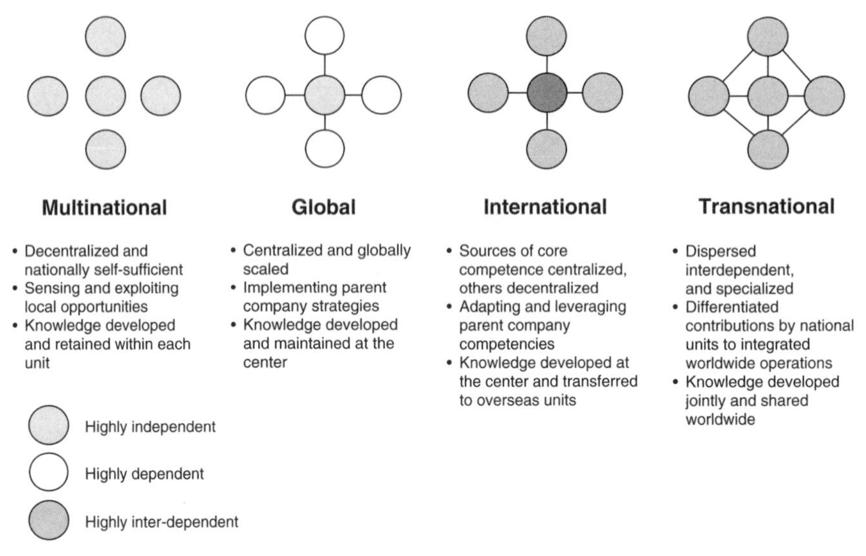

Figure 2.7 International Structures
Source: Adapted from Bartlett and Ghoshal (1998).

and the PMO who works for the VP IT in New York shown in the figure. Many organizations use a similar organizational approach, at least those interested in the benefits of project management and process improvement.

Imagine that one participant in the project has an organizational structure like that described and another has an organizational structure with no VP Projects and no PMO. Project management is an ad hoc affair and is a part of the normal workday rather than a dedicated resource. Think about the lack of process in the second organization, and the conflicts that could occur when process meets ad hoc. Thus, organizational structures matter. The lead project manager must understand how the organizations of the different participants are structured so that she or he can anticipate where problems could occur. Then there is the structure for organizations that operate in multiple countries, as shown in Figure 2.7.

2.4 INTERNATIONAL ORGANIZATIONAL STRUCTURES

In international markets, an array of organizational structures has evolved over the past few decades, from classical functional and matrix formats, to hybrid mixes, weak to strong matrix organizations, and projectized organizations. The structure of an organization will determine the flexibility available on contract structures, project manager power, stakeholders, sponsors, communications, and much, much more. It is critical that international project managers understand the enterprise within which they function, that of the other project participants, and that of the value chain. There are, of course, many variations of the theme, but Bartlett and Ghoshal (1998) provide an excellent starting point. Figure 2.7 provides a summary view of their concepts and shows how each organization deals with power and knowledge management. Let us now look at one implication for a simple international project.

Imagine an international project manager located in the United Arab Emirates (UAE) office of a multinational firm headquartered in Berlin; a partner that is a transnational firm headquartered in Buenos Aires with an branch office in the UAE; and a governmental organization as a customer in the UAE. First, the relationships between the governmental organization and the local entity likely will be close and enduring, whereas the project relationships will be transactional. In the Buenos Aires firm, there will be a different approach to knowledge transfer from the firm in Berlin and many more communications conduits in place. The international project manager should plan the stakeholder management with an understanding of the differences bewteen how her firm and the Buenos Aires firm deal with information and knowledge. There are likely a greater number of key stakeholders who expect information directly in the Buenos Aires firm. The international project manager for the Berlin firm will have an advantage of local knowledge that is not expected to be shared with the Berlin headquarters. The stakeholders in this firm are likely to prefer a more hands-off approach to information and knowledge.

Understanding the structures and norms of the organizations participating in the project is critical, for it will tell the international project manager how communications function and, even more important, how power is distributed and used. The next step is to understand how the branch offices interpret the organizational policies and norms. On a recent project in India, the customer was a consortium that consisted of a local government agency, two global firms, and one transnational firm. The seller was a consortium that consisted of an international lender, two global firms, and one transnational firm. Then there was the value chain, the major participants of which included global, international, and transnational. On this project there were literally hundreds of firms with wildly different structures and norms. Obviously an international project manager cannot learn the details of each organization in such a complex value chain, but she or he must analyze the major participants and have the ability to read other participants quickly if the need arises.

An international lead project manager with high XLQ will leverage his or her skills. For example, knowing the problems that can be encountered due to the disparities in organizational structures, the lead project manager can take on the role of mentor and coach for the project managers in value chain organizations. In this way, the lead project manager can enhance the abilities of the other project managers. This role of mentor and coach binds people together and enables the lead project manager to leverage or amplify their skills through others. As it is said of successful negotiations, the result is a "win-win" proposition.

All of these international structures are, of course, overlaid onto the conventional heirarchical or matrix structures used by most organizations, as noted earlier. The international project manager must create a "micro" culture and organization to conduct the project. That is the next consideration.

2.4.1 Collaborative Project Enterprise

Experience shows that the initiating and planning for an international project requires approximately the same amount of time as does the actual execution and close-out. The structure of the temporary CPE and of the contract is established in the initiating and planning phases of the project life cycle, well before the execution of the project begins. Unfortunately, experience also shows that often the lead project manager is parachuted into the project, along with the performing organiztions, at the end of the planning phase. This forces the international project manager to design, build, and motivate the CPE concurrent with project execution. Such an approach is not recommended and will seriously hamper, if not kill, the lead project manager's ability to build a CPE and a team culture.

The contract structure and structure of the CPE are the seeds of open communications, team culture, and trust. These structures will determine how effective the participants can be in successfully completing the project. That is to say, the structures can limit the abilities of the parties to work together. The structure of a CPE is established, or approved, by the customer. On competitive publicly bid projects, the customer mandates in an autocratic manner the CPE in the general and special conditions. This structure may be implicit in the communication protocols described or explicit in the contractual relationships defined.

At the other extreme, on negotiated design-build projects, all of the participants will jointly engage in the design of the contract and CPE structure to some extent. That is not to imply that it will be well designed but rather that the approach is participatory rather than autocratic. The parties have an opportunity to decide what type of organization will be utilized on the project, how communications will be conducted, how knowledge will be shared, and what type of culture there will be for the project team.

There has been some work on CPEs, called temporary project organizations as they relate to project management including Mintzberg (1983); Hastings (1995); Brown and Duguid (1996); Toffler (1997); DeFillipi and Arthur (1998); Bigley and Roberts (2001); Turner and Müeller (2003); Jensen, Johansson et al. (2006); Winter, Smith et al. (2006); and Grisham and Srinivasan (2007). According to DeFillippi and Arthur, reputation, relationships, and heavy reliance on the value chain are essential needs for temporary project organizations. Grabher concluded: "The formation and operation of projects essentially relies on a societal infrastructure which is built on and around networks, localities, institutions and firms" (p. 211).

Brown and Duguid (1996) found that team members are enculturated by the telling of stories that are community appropriate. Turner and Müeller (2003) concluded that a project manager functions as the chief executive officer of a CPE and that the primary role of the lead project manager is to set goals and objectives and to motivate team members, not to focus on planning and execution. This view is more representative of international project mangement realities in the twenty-first century and argues for early involvement of the lead project manager in the project.

Jensen, Johansson et al. (2006) proposed a model for analyzing interactional uncertainty between organizations, which we have described as especially important. They found that "[i]f trust between project owner and project improves, the image of the project will certainly also improve. This may lead to changing conditions for and the position of the project" (p. 10). Trust is essential for the nurturing and growth of a CPE. Project team leadership is a fundamental requirement, as is the necessity for the international project manager to act as mentor and coach. The project manager must reinforce the culture of networking, imbue a culture of open dialogue and transparency, and sense problems that prevent organizations and people from operating effectively. Often, in our experience, the initiation and planning for a project is done without the participation of an experienced international project manager.

If a project is viewed as CPE, what are the considerations for developing a contract structure that will plant the seeds for the growth of a healthy team culture? We explore the answer to this question in Chapter 11. For now, recognize that the lead project manager needs to be involved at the initiating phase and needs to build trust.

There are enough anagrams out there for project managers, but we must add a couple in this book, though we promise to keep them to a minimum. A temporary (connotation is short term) project organization (could imply an organizational chart) is the terminology that has been used to describe the concept of how a project really functions. We would like to change the terms to better describe what we know works in practice. Thinking about a project as a temporary organization is an attitude and is an integral part of XLQ. To use a metaphor, it is the ability to cause the participants of a project to cry when the lead project manager reads the poetry of Rumi. We prefer then the term "collaborative project enterprise" (CPE) to "temporary project organization."

Common definitions for the terms will help frame the discussion. According to *Webster's Ninth New Collegiate Dictionary*:

Collaborate—to work jointly with others in an intellectual endeavor.

All projects, even those that are adversarial, require organizations to work jointly, and, arguably, all projects are intellectual endeavors. (Not all those who do IT projects may agree that this seems to be the case when it comes to the project charter scope description.) But the planning phase of a project is really building a virtual model of the product that the project will produce.

Collaborate—to cooperate with or willingly assist an enemy.

How many of us can move past our emotions and do this? It is difficult and a challenge for everyone. However, it is essential in leading international projects. For our purposes, think of "competitor" rather than "enemy" and the definition is a perfect fit. We devote much of the book to the discussion of this attitude and its facets.

Collaborate—to cooperate with an agency or instrumentality with which one is not immediately connected.

As we will see, most international projects cross over organizational boundaries, especially in this global economy and value chain environment. As with assisting competitors, this is a critical attitude on international projects.

Enterprise—a project or undertaking that is especially difficult, complicated or risky.

Not a bad description of most any international project and any international business.

Enterprise—readiness to engage in daring action.

We discuss this in detail during our discussions of XLQ as part of trust. By "daring action," think about assisting a competitor or cooperating across business boundaries. Imagine you go to your boss and say that you will be sharing information with a competitor—daring action.

Enterprise—a unit of economic organization or activity especially a business organization.

A project is certainly an economic entity from a customer's point of view, whether it is to increase revenues, reduce cost, or improve quality of life. Although each member of the organization may have a different time horizon and profit motive, each must attend to the costs incurred and make sure that they are reimbursed.

Enterprise—a systematic purposeful activity.

Project management is a profession that is based on process; this is one reason why it has become popular with international businesses. A process can be interntionalized, can be repeated, and can be enhanced. In this way, organizations can improve productivity and lower cost while keeping quality high. It is again a good definition of what a project should be.

Here is the *PMBOK* definition:

Project—A temporary endeavor undertaken to create a unique product, service, or result.

Here "temporary" does not mean short term but rather that the project has a start and a finish.

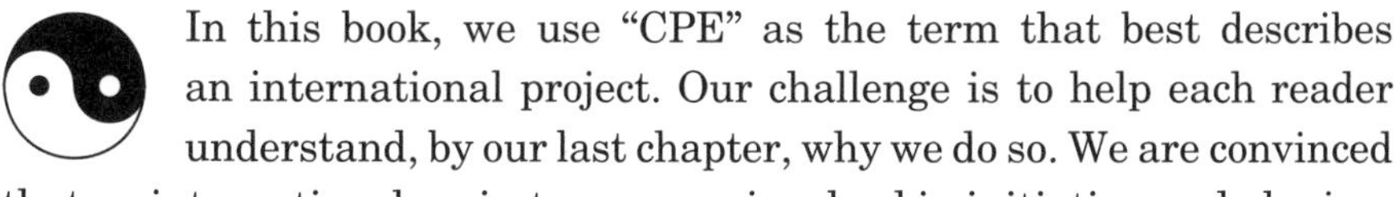

In this book, we use "CPE" as the term that best describes an international project. Our challenge is to help each reader understand, by our last chapter, why we do so. We are convinced that an international project manager, involved in initiating and planing, can double the chances for project success if she or he structures and leads the CPE to leverage the knowledge of the organizations and value chain in an open and trusting environment. An international project manager must inspire and engender trust in all participants.

Having created and nourished the creation of an effective CPE, the next challenge is to plan for change.

2.5 MANAGING CHANGE

One certainty on all projects is that they will change. Change comes from internal and external sources, from governmental regulations, from technology, from culture, and from force majeure, to mention a few possible sources. Imagine a topic, any topic, and it can likely be the source of change on a project. International projects are even more susceptible to change because of the complexity of the markets and the diversity in the politics, laws, customs and norms, culture, and so on. The project management literature focuses on the processes that should be in place to manage change. We discuss change in more detail in Chapter 6.

An international project manager must lead the CPE in embracing change. To do this, she or he must first gain the trust of the CPE. The lead project manager must provide the team with the opportunity to achieve and the confidence, sometimes, to fail. The protection felt by team members when they do fail creates the trust necessary to embrace change. If a team member fails, the project manager needs to help that person learn from the mistake, through mentoring and coaching. If the mistake impacts on the project, however, the project manager must take the responsibility for the error with the stakeholders.

Imagine a project manager who gathers her team and says that she wants, and encourages, each person to achieve beyond their expectations of themselves: strive for more, stretch your boundaries, take chances. These are the seeds of building a culture of change in a CPE. On international projects, members of different cultures will have different feelings about being offered such personal volition. In individualistic cultures—those where individuals prize their independence, such as the United States or parts of Europe—this will be well received. In collectivistic cultures, such as Japan or Korea, this may not be a welcome idea. However, there is far more to the story.

The GLOBE study (House, Hanges et al. 2004) asked people in different cultures what values and norms they actually had (as is) and what they thought they should have (should be), taking account of the globalization of attitudes. In this study, they asked people about "uncertainty avoidance"—the extent to which a people seek orderliness, consistency, structure, formalized procedures and laws in their daily lives. The study used a 7.0 scale with the higher the number, the greater the avoidance. For the as-is measurement, Switzerland scored 5.37 and Russia scored 2.88. For the should-be measurement, Switzerland scored 3.16 and Russia scored 5.07. Therefore, if you were the project manager with a team in each of these countries, you would have the Russians on board, for they are accustomed to dealing with uncertainty and expect it. They may have to show the way initially for the Swiss, who would likely push back, but underneath, both groups likely would think that a more central should-be position is preferable. The CPE, again, has its own culture that each participant can acknowledge and demonstrate to different degrees, at different times. The job of the project manager is to recognize the societal cultural differences and leverage the strength inherent in the team's diversity for the benefit of the CPE.

An international project manager must lead change, seek it, cherish it, manage it, celebrate it, and infect the project team with these attitudes. The Harvard Business School (2003) published a volume that addresses change from the perspective of business. The introduction to the book provides an excellent description:

> [C]hange is almost always disruptive and, at times, traumatic. Because of this, many people avoid it if they can. Nevertheless, change is part of organizational life and essential for progress. Those who know how to anticipate it, catalyze it, and manage it will find their careers, and their organizations, more satisfying and successful (p. 3).

The seven steps suggested for managing change from the Harvard book are as shown in Table 2.3.

The last chapter in the Harvard book addresses the issue of continuous change and whether people can accept it as a standard. The conclusion reached is that people should be able to handle regular change as long as it is explained properly, anticipated, handled in manageable doses, participatory rather than imposed, and made routine. Continuous change is normal in international project management, so it is certainly routine, and anticipated. The project manager must create a culture that accepts change. Then she or he must imbue this culture in the CPE, built on a foundation of trust.

Table 2.3 Change Management (Harvard Business School)

Author Step	Discussion
Mobilize energy and commitment through joint identification of business (PM) problems and their solutions.	According to Beers and Nohria (2000), the basic issue is to answer the question of "why" the change is being considered. For international project managers, the key is to demonstrate energy and commitment for change.
Develop a shared vision of how to organize and manage for competitiveness.	For the international project manager, the shared vision is critical for it builds buy-in as well as an understanding of why the change is needed.
Identify the leadership.	The identification process begins with the sponsor and the lead project manager, and includes champions for different aspects of the change when necessary.
Focus on results, not activities.	The activities will change themselves, so the project manager must keep the team focused on the "why" and the metrics for success.
Start change at the periphery, then let it spread to other units without pushing it from the top.	This can be done on large lengthy projects and is what we have called infection. Change can be led, not pushed, for at the end of the day, the team must come to accept and embrace the change. This goes to the issue of creating an environment where change is sought.
Institutionalize success through formal policies, systems, and structures.	The use of lessons learned, best practices, and building an environment that welcomes change. These help lead to success.
Monitor and adjust strategies in response to problems in the change process.	This is leading change.

The burden of manageable doses is the responsibility of the sponsor and of the project manager, and it should never be borne by the team members. One common mistake on international projects, and in many organizations, is to attempt to force team members to overcome changes without adjusting hours, days, or staffing levels—this is forced multi-tasking. The sponsor and the lead project manager must shoulder this burden, not the team. Imagine playing tennis with one opponent, then two, three, four.... At some point a person will be unable to hit all of the balls back. If each ball represents a change, you can see the idea. Grisham's second law of project management is "listen, question, think, then act." Even in a culture of change, people need time to do more than act; otherwise they will be unable to add their knowledge and value to the project. If there are six opponents hitting balls at one every day, this will lead to frustration, and worse.

One other view is useful in the discussion of change, that of Elizabeth Kübler-Ross (1969). She explored how people cope with losing a loved one and the five stages a person suffers through. Dr. Kübler-Ross provides a quotation for each stage from Tagore (2004):

1. Denial and isolation. "Man barricades against himself."
2. Anger. "We read the world wrong and say that it deceives us."
3. Bargaining. "The woodcutter's axe begged for its handle from the tree. The tree gave it."
4. Depression. "The world rushes on over the strings of the lingering heart making the music of sadness."
5. Acceptance. "I have my leave...a summons has come and I am ready for my journey."

Although the work of Dr. Kübler-Ross focused on terminally ill patients, her findings no doubt are familiar to all. Recognizing that these steps occur, a project manager can empathize with team members, and, when necessary, support them through difficult personal periods. If, for example, a person is stuck in the anger phase, the project manager may be able to coach him or her to see the options and perhaps move to the bargaining phase.

Look back at Figures 2.5, 2.6, and 2.7. Ignoring personalities and cultures, the potential for change is nearly infinite, and project managers cannot hope to avoid each one that might occur. Nor should she or he attempt to do so. The task for the international project manager is to consider the areas of greatest potential for change by understanding the structures of the organizations. In this way, she or he can demonstrate for the team that the changes are anticipated and normal and are not to be feared.

An international project manager must lead change by creating, again, the environment that enables people to cope with changes of all types. One of the key aspects of such an environment, and indeed of leadership itself, is trust. If the team sees constancy in the project manager's willingness to embrace change and understand that the project manager will involve and protect the team from the ravages of change, trust will develop and change will not be perceived as a threat. Creating such an environment is critical on international projects, and must be attended to each day by the project manager.

As we said, the project manager's ability to anticipate change is a key component of managing change. The ability to anticipate is based in large part on information and knowledge.

2.6 KNOWLEDGE MANAGEMENT

In our knowledge-based economy (Drucker 2000), businesses must harvest knowledge. There is an abundance of information available on the Web today, and this explicit information must be transformed into tacit knowledge before it becomes an asset for an organization or CPE (Wagner and Sternberg 1985; Lave and Wenger 1991; Spicer 1997; Earley and Ang 2003). The challenge is to first create an infrastructure that captures explicit information, and makes that information easily searchable—think of Google. Explicit information is just that, information, not knowledge. To create tacit knowledge, there must be a common context, and there must be an internalization of the explicit information—the "listen, question, think" portion of Grisham's second law of project management. Capturing and harvesting knowledge is a challenge; making knowledge easy to use is devilishly difficult. Buckminster Fuller, geodesic dome designer, once said that he could tell someone everything he knew given a number of days. That is knowing what you know.

In addition to the technical problems associated with knowledge management, there also the problem of willingness to share. Knowledge is valuable, and a competitive advantage. According to Sveiby (1998), one way to look at the value of knowledge is to compare the book value of an organization per share to its stock price; the difference is the organization's ability to translate knowledge into profit. We did a calculation for IBM in 2007, and using this approach, the value of knowledge came to over $630,000 per employee. That is a lot. Knowledge is a critical component for organizations and CPEs, regardless of how they attract, nurture, measure, harvest, and record it. The trouble is that people understand this linkage intuitively and are not always willing to give up their knowledge for free. We talk more about this throughout the book.

The early models of knowledge management proposed by Nonaka (Nonaka 1991; Nonaka and Takeuchi 1995) revolved around what has become to known as the SECI process. Knowledge management is seen in terms of a knowledge creating cycle that includes:

- Individuals sharing tacit knowledge through socialization (S)
- Articulating this either verbally or textually to make tacit knowledge explicit (E)
- Combining the explicit knowledge shared with existing explicit knowledge such as operating procedures, manuals, and information bases (C)
- Then, through reflection and embodying that reframed explicit knowledge, internalizing it so that it becomes refined tacit knowledge for many individuals across the organization (I)

Further, the need for a supportive environment created for the knowledge creation, transfer, and use was stressed and the concept of the importance of "ba," or a shared space where learning and knowledge work takes place was offered (Nonaka and Konno 1998; Nonaka 2001). This supportive environment is the responsibility of the lead project manager; earlier we called it a microculture.

Other concepts have been offered around the issue of providing a place, real or virtual, where people meet to create and share insights about knowledge. The community of practice (COP) was one such idea. It extended the original concepts of guilds and collections of workers that used these gatherings as a means of creating and sharing knowledge (Lave and Wenger 1991; Grisham and Walker 2005). This work started a new wave of knowledge management (KM) thought, and the COP ideas took hold with a widely cited book by Etienne Wenger (Wenger, McDermott et al. 2002) and others who also stressed the social side of KM (Sveiby 2001), the strategic side (Zack 1999), and the leadership side (Cavaleri and Seivert 2005). Interestingly, perceptions of the technology dominance of KM have been slipping as more and more KM thinkers explore how knowledge is created and used, with the view that KM initiatives should be 33% technology and 67% people oriented (Davenport and Prusak 1998). This fits back into the idea of structure for the CPE, trust in the project manager, and a CPE culture.

Information communication technology is seen primarily as an enabler, with knowledge management being driven by people (Walker 2005). The SECI model has been supplemented by refinements that show how individuals and groups attempt to manage knowledge. One such model, the 4-I's model put forward by Crossan, Lane, and White (1999), focuses on intuiting, interpreting, integrating, and institutionalizing in which knowledge flows forward from individuals to groups then to the entire firm to be recycled through feedback loops. This model was augmented by a better understanding of the role of power and influence and how organizational culture mediates this process (Lawrence, Mauws et al. 2005).

These views provide an internal perspective of KM in an organization and in our case a CPE. For internal projects, such as IT endeavors, the knowledge can be harvested as an extension of the KM systems that exist within an organization. The first level is knowledge that needs to be harvested as part of the life-blood of a organization—we refer to this as internal organizational knowledge. For projects that are performed externally to the organization, we call this project knowledge.

Project knowledge plays a key role in meeting the primary goals for an international project manager shown in Figure 2.8. In Chapter 10, we further define the considerations for customer satisfaction. For now,

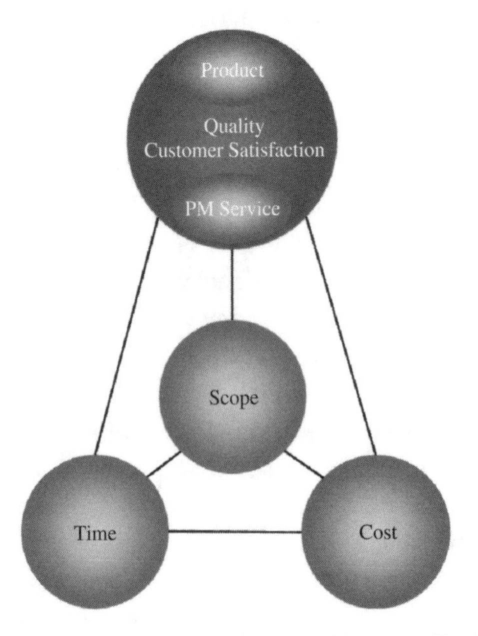

Figure 2.8 International Project Manager Goals

though, think about why a customer hires a consultant to provide services. Sometimes it is just outsourcing for people due to a temporary shortage of staff. More often, the reason is to procure expertise that the customer does not have. In such cases, knowledge transfer is a fundamental part of the project, although it is only implied. Consider this question: Would you prefer to have a customer who knows more about the technology and business than you, or a customer who knows little or nothing? If you chose the second case, you will certainly be in the knowledge transfer business, starting with the scope.

There is an interplay and integration between each of the goals shown in Figure 2.8, and knowledge is created as the project moves forward. The feedback loops and knowledge harvesting for internal organizational knowledge only require making the links from the project into the parent organization.

For the project knowledge, a knowledge management system must be developed that respects the privacy of each participating organization and protects intellectual property. Yet at the same time, the system must provide as much information and knowledge as is possible to the rest of the CPE. The project system must provide a way to enhance the project goals by the harvesting and sharing of project knowledge and knowledge that resides within each organization. We provide a few more references throughout the chapter to enable readers who are not familiar with the concept of KM a starting place to further your knowledge in this area. We also discuss one possible project architecture for KM in Chapter 12.

All of the XLQ attributes are necessary to harvest and use knowledge. Another critical consideration for KM is the structure of the contractual agreements, the subject of the next section.

2.7 INTERNATIONAL PROJECT STRUCTURES

As indicated in Section 2.3, globalization has changed how international projects are structured and implemented. As value chains become more pervasive and intricate, contract structures must consider the increased need for cross-cultural leadership, communications, and knowledge management. The selection of a contract structure is critical to the success of an international project, for it will affect the project's trust, communications, knowledge, and goals shown in Figure 2.8. We discuss this critical structure issue more fully in Chapter 11, but the next basic introduction will serve as a reference point.

Rigid contract structures—Fixed price affairs. They are rigid because the relationships between the parties are often fixed in advance, along with the communication channels. Fixed price contracts generally are used by public entities in a competitive bidding environment. At the extreme, they offer little communication between the bidder and the customer in advance of the tender. Ideally, contracts of this type are based on a design that is complete and accurate and on a contract that fully and accurately describes the customer's goals, objectives, and scope.

Flexible contract structures—Include cost plus, time and materials, and design-build. In these structures, communications begin early and continue through project close-out. On design-build contracts, the customer and seller (contracting entity or consultant) engage in dialogue from the product's conceptual phase into the operational phase. Flexible structures are often employed when the scope of the work is not sufficiently definable in advance of its execution. However, a common misconception is that flexible contracts diminish or eliminate the need for customer participation. In fact, they require more customer participation that does a rigid structure approach.

Collaborative structures—Include joint ventures, partnerships, consortiums, and alliances. In these structures, multiple parties perform the scope of the work and often can be both supplier and customer. The hallmark of collaborative structures is that they embody the need for more enduring and intimate relationships, with knowledge shared more openly. As with flexible structures, these structures are used when the scope of the work is not defined clearly in advance of the execution. In international markets, projects often require the blending of capabilities and assets that extend past the boundaries of a single organization.

For example, imagine that a nongovernmental organization (NGO) is to undertake an AIDS project. The customer might be a charitable organization, and the partners might be the local government, a pharmaceutical firm, and an NGO, such as like Médecins Sans Frontieres with people in country. Creating a contract scope to describe such an arrangement often requires as much time as does executing and closing out the project. More flexible and trusting contract structures such as this require more intimacy on the part of the various parties involved if the project is to be successful.

In our experience, this is the structure used more often in the international marketplace. Performing an international project that requires a convergence of needs, funding, technical expertise, value chain management, cost minimization, and local content concurrently requires more flexibility, and more participants. It also requires communications that are far more intimate than the other contract structures discussed.

Of course, there are innumerable variations on the general structures described. The key is to encourage transparency from the beginning, and a culture of knowledge sharing and appetite for change. Again, this requires all of the XLQ attributes.

2.8 FINANCING

The financing for international projects is accomplished in an array of ways. In international business and project management, there are an amazing variety of opportunities that range from venture capital to bonds. We discuss some of the categories of financing that is available to give our readers an appreciation of the options:

- Capital provided from cash on hand
- Bonds
- International financial organizations
- Conventional banks
- Investment banks
- Sovereign wealth funds

Capital provided by the customer from cash on hand is a source of funding. The customer may be a private organization with a strong balance sheet, a wealthy individual, a charitable trust, or an NGO. This is the easiest source of funding, as it involves funds that exist, are earmarked for the project, and reside in a fungible state. We once did a project for the government of Sir Bolkiah Hassanal, the Sultan of Brunei. This is a prime example of an organization with plenty of cash on hand.

Bonds can be issued by a government organization, or a private sector organization. Bonds are obligations denominated in a currency that have a risk rating, such as Moody's or Standard & Poor's, and are backed by a promise to repay with a designated amount of interest in a time frame. On international projects, the currency of the bond is quite important. For example, for a bond issued in Venezuelan Bolivars (VEB), the customer, assume they are Venezuelan, would pay the loan and interest in VEB. If the bond holders are located in Argentina, they will likely need to convert VEB into Argentinean pesos (ARS). If the VEB becomes weaker against the ARS, the bond holders will suffer. The reverse could put a financial burden on the Venezuelan customer. Therefore, a risk must be factored into the sale and purchase of bonds. Numerous organizations can arrange to float bonds.

International financial organizations are another source of funding. These come in a variety of types and finance a variety of projects globally. Here are a few:

- According to its Web page (www.ifc.org), the International Finance Corporation (IFC) "fosters sustainable economic growth in developing countries by financing private sector investment, mobilizing capital in the international financial markets, and providing advisory services to businesses and governments." It is are a part of the World Bank Group.
- According to its Web site (www.imf.org/external), the International Monetary Fund (IMF) "also lends to countries with balance of payments difficulties, to provide temporary financing and to support policies aimed at correcting the underlying problems; loans to low-income countries are also aimed especially at poverty reduction."
- According to its Web site (www.worldbank.org), the World Bank:

  > [I]s a vital source of financial and technical assistance to developing countries around the world. We are not a bank in the common sense. We are made up of two unique development institutions owned by 185 member countries—the International Bank for Reconstruction and Development (IBRD) and the International Development Association (IDA). Each institution plays a different but collaborative role to advance the vision of an inclusive and sustainable globalization. The IBRD focuses on middle income and creditworthy poor countries, while IDA focuses on the poorest countries in the world. Together we provide low-interest loans, interest-free credits and grants to developing countries for a wide array of purposes that include investments in education, health, public administration, infrastructure, financial and private sector development, agriculture, and environmental and natural resource management.

- The Asian Development Bank (ADB) "is an international development finance institution whose mission is to help its developing member countries reduce poverty and improve the quality of life of their people. Headquartered in Manila, and established in 1966, ADB is owned and financed by its 67 members, of which 48 are from the region and 19 are from other parts of the globe" (www.adb.org).

- The Inter-American Development Bank (IDB) Web site (www.iadb.org) explains:

  The IDB, established in 1959 to support the process of economic and social development in Latin America and the Caribbean, is the main source of multilateral financing in the region. The IDB Group provides solutions to development challenges by partnering with governments, companies and civil society organizations, thus reaching its clients ranging from central governments to city authorities and businesses. The IDB lends money and provides grants. With a triple-A rating, the Bank borrows in international markets at competitive rates. Hence, it can structure loans at competitive conditions for its clients in its 26 borrowing member countries.

- The African Development Bank (AfDB) Web site (www.afdb.org/en/home) explains:

  The African Development Bank (AfDB) Group's mission is to help reduce poverty, improve living conditions for Africans and mobilize resources for the continent's economic and social development. With this objective in mind, the institution aims at assisting African countries—individually and collectively—in their efforts to achieve sustainable economic development and social progress. Combating poverty is at the heart of the continent's efforts to attain sustainable economic growth. To this end, the Bank seeks to stimulate and mobilize internal and external resources to promote investments as well as provide its regional member countries with technical and financial assistance.

- According to the Web site (www.ebrd.com) of the European Bank for Reconstruction and Development (EBRD):

  The EBRD is an international financial institution that supports projects in 30 countries from central Europe to central Asia. Investing primarily in private sector clients whose needs cannot be fully met by the market, the Bank promotes entrepreneurship and fosters transition towards open and democratic market economies. The EBRD is the largest single investor in the region and also mobilizes significant foreign direct investment into its countries of operations. The Bank invests mainly in private enterprises, usually together with commercial partners. It provides project financing for the financial sector and the real economy, both new ventures and investments in existing companies. It also works with publicly-owned companies to support privatization, restructuring of state-owned firms and improvement of municipal services.

- According to its Web site (www.exim.gov):

  The Export-Import Bank of the United States (Ex-Im Bank) is the official export credit agency of the United States. Ex-Im Bank's mission is to assist in financing the export of U.S. goods and services to international markets. Ex-Im Bank enables U.S. companies—large and small—to turn export opportunities into real sales that help to maintain and create U.S. jobs and contribute to a stronger national economy. Ex-Im Bank does not compete with private sector lenders but provides export financing products that fill gaps in trade financing. We assume credit and country risks that the private sector is unable or unwilling to accept. We also help to level the playing field for U.S. exporters by matching the financing that other governments provide to their exporters. Ex-Im Bank provides working capital guarantees (pre-export financing); export credit insurance; and loan guarantees and direct loans (buyer financing). No transaction is too large or too small. On average, 85% of our transactions directly benefit U.S. small businesses.

- According to its Web site (www.jbic.go.jp/en), the mission of the Japan Bank for International Cooperation is:

  [T]o contribute to the sound development of the Japanese and international economy by conducting international finance operation in the following three fields: promoting overseas development and acquisition of strategically important natural resources to Japan, maintaining and improving the international competitiveness of Japanese industries, and responding to disruptions in financial order in the international economy.

- "Conventional" banks also provide financing. The following are the approximate 10 largest banks in 2009 with the estimated market values in billions of U.S. dollars (in parentheses). The financial crisis has caused some variation in the values:
  1. ICBC—China ($277.5)
  2. Construction—China ($136.7)
  3. Bank of China—China ($113.7)

4. HSBC—United Kingdom ($112.8)
5. JP Morgan—United States ($111.5)
6. Wells Fargo—United States ($91.5)
7. Bank of America—United States ($71.6)
8. Mitsubishi UFJ—Japan ($67.7)
9. Banco Santander—Spain ($53.0)
10. Citigroup—United States ($40.6)

Investment banks are another source of funding. The link to Wikipedia (http://en.wikipedia.org/wiki/List_of_investment_banks) provides a good listing of such institutions, and there are quite a number. Venture capital and private equity organizations (www.entrepreneur.com/vc100) also provide funding. Private equity organizations include such groups as the Carlyle Group, Kohlberg Kravis Roberts, Goldman Sachs Principal Investment Group, the Blackstone Group, and TPG Capital. Such organizations typically invest in companies and focus primarily on leveraged buyouts rather than venture capital—but not always.

Sovereign wealth funds (SWFs) represent excess liquidity in the governments of the respective countries. The largest SWFs are designated the super seven funds. As before, the amounts in parentheses are approximate market values in billions of U.S. dollars, as of 2009:

1. Abu Dhabi Investment Authority (ADIA) ($875)
2. Government Pension Fund of Norway ($350)
3. Government of Singapore Investment Corporation ($330)
4. Kuwait Investment Authority ($250)
5. China Investment Corporation ($200)
6. Singapore's Temasek Holdings ($159.2)
7. Stabilization Fund of the Russian Federation ($158)

These sources of financial assistance are by no means complete but will give readers a feeling for the types of options available internationally.

There are thousands of different structures for financial packages, and each is adjusted to the customer, the risk, the market conditions, and more. To give a sense of the creativity that is part of an international project, consider this scenario. A project is being planned for a new public health care system in Nigeria. The consortium that will operate the system is comprised of a U.K. multinational organization, a U.S. global organization, the Nigerian government, and a Chinese organization. The consortium that will design and execute the project is comprised of different divisions of the same group that will operate the facility. But this consortium has a completely different contract agreement for the design and execution phases.

The consortium for finance has the same organizations, with the addition of the Japan Bank for International Cooperation (JBIC), The JBIC is providing financing to promote Japanese products with a loan denominated in yen and paid in naira. The JBIC's share is 30%, and it is taking the currency risk for its portion of the funding. To satisfy the JBIC's need for security, the Nigerian government is guaranteeing the loan and is providing 10% of the funding. The loan likely would be paid from the revenues generated from operations, but it also could be paid from a public stock offering if the operations consortium were to go public.

In the design and execution consortium, the U.K. organization is providing 40% of the funding through a subsidiary group of banks and venture capitalists comprised of six other organizations. This group is providing financing in their respective currencies, and it is up to the U.K. organization to take the currency risk. The U.S. organization is providing 10% of the funding, which is comprised of 5% cash and 5% value added by discounts provided on equipment. The currency risk for this portion resides with the U.K. consortium partner. The Chinese organization is providing the remaining 10% of funding. Its portion is 5% from a Chinese bank, 2% cash from the organization, and the remaining 3% discounts on services performed. The currency risks are to be taken by the Nigerian government for this portion.

This idea of "sweat equity," or trading services for shares, is not uncommon, particularly in economic conditions where credit is in short supply.

Finance is an extremely interesting portion of international projects. It is diverse, creative, and risky. Lead international project managers need to have at least a rudimentary understanding of finance and markets so that they can anticipate potential problems.

2.9 SUMMARY

We have described the international environment and some of the reasons that international project managers must have a large set of skills and abilities. We described an international project manager's need for all attributes of cross-cultural leadership intelligence (XLQ). Then we discussed the international environment issues of globalization, sustainability, ethics and laws, value chains, virtual teams, CPEs, change management, knowledge management, and the structures of international projects.

An international project manager must be familiar with these topics and must have a hunger to learn for there is always more to know. In Chapter 3, we focus on the participants, life cycle, and contracting methods for international projects.

Chapter 3

Project Basics

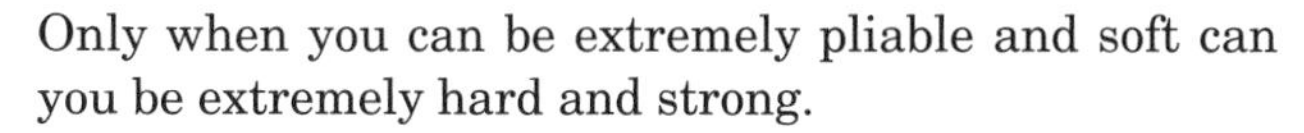

> Only when you can be extremely pliable and soft can you be extremely hard and strong.
>
> —*Zen proverb*

> Differences are not deficits to be changed and corrected, but gifts to be cherished and enjoyed.
>
> —*Samuel Paul*

In this chapter, we define some of the basic terms and definitions for leading international projects. We also discuss the project life cycle and the relationship between it and the different contract structures. First, however we need to explore the participants on a project and to introduce some basic terminology.

3.1 PARTICIPANTS

Based on project management standards, the definition of what a project is varies. Here are excerpts from the recognized international standards:

- According to the *Project Management Body of Knowledge* (*PMBOK*), Section 1.2.1, a project is a temporary endeavor undertaken to create a unique product, service, or result.
- According to the International Project Management Association (IPMA), Section 2.7.1, a project is a time and cost constrained operation to realize a set of defined deliverables up to quality standards and requirements.
- According to the Association of Project Managers (APM), Section 1.1, a project is a unique, transient endeavor undertaken to achieve a desired outcome.

This book uses a hybrid definition for an international project:

> International project—A unique, transient endeavor undertaken to create a unique product or service that utilizes resources from, or provides product or services in, more than one country.

The customer is an organization, or group of organizations, that determines or approves the time, budget, scope, quality and structure of the collaborative project enterprise (CPE). This determination or approval, of course, may be done in collaboration with the group providing the services for the project (more on that later), or it may be done unilaterally. The key goal on international projects is to ensure that the customer is satisfied with the product and the service. This book emphasizes repeatedly the critical importance of customer satisfaction; if the customer is not satisfied, the project will likely not be successful. It is also important to stress that a project is made up of two components: the product of the project (e.g., new power plant) and the project management service (e.g., scheduling, estimating, communications, etc.). Too often, the focus is on the product rather than on the service, to the detriment of the project.

Projects are unique endeavors in that they provide products that are physically different, utilize teams that have not worked together before, utilize new technology, or are located in a new geographical or political area. Firms may undertake multiple projects for a single customer that are similar to previous projects, but such projects still are unique because most often they include individuals who have not worked together before.

The transient nature of projects is what distinguishes them from ongoing business operations—projects have a beginning and an end. Information technology (IT) projects can include support for new Web platforms that have been designed, implemented, and tested as a project. The question is whether the maintenance (or warranty) is part of the project. Certainly, the warranty is part of the agreement with the customer, but often a group separate from the team that executes the project fulfills the warranty obligations. Utilizing our definition of "project," the warranty generally is considered part of the project, certainly from the customer's point of view. The seller, however, may have a different project manager for the warranty phase; this person may be part of operations, not projects. Some organizations use the project manager title to make the responsibilities of the position clearer to the customer. In fact, often the project manager for initiation is different from the project manager who executes the project.

Imagine you are a customer who wants to develop a new rail link from the Pacific to the Atlantic coast of South America. It is a long-term project and requires building and nurturing a multitude of relationships. You select a large global firm with a strong knowledge base and begin. The first project manager spends a year guiding the development of the financials, high-level planning for the project, building relationships and political contacts, structuring the contract(s), and learning the implied needs of the customer. Then comes the second project manager, who spends a year engaged in the detail planning. Then, when the project begins, the third project manager implements the project; last, the fourth attends to the operation and maintenance of the new rail link. You, as the customer, have been forced to educate four project managers and work through all of the interpersonal and political issues with each one. Most large customers will not tolerate such a revolving door, it is simply too much work and too disruptive. More on this shortly.

If customers are vague about their expectations for services, their desire for ongoing operations or warranty, or their desires about number of project managers in their scope document, the project manager should work to clarify these points with the customer. In Chapter 6 we discuss the issue of scope creep and suggest that one way to mitigate it is to ensure that everyone has a common understanding of what the customer expects.

3.1.1 International Project Manager

An international project manager is directly responsible to the sponsor for the successful completion of the organizations portion of a project, internally. An international lead project manager is responsible for the successful completion of the project to the customer and the participants. The standards define a project manager in these ways:

APM: Project manager—"the individual responsible and accountable for the successful delivery of the project" (p. 15). The focus is from within one organization.

IPMA: Project manager—"the professional specialist who plans and controls a project" (p. v). Again, the focus is within one organization.

PMBOK: Project manager—"the person assigned by the performing organization to achieve the project objectives" (p. 369). This is the same view as the IPMA.

Figure 3.1 shows one basic international CPE structure, with a project manager in each organization. As you can see, this structure

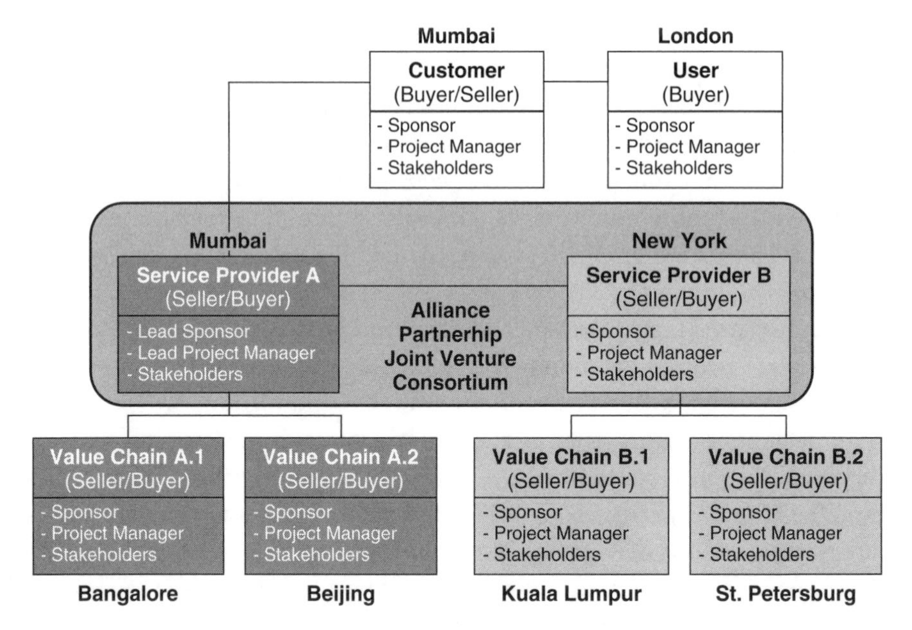

Figure 3.1 Basic International CPE

is problematic for the definitions provided in the standards. What does one call the lead project manager in the alliance organization if you use the *PMBOK*, APM, or the IPMA? We use the modifier "lead" to describe the project manager who has ultimate responsibility for the execution and successful completion of the project, not just for her organization, but for the CPE. That is because the "performing organization" has control over means and methods. The standard definitions work quite well in the context of each organization but not so well in the context of a typical international project. Therefore, the definition we use in this book is similar to the standards with regard to the project manager but expanded for the lead project manager. Our definitions are:

> International project manager—The person responsible to the sponsor for the successful completion of a project within one organization.
>
> International lead project manager—The person responsible to the customer for the successful completion of the project for the entire CPE.

The importance of this distinction on international projects is more than semantics. Proverbs offer a different way to consider the differences. In German, *Viele Hände bringt's gleich zu Ende* means "Many hands bring it quickly to a conclusion," or the more people involved, the quicker the work gets accomplished —as in building a wall. Leading a CPE also relates to the English proverb "Too many cooks spoil the broth," or multiple leaders will lead in multiple directions. It is our view that there must be a single person responsible for international CPEs. The lead international project manager has an obvious duty to her or his organization, a duty to the other organizations in the CPE, and overall responsibility to achieve the goals and objectives of the CPE. That does not mean that a customer or user must cede full authority to the lead project manager; rather, enough authority must be ceded to enable him or her to lead the CPE.

On international projects, the lead project manager must take, and demonstrate, responsibility for all of the organizations that are part of the CPE. This will result in conflicts of interest between what is good for the project manager's own firm and what is good for the CPE. It is an inescapable conundrum but one that can offer opportunities to leverage the diversity. Fortunately, if properly managed, it can increase the leadership stature of the lead project manager through the application of cross-cultural leadership intelligence (XLQ).

The lead project manager concept is a critical point, so allow us to repeat it using a metaphor. Picture a group of people standing on a knife-edge ridge in the Dolomites with a 300-meter vertical drop. Each person is tied to the adjacent person with two meters of slack in the rope. Each person has a razor and can cut the rope if he or she chooses to do so (think of the book *Into Thin Air*). If everyone keeps the rope connected, there is enough mass to catch the person who falls. If the lead project manager removes the razors, the people are safe, but the few near the person who falls may be frightened, and perhaps scraped up a bit, for they may have gone over the edge as well. The lead project manager could take the approach of focusing only on his or her company and seeing that there will be plenty of time to sever the rope if someone slips. Imagine the difference in attitude people would have toward the lead project manager at each extreme. Removing the razors builds trust, demonstrates empathy, and shows XLQ.

Having defined an international project and a lead international project, let us now discuss the other participants.

3.1.2 Stakeholders

As discussed, CPEs on international projects usually are made up of multiple organizations. Figure 3.1 shows that each participant has a sponsor, project manager, and stakeholders, although the titles may be different. It is important to keep this structure in mind for the definitions that follow. As noted in Chapter 2, many international projects are performed by an alliance, partnership, joint venture, or consortium. Often customers prefer to communicate contractually with only one lead firm, although in some cases all firms in the alliance, partnership, joint venture, or consortium may be required to sign the contract with customers. Sometimes the user may not desire an active role in the project and will communicate only through customers. It is within this context that we begin with a review of the definitions provided in the standards, starting with the sponsor.

APM: Sponsor—"The individual or body for whom the project is undertaken. The sponsor is the primary risk taker is, is responsible for the business case, and is responsible for the project" (p. 159).

IPMA: Sponsor—"The person responsible for the business case" (Section 3.06).

PMBOK: Sponsor—"The person or group that provides the financial resources in cash or in kind for the project and can turn to for support" (Section 2.2).

The definitions address internal sponsors, and all point to a higher level of authority within the organization. The sponsor must be in a position to help relocate political barricades, mobilize scarce resources, and make profit and loss decisions. Externally, the sponsor is the person who communicates with the sponsors from other organizations who have a similar status within their organizations. Our definition of a sponsor, which relies heavily on the APM definition, is presented next.

> Sponsor—A person in top-level management who is responsible for the project, who has the ability to commit organizational resources, to alter profit and loss metrics, and to remove obstacles.

The difficult issues on all projects involve money, and most international project managers simply do not have enough authority to resolve most of such issues that arise. On large complex international projects, it is extremely important to have a management-level sponsor who can make sizable financial commitments personally. As many who read this book will know, on a large project, the cost of delays in making a decision frequently dwarfs the actual cost of the change and, worse yet, ripples through the value chain. Lead project managers, for example, might have a limit of authority of perhaps $10,000. On a $400 million project, a change of $20,000 could easily result in delay costs of five times this amount. It is then necessary for the sponsor to provide higher levels of authority or to be available more frequently, and on short notice.

Stakeholders on projects are people who have a stake in the outcome. So certainly, a sponsor would be a stakeholder, as would be the project manager, the project team, and the rest of the value chain. Often there are key stakeholders that are not a part of the CPE. For example, a power plant project may cause noise, pollution, or property complaints from local citizens. These citizens have an interest in the project, because it is their community, but they may not have a spokesperson until a complaint arises. Experience has shown that there can be thousands of stakeholders on a project; one project in India comes to mind. In Chapter 5, we discuss the importance of determining the key stakeholders on a project and how to figure out their needs. For now, the standards define stakeholders in these ways:

APM: Stakeholder—"The organization of people who have an interest or role in the project or impacted by it" (p. 159).
IPMA: Stakeholder—"A party interested in the performance and/or success of the project, or who is constrained by it" (Section 1.2).
PMBOK: Stakeholder—"A person or organization that is actively involved in the project, or whose interests may be affected by the execution or completion of the project" (p. 376).

Next is the definition we use in this book.

> Stakeholder—A person or organization that has an interest in the project, or who could be impacted by it.

A ferry terminal constructed in the United States required dredging for the entrance channel. The spoils were discharged on land, and an endangered species was found in the area. A number of previously unidentified stakeholders shut the project down for quite some time until the issue was resolved to their satisfaction. Again, it is important to emphasize that there may be many stakeholders on a project, including some that are unknown to the CPE, until problems develop. An international project manager must consider the larger world within which we all work and the pressures that globalization brings. Organizations that strive for sustainable operations will be better trained to look at the world from a macro view and thus better able to adapt to these undefined stakeholders.

User is a term that is used in a variety of ways depending on the industry. Let us start with the definitions established by the standards:

APM: User—"The group of people who are intended to benefit from the project or operate the deliverable" (p. 163).

IPMA: User—"The group or person who will work with the results of a project and deliver the benefits" (Section 3.06).

PMBOK: User—"The person or organization that will use the project's product or service" (p. 378).

For example, consider an IT project for a bank that entails the creation of a new online banking service platform. Internal to the bank there is a sponsor, project manager, stakeholders, and users. The internal user could be the functional organizational unit that will generate revenues from the new IT business asset. There are also external users, sometimes also called customers and stakeholders, who will make use of the new service. The external users are likely to be polled to determine basic preferences. Although they do not participate in the project, they are stakeholders as well. The definition for user that we will use in this book is:

> User—A person or organization that will use the product of the project and that may be internal or external to project.

Last, but not least, is the customer. Customers can be internal or external, can be users as well, and are always the primary stakeholders—key stakeholders. Key stakeholders are the people or organizations that can significantly help or hinder the project. In Figure 3.1, the key stakeholders would be the sponsors of each organization. On a project that involves the environment, a key stakeholder might be the spokesperson of an environmental watch group, or a politician, that may or may not be a contractual party to the project.

As shown in Figure 2.8, customer satisfaction is the critical metric for a successful project. In practice, a project is seldom, if ever, successful where the customer was not satisfied with the product and the service. Many firms have adopted a customer-centric approach in all their activities, and they consider the customer to be *the* key stakeholder. Here are the definitions proposed by the standards for customer.

APM: Uses owner rather than customer. Owner—"Person or organization for which the project is ultimately undertaken and who will own, operate and benefit from the facility in the long term" (p. 146). As you can see, this definition is very close to that of user.

IPMA: A customer is an interested party (Section 1.2). IPMA emphasizes the importance of customer satisfaction, but it does not define what a customer is.

PMBOK: Customer—"The person or organization that will use the project's product or service or result" (p. 358). This definition also could describe a user.

A customer is most often the organization that contracts for the project. This contract can range from an informal internal agreement to a lengthy external written contract. The customer is the entity that requests the project from the performing organization(s), pays for the costs incurred, and accepts the project when completed. The definition of customer that we use in this book is:

> Customer—A person or organization that contracts for, pays for, and accepts the product of the project.

Experience has shown that the user may in fact provide the funding itself, in the form of cash or loans. However, the customer normally is the entity that actually pays for the services performed. Acceptance is an equally messy affair on many projects. The acceptance of the product sometimes is contingent on approval by both the customer and user. Even in such circumstances, the contractual relationship is usually between the customer and the service provider, without the user being a party to the agreement. In our experience, the user most often has a far more passive role in the project than does the customer, until acceptance.

Be cautious of users that have a say in the acceptance and are not direct signatories to the contracts. Scrutinize the contract language, and verify with the customer what exactly the role of the user is to be. Will the user review the scope and design? Will the user be involved in the testing or inspection of the services prior to acceptance? And perhaps most important, does the user provide any of the funding for the project? We do not mean to imply that users are trouble but rather that the lead project manager needs to engage them in the process, not wait until the end.

3.1.3 Society and Government

International projects can affect the lives of people in multiple countries and their governments. These groups may or may not be part of the CPE and thus can be overlooked in the planning process. One major exception to this is public-private partnerships (PPPs). PPPs are endeavors undertaken by government agencies in concert with the private sector; in these cases, the government is an integral part of the CPE. For example, if the government of Ecuador needs to construct a new water distribution system, it may choose to engage in a PPP. The PPP could provide the financing, design, construction, and operation of the facility for the benefit of the government and its people. The same project could be undertaken by the government by outsourcing each phase, but a PPP arrangement can provide a more integrated and efficient solution.

If an organization is rolling out a new cross-country e-business platform, planning must address local governmental rules and regulations. If the lead organization has a reputation of being predatory or of being responsible for worker abuses in other countries, citizen groups can protest the project. The citizens are stakeholders, always, but they may not have a spokesperson. If this is the case, the international lead project manager should anticipate problems, but the challenge of how to manage them if they occur remains.

In India, a consortium organization undertook to plan and build a small city to support the construction and operation of a large new utility project. The lead member of the organizational consortium had a reputation for being aggressive and bottom-line focused. The consortium undertook to engage the local communities, who had lived in the area for millennia but who were not organized. Help was provided to create a citizens' group and to engage them in the planning process so that they had a voice, albeit limited, in the effect the project had on their lives. Despite this, a scathing Human Rights Watch report was issued condemning the consortium's practices. Ultimately the facts mostly vindicated the consortium, but it took months of investigation, caused delays in the project, and contaminated the reputations of all consortium members. The consortium was proactive in creating a stakeholder spokesperson, yet the underlying culture of the lead member overshadowed the good intentions in the implementation. Organizations should always be good corporate citizens and treat the local people with the same respect and consideration that they would like to receive. However, that may not be enough.

Governments can provide a different type of challenge. Often, finding a single point of contact within a governmental organization is difficult if not impossible. Determining who has jurisdiction can be a daunting experience with no clear answers forthcoming. To mitigate this issue, CPEs often include a local expert who knows how the bureaucracy functions. More difficult yet is attempting to predict political changes. When elections occur, the stakeholders can change, and with them their attitudes and policies. A recent case was in Bolivia, when Juan Avo Morales was elected president on a popular platform of regaining the natural gas reserves of the country. After taking office, he fulfilled his promise. As a result, contracts made with global firms were "renegotiated." If you had been the international project manager on a gas pipeline project in Bolivia that started under the presidency of Gonzalo Sánchez de Lozada, you would have identified the key governmental stakeholders and planned accordingly for their participation. With President Morales, you must begin again.

Political risks must be managed, and the ongoing identification and participation of governmental stakeholders is essential on international projects. Having identified and defined the major participants in international projects, we now turn our attention to the project life cycle and contracting methods.

3.2 PROJECT LIFE CYCLE

The life cycle of a project starts when the project begins and ends when all of the obligations have been completed and the project has been accepted. The challenge, in practice, is to know when these two events actually have occurred. Before addressing that question, we first need to discuss the project life cycle from the perspectives of the standards. Figure 3.2

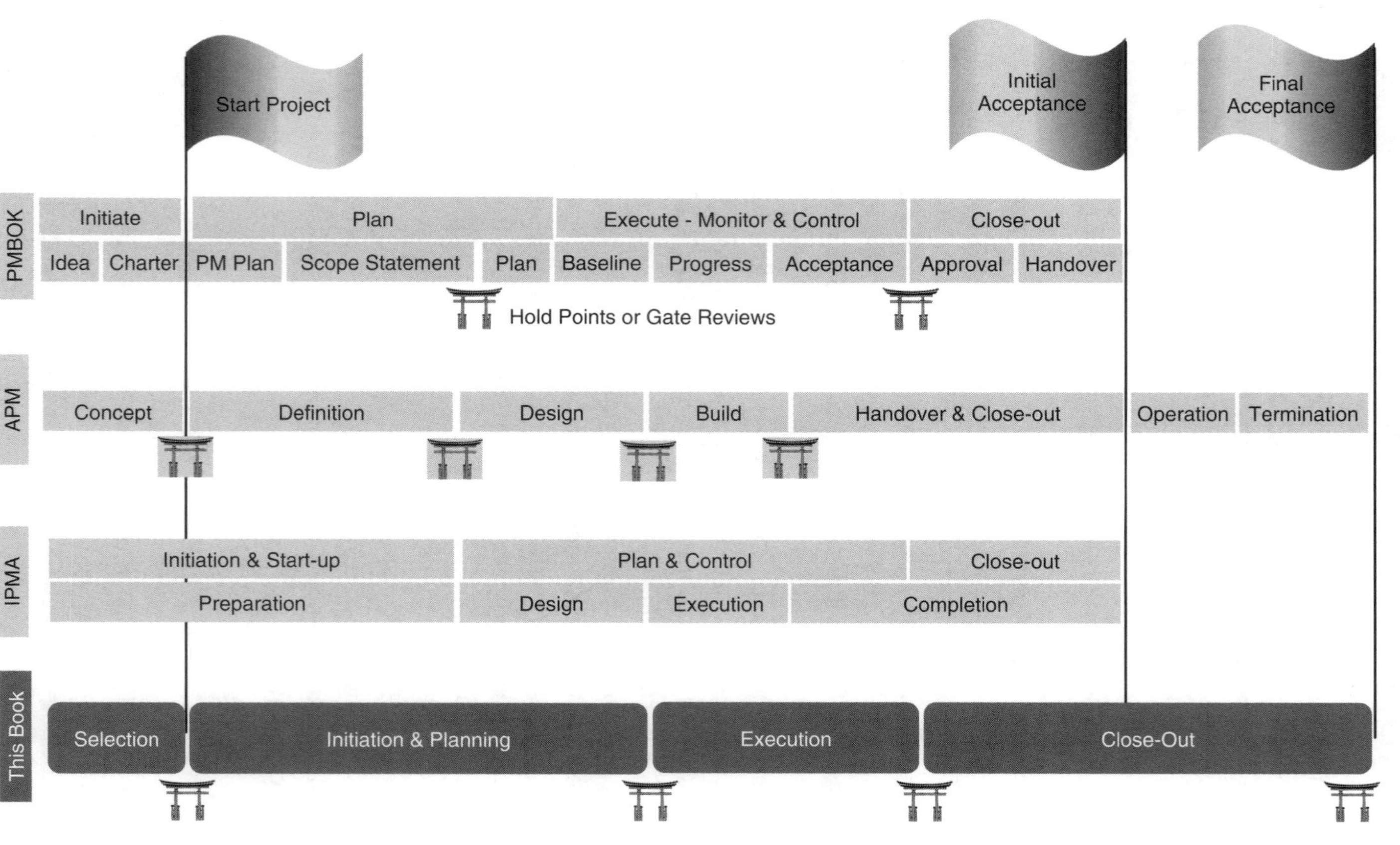

Figure 3.2 Project Life Cycle

provides a summary view of the standards and the definitions that are used in this book for easy comparison.

3.2.1 Selection

Selection of the project to be performed occurs in many ways in practice. First, consider an internal project undertaken for one office of a multinational firm to provide an IT product for another office. The need and service to be provided may be uncertain, and the actual project may in fact be to conduct a feasibility study. If the results of that study confirm the viability of the effort, then another project could be undertaken to perform the work. Each project would progress, using the *PMBOK* phases from Figure 3.2, through the selection, initiating and planning, execution and control, and close-out phases. The selection of the feasibility study would come from a business need; the selection for the IT project would come from the results of the feasibility study project.

What if the organization prefers to have continuity of the project manager participating in both and so elects to perform the feasibility study as the initial part of the IT project? The same terms still apply, with one slight clarification. The *PMBOK* utilizes the project charter as the vehicle for formally authorizing the project. In this case, the project charter would need to be finished at the beginning of the selection phase (i.e., the feasibility study) rather than at the end of the selection phase. The reason is funding. Even our example of a feasibility study, few projects are launched unless funding is available, and in most organizations that requires a bank account number, which requires a document like a project charter.

Now consider a different case. Imagine that you represent the lead global firm in the partnership shown in Figure 3.1. Your organization is trying to decide if it will tender for a power project in Spain that has a budget of US$100 million and will likely require partnering with another organization for work you cannot perform. This organization is not yet part of the team. Assume it will cost your firm $100,000 to tender and your success rate is 20%, but the margins on the project look promising. The sales force in the Madrid office will negotiate the contract, and its members look to you, the international project manager in the Tokyo office, to find a Japanese manufacturing firm to provide equipment and a Japanese export loan facility to finance a portion of the project. Many firms charge the costs of the tendering against a sales account and the time that an international project manager spends the helping with the bidding part of the tendering project. If your firm is successful in winning the project, the project is selected, a project charter created, and initiation and planning begins. There is often a hand-off at this point from the sales group to the projects group. In our experience, there can be a sales project manager and an execution project manager.

Look at this last example from the customer's viewpoint. Let us assume the customer performs a dozen projects per year, and the creation of the tender documents is a sign that the project has already been selected and moved into the planning phase, especially if it is to be tendered using a public competitive bid. If it is to be tendered using a negotiated design-build platform (more on this shortly), then it is likely to be moving into the initiation and planning phase.

The reason for providing these examples is to show that the terminology used by the different standards requires some practicality when applying it. Organizations use different terminology to describe their own internal business functions and the project life cycle. In Chapter 2, we introduced the idea of the CPE. As the lead international project manager, you must understand the concept of the generic project life cycle so that you can understand where other organizations perceive themselves to be in the cycle. This information provides invaluable insight into how each organization views the project, and it contributes to the XLQ dimensions of empathy, transformation, and trust.

Based on our experience, the selection phase is not part of the project itself, is likely not funded by project funds, and normally there has not been a project charter issued. For each organization shown in Figure 3.1, there would be a project charter, and it would open the gate to the next phase of initiation and planning. The Project Charter Format in Chapter 5 provides an outline of the information needed for a project charter. As the template indicates, the scope definition is very high level, and is not adequate for planning a project. The project charter opens the bank account for accumulating project costs.

We recommend that international projects have a collaborative structure and are negotiated with the involvement of all the major parties. It is a contract that is negotiated by the parties, all if possible, beginning with the formulation of the preliminary scope statement and ending with the completion of all obligations. This negotiation period is called "Selection" under the "This Book" view at the bottom of Figure 3.2. In a perfect world, the cost of the meetings and negotiations are subsumed into the project costs. In this way, there are no risks taken by the participant organizations; nor is there a need to hide or bury the costs in another account.

In such a case, it is recommended that the lead project manager issue a project charter for the entire CPE. This provides a formal beginning for the project, demonstrates a collaborative intent, starts building trust, initiates a team culture, and demonstrates leadership. This project charter would include input from each organization in the CPE. Then, after discussion, the lead project manager can codify this information into a single CPE project charter and distribute it to each organization. An organization's internal project charter can provide any profit metrics or other proprietary information required by the individual enterprises for their own use, but transparency should be the goal for all of the organizations. Boundaries for legal requirements, proprietary technology, and competitive information must be protected, of course, but the goal should be to maximize the knowledge shared. The structure of the CPE and the contract can describe what information is to be shared.

We describe the adversarial approach later in the book, as it is frequently used although we do not recommend it. There are numerous structures between the collaborative and adversarial extremes, and a discussion of all is not possible. For our purposes, here think of the project life cycle phases as time zones. Depending on the structure of the project, organizations can be in different time zones at different times. Our recommended approach places most of the organizations in the same time zone.

This phase of the life cycle creates a high-level view of the project; defines key stakeholders, the lead project manager, and other project managers; and creates the bank accounts to fund the project work. There is always some overlap in the phases. The point here is that to create a project charter, the CPE must have done some planning to describe the boundaries of the project.

3.2.2 Initiation and Planning

According to studies of IT projects, project failure rates range from 60% to a staggering 84%. Clearly, a business leader can get better odds at Monte Carlo. In our experience, the majority of projects that fail suffer from a lack of attention and commitment during the initiation and planning phases. Worse, they require the planning phase to be concurrent with

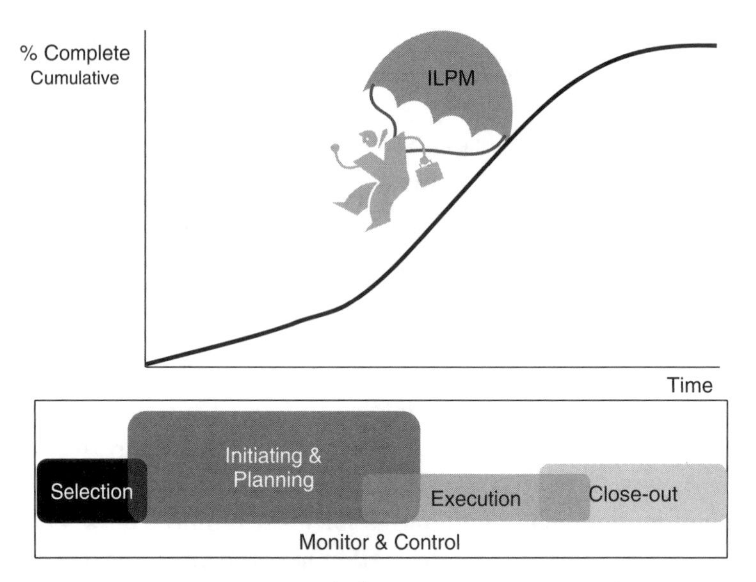

Figure 3.3 S Curve Excluding Warranty or Operations

the execution of the project—such as the lead project manager being named, as shown in Figure 3.3. We have informally polled hundreds of international students experienced in project management to check our experience with that in the industry. Asked what percent of a project manager's total effort on a project should go into initiation and planning phase, the response averaged approximately 70%. When asked how much participation they actually have during these phases, the answer is 30%.[1] In our experience, 70% to 80% of the success of a project is attributable to the work during the initiating and planning phase, and we strongly recommend that all international project managers focus accordingly. This is where international project managers can leverage their experience, add significant value, and realize a large return on their investment of time.

Certainly failures can and do occur in all phases due to resource constraints, force majeure, and the like. but the single most important phase is that of initiation and planning. Scope ambiguity is one of the largest causes of project failure, and we argue throughout the book that a thorough, professional job of building a work breakdown structure (WBS)

[1]KMPG Canada Survey (1997): 61% failure; the Chaos Report, Standish Group (1995): 16% of projects completed on time and under budget; the OASIG Study (1995): 70% of projects fail.

with the participation of the entire CPE, can minimize this problem. The phase where scope certainty is constructed is during initiation and planning.

The initiation and planning phase of the project could begin with a breakdown of a one-paragraph scope statement, or it could begin with the review and organization of detailed WBS. In either case, it will end with a battery of plans that will chart a course for the project. This book groups these two activities together for it is often impossible to know where one finishes and the other begins. Unless they represent gates for your internal processes, it is really not critical to distinguish between them in practice. We illustrate by walking through a common process.

Imagine that you are the lead project manager for a Swedish-based charitable nongovernmental organization that has decided to undertake a project for AIDS prevention and treatment project in Botswana. You have a project charter. Assume it is the first project of its kind on this continent, and few details are known. As customer satisfaction is the hallmark of a successful project, the lead project manager for the Swedish firm wants to begin with a well-defined, mutually agreed-upon scope of work. At the start of the initiation phase, the lead project manager should detail the components of the project and its deliverables. A working group can be assembled and a brainstorming session conducted with members of the CPE. Depending on the Swedish organization's experience base, this likely will produce a number of questions where expert advice is necessary. There will be questions about the country, the information and medicines, the logistics, metrics, and the like that require further information. The lead project manager can create a preliminary list of experts that are required, and begin soliciting cost proposals for the subject matter experts. This initial brainstorming session will also produce preliminary ideas and questions about quality, cost, time, risk, procurement, communications, resources, stakeholder expectations, and the attitude of the customer toward these dimensions.

The definition of scope can be addressed in a multitude of ways, from having a detailed technical specification provided by the customer, to having a performance specification that requires the contractor (the Swedish firm in our case) to detail out the technical requirements. The scope can be unchangeable if competitively bid via a lump-sum or fixed price contract (adversarial), or it can be negotiated in a cost-plus contract structure or a design-build approach (collaborative). Regardless of where the project falls on this continuum, the lead project manager must reconfirm the scope with the customer early, and often, to make certain a common understanding exists. Our experience, from the customer's perspective, is that having the contractor or seller question the scope early and often improves the probability of success. After all, frequently the reason for hiring a service provider is that the customer does not have the expertise internally to plan and execute the project, so an implied part of the effort is to have the contractor—the professional—scrutinize the scope for ambiguities and missing components. A more detailed discussion of scope is provided in Chapter 6.

To use a sailing metaphor, the initiation and planning are charting the course and laying in the provisions that will be needed for the voyage. Once you push away from the dock, the changes begin. As Charles Darwin said, "It is not the strongest of the species that survives, nor the most intelligent, but the one most responsive to change." The lead project manager can build a culture of change in the CPE, as we discuss especially in Chapter 4. It is the acceptance, perhaps embracing, of change that enables the CPE to handle whatever comes gracefully.

Now it is time to measure actual progress, compare it to the course (baseline) you have plotted, then adjust and plan again during the execution.

3.2.3 Execution and Control

One of the joys that all international project managers should experience is being parachuted into a project at the point of execution, as shown in Figure 3.3. The more a project manager knows about the goals and objectives of the project, the more time she or he has to develop the CPE and build trust and relationships, the greater the probability of success on a project. If a person with no experience and training in international project management does the initiation and planning, the risk of failure can easily exceed 50%. Even part-time involvement by an experienced international project manager is far preferable. On our ideal international project, the project managers are assigned at the time the CPE is created, at the time of the project charter.

To continue our sailing metaphor, once at sea, the lead project manager must take location readings, check them to the plan, and then decide if course adjustments are required. As with sailing, small deviations from the plan early in a project can lead to large mistakes at the end—such as planning to sail from San Francisco to Hong Kong and ending up in Singapore.

In Figure 3.3 the phases used in this book are shown at the bottom. The importance of each phase is reflected by the height of the boxes and the fact that the phases overlap unless a strict gate process (hold points) is utilized. Also please notice that monitoring and controlling is necessary during all phases of the project, even during the selection and close-out phases. Global markets change rapidly, so a project selected today may need to be placed on hold due to changing political or financial realities, then brought back to life later. International project managers should keep good records of the state of the planning and of the work done. It is simply too wasteful and inefficient not to. Record keeping is an essential part of any professional monitoring and control system, as are lessons learned, beginning at selection and running through close-out.

During the initiation and planning phase, the forming, storming, and norming (Tuckman and Jensen 1977) occur. This is the phase during which the lead project manager needs to create the CPE culture and to build long-term relationships, which are critically important to project success. To parachute a project manager—worse yet the lead project manager—into a project at the start of execution is a near-insurmountable challenge. Overcoming such a deficit would require a lead project manager with extremely high XLQ and many years of experience. It will also require luck.

A heuristic that we have used for years is that during the first 10% of a project, 15% of the work gets accomplished, and during the last 15% of the project, 10% of the work gets accomplished. Frequently we see a lead project manager, like the one in Figure 3.3, set the course after the initiation and planning and start measuring from there. In either case, the early period of measurement is critical, because often small deviations can be considered acceptable. For example, would a cumulative plan of 2.5% against an actual of 2.0% be considered acceptable? If one is using a control chart with limits of +/–10%, this is a deviation and would need to be investigated. Of course, this means that the organization must have a control system to determine costs within 90% accuracy.

The distance from San Francisco to Hong Kong is 11,116 kilometers (6907 miles). If the 0.5% shortfall goes uncorrected for four months, then the crew will end up in Singapore instead of Hong Kong. Early in a project, the slope of the S curve is increasing at an increasing rate, for those of you who like math. This means that the shortfalls have a magnification factor. To recover the loss of 2%, using our example, may be the equivalent of a month of progress. Early in the project, only so much work can be performed. We have seen this occur many times on projects. If, as is often the case, other organizations are introducing their project managers as well, as in Figure 3.3, it is easy to get off course before you know you are. For this reason, we cannot overemphasize the importance of having project managers involved in the planning and initiation phase. Our recommendation is to undertake these tasks in initiating and planning, *not* during execution:

- Creating the CPE culture
- Team building (forming, storming, norming)
- Designing and structuring the CPE
- Completing the learning curve for relationships, cultures, politics, financials, technology, and so on
- Completing the stakeholder analysis
- Completing and testing the project management plan

One actual example of Figure 3.3 is a project in Argentina. In order for the sales team to hand off the project to the implementation team, a kick-off meeting was held. The project manager for the sales portion (initiation and planning) of the project was transferring the project to the project manager for the execution portion. Later in the project, there was another hand-off at the start of operations to the warranty project manager early in the close-out phase. The project start date and finish date were included in the contract documents and were not negotiable. Clearly, the focus of the project manager for the sales portion was on getting the contract signed. This led to an unhappy customer who had to repeat all discussions with the new execution project manager and then with the warranty project manager. Worse yet, the project required a consortium partner to perform a significant portion of the local scope; this partner did not yet even know of the project. Imagine if a lead project manager had been involved in the sales effort, execution, and warranty. In fact, we actually changed our business model to provide cradle-to-grave services to select customers.

Establishing the course, or the baseline plan, is perhaps the easiest of the endeavors for an international project manager. The challenging work lies in the monitoring and control. Building a project plan for the CPE requires a lot of hard work. The monitoring and controlling extends over the life of the project and requires knowing where one is supposed to be, where one is, and where one is going. It is the plan-do-check-act concept that was originated by Dr. Walter Shewart and made popular by Dr. W. Edwards Deming. Later Dr. Deming changed the term to the plan-do-study-act cycle to better describe the process. It is also the basis of the *PMBOK*, the Japanese kaizen process, six sigma, and quality programs. We recommend that each and every reporting period be considered to be miniproject with a close-out. It is an opportunity to plan the project again with better information than what was known the previous period.

Returning to our sailing metaphor, knowing that the plan was to be at a certain latitude and longitude on a given day, having consumed so many liters of water and so many kilos of food—and of course there is the wine at one bottle per day. Imagine the trip to Hong Kong requires 30 days with an average of 370 kilometers per day. At the end of 10 days, a progress measurement is taken. The crew is not 3,700 kilometers into the journey but rather 2,700, and off course by 370 kilometers. The wine consumption has been 20 bottles, and there are only 15 days of food and water remaining. That is the monitoring part, or the "do" part of the Deming cycle.

For control, the "study" part of the Deming cycle, the captain must return to look at the knowledge gained. Did the plan anticipate such heavy winds and tides, did the crew consume too much food, why did the ship get so far off course, were our original estimates flawed, and so on? The captain must now plot a new course, ration the supplies, and lock the wine away. It is a return to the initiating and planning phase to rethink the journey. The reason we say this portion is the most challenging is because of the number of variables. On international projects, this part is complicated by the number of organizations, by those who eschew transparency, by the lag in determining the actual costs, and so much more.

In informal polling of global project management professional (PMP) candidates from a variety of industries and countries, there is a consensus that estimates and durations at the start of execution prove to be about 80% accurate at completion. This is about 5% more conservative than the 75% accuracy that field research done on North Sea platforms has shown (Emhjellen, Emhjellen et al. 2003). Project managers know this intuitively and plan by including contingencies in their project baselines. The problem in execution and control is that a lead project manager probably does not know which specific activities contain the contingency.

(More on this in Chapters 5 and 6.) Compasses must be adjusted for errors in true north before they are used. Unfortunately, an international project manager has numerous compasses, none of which can be adjusted for true north unless there is transparency. Therefore, on any project progress report, there is uncertainty as to whether the measurement itself is inaccurate or the baseline was imperfectly estimated. Using the contingency is easy enough, if there is one, but then what about the rest of the project? It is a problem on many projects knowing the actual location or progress achieved, it is an even larger problem projecting what changes need to be made to recover your course or project goal.

Experienced project managers are always skeptical of progress measurements, which are subject to the inherent errors in the baseline described earlier and to the accuracy of the value of progress achieved. For example, if an IT project requires half a million lines of code over a period of five months, how would one determine the percent complete at the end of the third week? One can apportion time or costs, use a statistical assessment, or make a subjective assessment. However, likely it is not possible to know if the lines are accurate or produce the desired result. Perfection is expensive; it is what economists call the cost of perfect information. Errors could be reduced if, for example, a $100 million project that is to be completed in two years could be planned with the longest work package being one hour and progress measurements taken every hour, but the cost of taking measurements every hour would be astronomical. On international projects, balance is important, as is the recognition that baselines are not perfect. There is another problem, of course: projecting completion.

We call the return on investment measurement the burn rate. By "burn rate," we mean the resources used to generate the progress achieved. So, for example, if the project spent US$100,000 but achieved only US$90,000 of value, the burn rate would be 0.90; said another way, you are 90% effective. This idea is the same as applied in the earned value (EV) method, but it is not without its challenges. To use the EV method properly requires knowing the actual cost of the work, and some physical measure of work units. Chapter 7 presents a full discussion of the EV method and the attendant challenges. Returning to our sailing metaphor, the challenge again is to determine where you are, what resources (food, water, wine) you consumed to get there, and of course where you are going. Recognizing that baselines are not perfect and that the burn rate experienced is knowledge about your specific project are both critical components of knowing the status of the project. At each progress update, you need to determine if the baseline needs to be changed or if course corrections will enable you to complete the project successfully. We used to say of projects in Saudi Arabia that the negotiations began once the contract was signed. Similarly, the difficult work for an international project manager is in monitoring and controlling, especially after the boat leaves the dock.

The execution phase is not the time for establishing project baselines, as we indicated. This is the time to cater to the customer, to nurture the CPE, and to anticipate the changes. It is a time to measure the pulse of the project team, to devote attention to the CPE members, to diligently test and check that the communications channels are open and effective, and to address the changes enthusiastically. This is the time to use the XLQ skills to their maximum advantage. It is also a time to look for opportunities to engender a routine of acceptance or partial close-outs.

3.2.4 Close-out

On a project in Brunei where we were the service provider, we had been dispatched to close out the project and to retrieve the retention withheld. We were the one of numerous project managers assigned to the project, which had been placed into operation eight years earlier and on which the warranty had expired. Suffice it to say that there were more than a few lessons learned on this project! It is a stark example of how things can go wrong when a routine of acceptance is not built up over the course of a project. The customer was incredibly patient and understanding, despite the frustration felt. We had in fact misplaced the project records and had to rely on the customer's files to research issues. How much confidence would that inspire?

When we were the project manager for the customer on other projects, we always wanted to be involved in the project and to be educated and consulted as the work progressed. We preferred to accept work in packets. Each reporting period was like a mini–close-out, and at the end of the project, completion was only one reporting period away. We wanted to be involved enough to monitor our investment and to feel that we could trust the service providers. However, we did not want our trust abused. Some service providers attempt to extrapolate a partial acceptance into implied acceptance of all associated work and scope. Fortunately, only a minority of service providers do so.

On another project in the Philippines, the customer had developed a specification where the definition of overall project acceptance was simply unintelligible. It was clear that the customer did not understand the end product, was a first-time buyer, and intended to engage in a contentious relationship. On this project, we were involved during the selection phase to support the negotiations. As discussions ensued, it became apparent that the customer had an end user that was an investor, not an operator. Knowing this, we were able to suggest alternative contract language that would hit the user's financial concerns while defining what acceptance meant in a technically cogent manner. It is extremely important to have a clear and unambiguous definition of what completion means.

Another major concern for international project managers and arguments for routine of acceptance are punch lists or deficiency lists. So-called scope creep or, as we like to call it, death by a thousand cuts[2] may occur anytime during the project life cycle, including close-out. On a large prison project, the deficiency list ran over 2,000 items. About 30% of these were legitimate items that needed some corrective work, about 40% were items where the work was not yet completed, and the remaining were a mixture of enhancements and added scope. We encouraged the use of routine acceptance and a rolling deficiency list, so the 40% category could be addressed on a periodic basis. The last 30% of "scope creep" was, and often is, a matter of a customer not understanding enough about the details of the project until the work is actually nearing completion. It is easier to see what is missing and what was not anticipated when the product is visible. A lead international project manager must continually try to flush out scope issues, so that such items can be addressed during the project rather than at close-out. The 30% of legitimate items that need to be corrected is a blessing for international project managers who wish to use mini–close-outs, for it provides critical feedback and knowledge of what is working and what is not. On international projects, lessons learned should not, and cannot, wait until acceptance.

On a large project in Thailand, we had a planned duration of five years and monthly reporting requirements. We had numerous occasions where firms in our value chain were simply not performing, and other occasions where the customer was not fulfilling its obligations. Getting this information back into our organization and back into the CPE was essential for course adjustments. Such lessons learned are too valuable

[2]Courtesy of Dr. William Ma.

to wait until the end of a project; they must be harvested and shared with each progress report. Waiting until the end of a project for lessons learned is not an approach that will provide success on international projects, especially at the program level. Lessons learned represent knowledge, and knowledge is the second greatest asset of an organization; people who have it are the first.

3.2.5 Warranty and Operations

On most projects, warranty obligations will last for months or years after the customer has accepted the project. On PPP projects, there are added obligations such as operation and maintenance of the product. As we indicated earlier, there is normally a hand-off (transfer of responsibility) from the lead project manager to an operations manager for the long-term use of the product. Although we prefer the continuity of a single lead project manager, many international project managers do not have the necessary training or expertise to operate the product. If this is the case, we recommend that the transition is a long one, perhaps six months. This permits the customer to learn to work with a new person but not be responsible for educating her or him about the project history. It also provides time to make certain everything is in good working order and that all issues have been closed out to the customer's satisfaction

Having explored the project life cycle, we now turn our attention to the structure of contracts and what impact they have on the CPE.

3.3 CONTRACTING ENVIRONMENTS

Customers utilize a multitude of contracting methods on international projects to allocate risks, establish their level of perceived participation, and seek out the most efficient use of their resources. Figure 3.4 shows four of the basic contract environments employed. Each of these environments includes a pricing methodology that ranges from firm fixed price to cost plus. As discussed earlier, the method selected will dictate the structure of the CPE, the communication networks created, the culture of the CPE, and the level of trust between the participants. While each method has its advantages and disadvantages, in our experience, the CPE structure is one of the critical considerations in structuring a successful project. We discuss this issue in detail in Chapter 11.

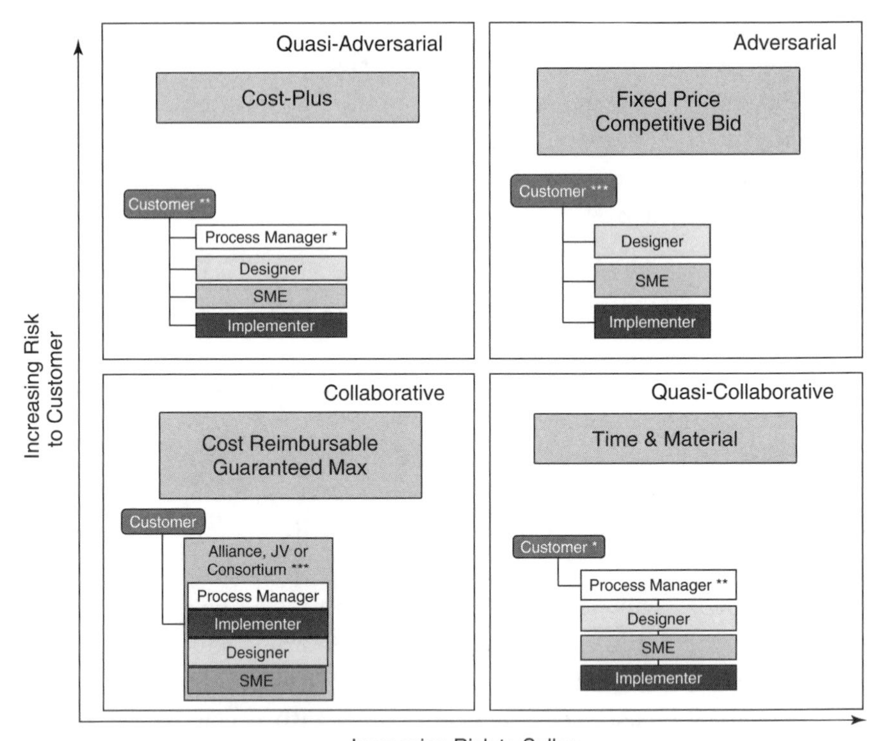

Figure 3.4 Contract Environments

In Figure 3.4, the participants are the customer, which is the entity that is funding the project and may be the end user of the product of the project. The process manager, usually a project management organization, is the organization that has been hired to manage the project. The difference between quasi-adversarial and quasi-collaborative is the amount of responsibility, authority, and risk that this managing organization assumes. The designer may be an IT consultant for system architecture, an architect for a building project, or the like. The subject matter expert (SME) is a professional whose knowledge is needed to design, manage, or implement the project—a country expert, for example. The implementer is the organization that performs the work itself. Of course, any of these participants can be individual organizations, alliances, joint ventures (JVs), consortiums, or value chains.

The conventional method used is the adversarial environment, which relies on direct individual contract relationships between the customer and the major participants. In this case, customers should accept the fact that they will have to devote consistent time and effort to the project and will have to serve as the de facto lead project manager for the CPE. (All of the participants together form the CPE in all cases.) The asterisks (*) indicate the proportion of the lead that the participants will in fact shoulder, in general. Customers who do not have the staff, the expertise, or the time but who wish to play an active role often choose the quasi-adversarial environment. Customers who want to shed most of the risk and have a less active role in the project sometimes use the quasi-collaborative environment. As the figure shows, there is a relationship between the risk and the structure of the contract; we delve into this beginning in Chapter 8.

Customers who want little or no involvement in the project and who want to take little risk frequently select the collaborative option—this sometimes is called design build. This method is used often when customers have only a vague idea of the scope of the work but needs for product to be completed quickly. In effect, customers contract out the scope of the initiation and planning that they would perform for the conventional structured project. Some customers consider the structure in the lower left quadrant to be risk free, but this is simply not the reality. The alliance, JV, or consortium does take most of the project risks, but the risk of this group misunderstanding the scope can be significant. If this approach is used, customers must devote time to the project to ensure that there is a common understanding of the scope. However, here the customers' role is as scope advisor rather than of lead project manager, as is the case with the approach in the upper right quadrant.

Trust, effective and open communications, and the willingness to share knowledge are critical to the health and success of all projects. Each of the methods shown in Figure 3.4 comes with inherent barriers (Grisham and Srinivasan 2007) to each of these critical components. Although the wording of the general and special conditions of the contract can mitigate the challenges to some extent, the structure itself establishes the barriers. To illustrate the point, look at the approach in the collaborative environment of the figure. There is only one communication channel external to the alliance, JV, or consortium: the customer. Now compare that to the adversarial environment, where there are three participants with direct responsibility and lines of communications with the customer but none with the other participants. Many international customers prefer to take have a single entity responsible for a project and thus often vie for some version of the collaborative environment. In our experience, this is the best environment to use as a foundation for international projects. When the customer is a governmental agency, there are often legal restrictions to using this approach, but that does not mean that they are mutually exclusive. We discuss this issue in more detail in Chapter 11.

In our experience, the structure of the CPE dictated by the contracting approach is an aspect of international project management that requires far more attention than it has received in practice, in the literature, or in the standards. We discuss in more detail the communication and knowledge-sharing aspects in Chapter 4, the procurement techniques in Chapter 11, and the risk in Chapter 8.

3.4 PUBLIC-PRIVATE PARTNERSHIPS

Imagine that the government of Pakistan needs to improve its infrastructure, schools, and hospitals and that public funding is not available. The government decides to approach the Asian Development Bank (ADB) for financial assistance. The government and bank might come to a preliminary agreement on funding and prepare a high-level plan to sketch out the time horizon, financial requirements, and technical support that will be needed. They then seek out other partners who can provide additional funding, expertise, and risk management. Generally these three things—risk transfer, funding, and technical expertise—are the reasons for PPPs.

The Pakistani government and the ADB bring on an additional organization to help with the program's overall planning and organization and to parcel out the projects within the program. Imagine the program includes 10 schools, 10 hospitals, and 100 kilometers of toll roads to connect a few smaller cities. The schools and hospitals are to be designed and constructed by a joint venture, formed for this specific program, made up of a Singaporean design and project management organization, a Chinese construction organization, and a local labor broker and material organization in Pakistan. The program requires additional funding beyond the ADB commitment, so a consortium of financial organizations comes together from London, Dubai, and Tokyo. Then an alliance is formed between a Pakistani, a Portuguese, and a Swiss organization to operate the assets before turning them over to the government. This amalgamation of organizations is then bound together with a PPP agreement.

The example is just one of the infinite number of ways to construct an international PPP. Historically most PPPs have been used in the transportation sector, but today development banks, such as the ADB, are also using the technique. At its best, a PPP is constructed through negotiations among the primary participants. Primary participants in our example would be the Pakistani government, the ADB, the joint venture, the consortium, and the alliance. Most public entities and development banks have strict policies about negotiated agreements, but the transparency and responsibility for public funds can be attended to in a number of ways. The critical component is that the potential partnership organizations must discuss the project, or in our case program, in sufficient detail to lead and manage it effectively.

As you can see from our example, there is so much diversity and so much complexity, and there are so many risks that organizations are not well served by attempting to create a PPP in an adversarial environment and by relying on competitive bidding, or low price wins. The devil, as they say, is in the details, and those require discussion.

If PPPs are formed in a collaborative environment, they represent perhaps the purest form of a CPE, which we describe in the next chapter. Ideally, they are formed by a group of organizations that can bring expertise, politics, and finance to bear on projects needed by the public.

There is a great deal of literature available on PPPs. A few works from 2009 include those by: Blaken and Dewulf; Fitzgerald and Duffield; Grisham and Srinivasan; Ismail, Takim et al.; Roumboutsos and Chiara; and Soliño and Vassallo. The paper by Ismail, Takim et al. provides a good tabular comparison of value for money (VFM) variables considered in the United Kingdom, Australia, and Japan. The authors describe VFM as "the optimum combination of whole-life costs, benefits, risks, and quality to meet user requirements at the lowest possible price" (p. 57).

The Romboutsos and Chiara paper provides a method that is very similar to the one presented later in this book for building a CPE plan. They base their approach on the well-known strength, opportunity, weakness, and threat (SWOT) model. In their approach, the public organization does a SWOT analysis on its portion of the project; private organizations do a SWOT analysis on their portion; and then these are combined into a single SWOT for the PPP. We take a similar approach by building a CPE project plan from the bottom of the value chain up. Romboutos and Chiara also describe what they call a political, economic, social, and technological analysis (PEST) and provide some useful metrics. We recommend that such an analysis be done on all international projects, regardless if they are collaborative PPP or adversarial. Here are a few dimensions to provide the concept:

- Political factor: government support, political stability, environmental concerns
- Economic factor: public budgetary constraints, low cost of money, investors seeking safe returns
- Technology factor: innovation, performance output, product life cycle
- Social factor: transparency, employment, public sector entrepreneurship.

For those of you looking for a case study, the Soliño and Vassallo work discusses transportation for the Madrid-Barajas airport.

The approach, environment, and metrics for PPPs provide useful lessons in developing and leading international projects that do not involve governmental organizations. In a collaborative environment, organizations that have the expertise necessary discuss the customer's long-term needs with the customer in active participation. These goals are then translated into more detailed plans and processes, and all of the primary organizations involved develop the plan for the project as they make the agreements necessary to create it. In other words, the lead project manager presides over the team building, CPE culture, and planning. This is the perfect-world scenario, but it is the one to strive for whenever the opportunity presents itself.

In this book, when we point to PPPs, we are suggesting this perfect-world goal as a benchmark. Now let us look at the most important aspect of international project management: people. This aspect of international project management is absolutely critical to success. It begins on the first day. If done properly, the benefits will last long after the project has been completed. Leadership builds long-term respect and relationships.

Chapter 4

Leading Diversity (Human Relations and Communications)

When we see men of worth, we should think of equaling them; when we see men of contrary character, we should turn inwards and examine ourselves.

—*Confucius*

Each man is good in the sight of the Great Spirit. It is not necessary for eagles to be crows.

—*Sitting Bull*

To be good is noble; but to show others how to be good is nobler and no trouble.

—*Mark Twain*

If you are neutral in situations of injustice, you have chosen the side of the oppressor. If an elephant has its foot on the tail of a mouse and you say that you are neutral, the mouse will not appreciate your neutrality.

—*Bishop Desmond Tutu*

Experience is the child of Thought, and Thought is the child of Action. We cannot learn men from books.

—*Benjamin Disraeli*

4.1 CROSS-CULTURAL LEADERSHIP INTELLIGENCE

Most projects today involve cultural diversity—personal, societal, group, and business—even those conducted within the borders of a single country. Immigration has seen to this. Projects that cross country boundaries, or international projects, certainly do. In a globalized world of outsourced goods and services, many projects take full advantage of competition on price and quality. The demand for project managers with cross-cultural leadership skills is increasing rapidly, and we saw a need for research to harvest the knowledge that had accumulated over the years on culture and on leadership into an integrated model. Our interest was to explore connections between theory and practical experience in numerous countries and cultures. This became the foundation for our thesis. The hypothesis is that there are cross-cultural leadership intelligence (XLQ) attributes that are effective, regardless of culture. The next sections provide a brief overview of the findings (Grisham 2009).

4.1.1 XLQ Model

The XLQ model is depicted as a wheel (see Figure 4.1), to imply movement, change, and relationship among the various components. The model should be understood to point in a direction, like the historical *quibla* compasses that people use to face in the direction of Mecca to pray. In early history, the compasses were not as precise as the global positioning systems in use today, but they were accurate. Likewise, the blend of attributes shown in the wheel is so complex, and the cultures so diverse, that they provide guidance, not strict standards. The complexity of the world in which we live demands that lead project managers not only have these skills but, more important, know when to use which ones and in what measures. Lead project managers *must* be able to adapt to their environment.

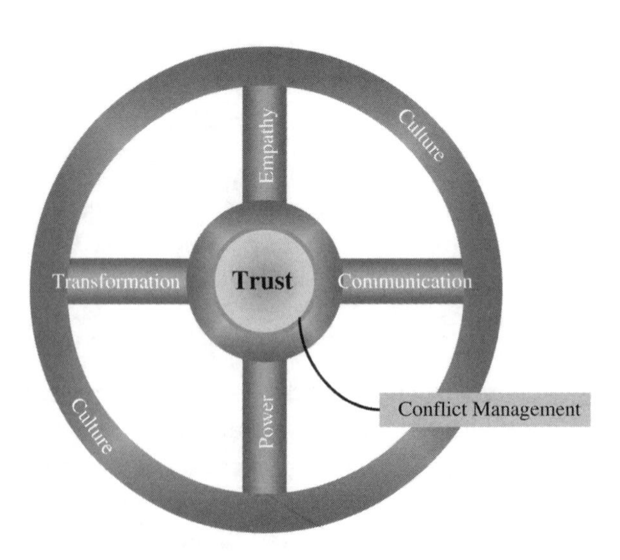

Figure 4.1 XLQ

Leadership can, we believe, be learned, and the model proposes that leadership can be displayed at different levels. For example, if one is weak in communications, the other spokes of the wheel and the rim of culture can offset the weakness. So leadership is not an either/or proposition but rather a spectrum of possibilities. Similarly, some attributes of the model will need to be emphasized, some deemphasized, depending on the personalities, and cultures present in the collaborative project enterprise (CPE). However, a wheel without a rim will not go far, so the need for cultural intelligence (CQ, as it is called) is essential. Trust is the hub of the wheel, and the lubricant for the wheel is conflict management. The spokes are empathy, communication, power, and transformation. The rim of the wheel is culture.

First, we need to define what we mean by "leadership." Our definition is "the ability to inspire the desire to follow and to inspire achievement beyond expectations." Thus, the people who follow define a leader. The second part of the definition is a combination of the expectations of the people who choose to follow about themselves and the expectations of the leader about the followers. International projects change, and a leader must create a CPE culture that enables people to strive to achieve beyond their own expectations of themselves, and fail safely. We are not suggesting that the lead project manager should accept failure lightly; what we are saying is that people must feel safe enough to stretch beyond their capabilities, knowing failure means knowledge not unemployment. This is one way to inculcate the acceptance of change into a CPE. However, permitting people to fail safely does much more; it also builds trust in the leader, demonstrates empathy for others, and displays transformation skills.

For culture, we prefer the definition of anthropologist Margaret Mead (1955), who defined it as "a body of learned behavior, a collection of beliefs, habits and traditions, shared by a group of people and successively learned by people who enter the society." For the word "society" in this definition, substitute "CPE." As we said earlier, there are personal, societal, business, and CPE cultures. The CPE culture is an organism. The lead project manager plants the seeds and nurtures the plant. In our view, cultures are organic in nature and, like a Japanese garden, require pruning and respectful attention. A wealth of literature is available on cultural characteristics; two works that we strongly recommend for further reading are Hofstede (1984) and the GLOBE Survey (House, Hanges et al. 2004). In our work on XLQ, we connected the hypothesis back to the GLOBE survey in our testing.

Our testing of the hypothesis utilized the Delphi technique and was accomplished with a group of international experts from the private sector, the public sector, nongovernmental organizations, and academia. For purposes of the testing, an expert was defined as (1) a person that has at least 20 years of practical experience working in an international/multicultural environment, in any industry; or (2) a person who has an advanced degree in leadership or cross-cultural studies with over 20 years of research, teaching, publication experience, or a combination of all. The demographics of the Delphi panel are provided in Table 4.1.

The cultural "clusters" shown in the table provided a consistent means of connecting our research back to the Global Leadership and Organizational Behavior Effectiveness Research Program (GLOBE) study clusters. Our research first looked at leadership, then at culture. We cover the XLQ dimensions soon, but first let us look at culture. The GLOBE survey provides nine dimensions of culture:

1. *Uncertainty avoidance.* Extent to which people strive to avoid uncertainty by relying on social norms, rituals, and bureaucratic practices
2. *Power distance.* Degree to which people expect and agree that power would be stratified and concentrated at high levels of organizations
3. *Institutional collectivism.* Degree to which society and organizations encourage and reward collective distribution of resources
4. *Group collectivism.* Degree to which individuals express pride, loyalty and cohesiveness in their organizations and families
5. *Gender egalitarianism.* Degree to which societies and organizations promote gender equality
6. *Assertiveness.* Degree to which an individual expresses assertive, confrontational, or aggressive behavior in organizations and society
7. *Future orientation.* Degree to which individuals engage in future activities, such as planning and postponing collective gratification
8. *Performance orientation.* Degree to which the society or organization rewards performance and excellence
9. *Humane orientation.* Degree to which societies and organizations reward fairness, altruism, friendliness, generosity, and caring for others

The GLOBE study was published in 2004, and it surveyed 17,300 midlevel managers representing 951 organizations including financial services, food processing, and telecommunications in 62 cultures. It also

Table 4.1 Delphi Demographics

Panel Demographics		
Cultures and Experience	**Panel Culture**	**Panel Experience Years**
Cultures		
Eastern Europe (Albania, Georgia, Greece, Hungary, Kazakhstan, Poland, Russia, Slovenia)	0	17
Nordic Europe (Denmark, Finland, Sweden)	0	11
Germanic Europe (Austria, Germany East, Germany West, Netherlands, Switzerland)	1	13
Latin Europe (France, Israel, Italy, Portugal, Spain)	2	59
Latin America (Argentina, Bolivia, Brazil, Colombia, Costa Rica, Ecuador, El Salvador, Guatemala, Mexico, Venezuela	1	21
Confusian Asia (China, Hong Kong, Japan, South Korea, Singapore, Taiwan)	3	95
Southern Asia (India, Indonesia, Iran, Malaysia, Philippines, Thailand)	3	71
Sub-Saharan Africa (Namibia, Nigeria, South Africa, Zambia, Zimbabwe)	0	14
Middle East (Egypt, Kuwait, Morocco, Qatar, Turkey)	1	49
Anglo (Australia, Canada, Ireland, New Zealand, South Africa, United Kingdom, United States)	13	349
English Caribbean	1	
Years of Experience		
Academia: Number of years		206
Business: Number of years		376
Government: Number of years		69
Nonprofit: Number of years		9

asked the participants the "as is" condition—How are things actually done in your country?—and a "should be" question: How do you think things should be in your country? The participants were quizzed on each of the nine dimensions listed. In asking "as is" and "should be" questions, we see patterns of behavior, from practice, that represent at least in part the globalization of societies, especially in the younger generation.

For our purposes here, the results of the Delphi panel are useful to describe the correlation between the attributes of XLQ and the nine cultural dimensions of the GLOBE survey. The results are shown in Figures 4.2 and 4.3. Any scores above 4.0 were considered correlated.

What we hypothesized, and found, was that there are different levels of correlation between cultural dimensions and leadership dimensions. In practice, one emphasizes certain attributes when working with passport holders from Russia and from Indonesia, for example. The bandwidth of correlations was tighter than we had thought it would be based on the diversity of panelist backgrounds.

The leadership dimensions were built from the exegetical research of leadership theory. Each leadership dimension is comprised of descriptors that further define the aspects of the particular dimension. That is the basic framework of the research testing and results for the model. Now let us examine each leadership attributes individually, beginning with the hub of trust.

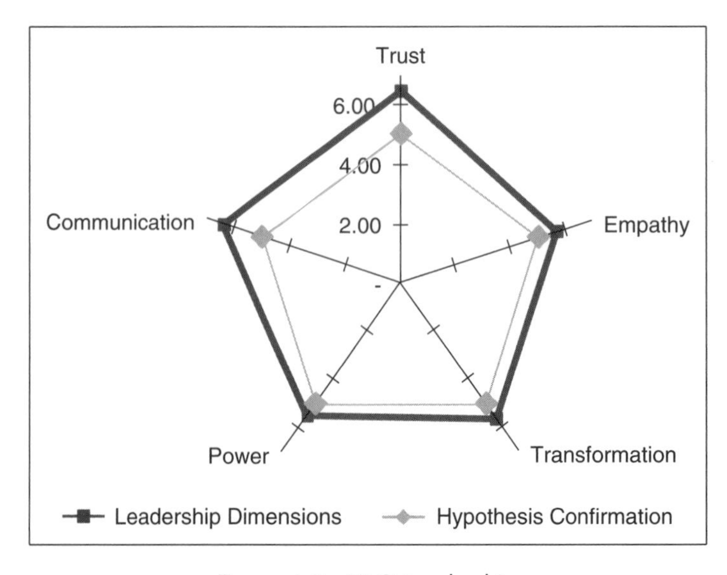

Figure 4.2 XLQ Leadership

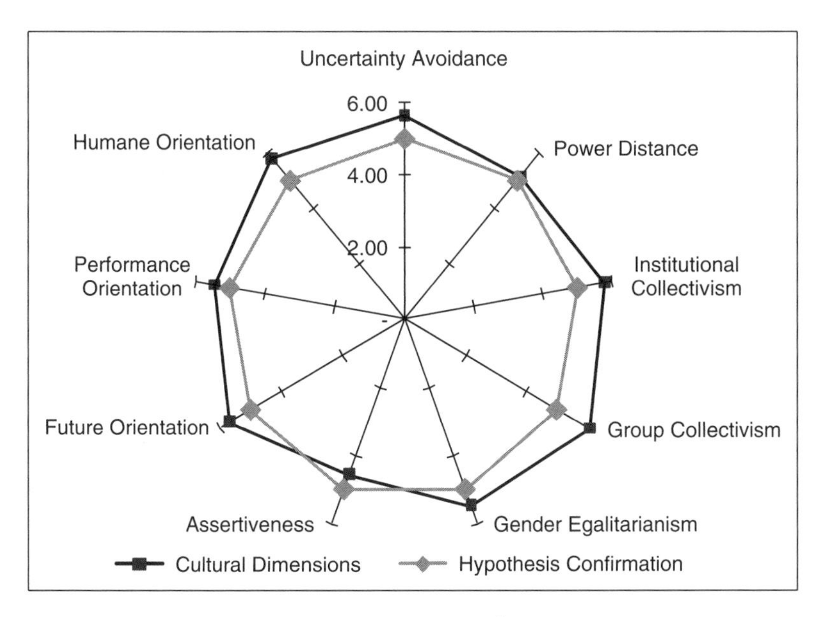

Figure 4.3 XLQ Culture

4.1.2 Trust

The definition of trust is based on the work of Mayer, Davis et al. (1995). They define trust as:

> [T]he willingness of a party to be vulnerable to the actions of another party based on the expectation that the other will perform a particular action important to the trustor, irrespective of the ability to monitor that other party (p. 712).

Imagine you are in Cape Town and you have project team in Kuala Lumpur that you have never met personally, and never will. How will you earn their trust? Trust is absolutely critical for international projects and CPEs. Trust is also a fragile commodity that takes time to build but can be lost very quickly.

Trust is like love; everyone knows it when they experience it, but describing it is sometimes quite difficult. The descriptors of trust may help to explain:

- Care and concern—also esteem, face
- Character—also honesty and integrity, duty and loyalty, admiration
- Competence—also technical ability, judgment

- Dependability—also predictability, keeping commitments
- Fearlessness—also confidence, self-sacrifice
- Humaneness—also tolerance, respect
- Integrator—also goal clarity, cohesiveness
- Integrity and ethics—also values, ethics
- Truth and justice—also fairness, candor

As you can see from the descriptors, many of these are skills that can be learned through a combination of training and experience, and some, such as humaneness or face, are a function of the culture in which people were raised. Some, such as keeping commitments, are a matter of discipline as well. Think of a person you trust and compare him or her to each of these descriptors. When we ask businesspeople around the globe who they think of as a leader, a few names always pop up: Mahatma Gandhi, Mother Teresa, and Nelson Mandela. For these three, think just about fearlessness. They are known for patient, unwavering confidence in the face of impossible odds and lives devoted to self-sacrifice.

If you are the lead project manager for a CPE that has an adversarial structure, which of these descriptors might you be able to emphasize? In our experience, most actually can be emphasized. In adversarial structures, the flow of information normally is restricted, so intended candor may be misinterpreted as feigned candor. If this is the case, it can have an effect on care and concern, character, fearlessness, humaneness, integrator, integrity and ethics, and truth and justice. Transparency is one way to defuse the negative effects of withholding information, but it is easier said than done in adversarial contract structures. If, however, transparency is the course selected, then imagine the benefits from the display of fearlessness. It adds greatly to the development of a CPE culture of transparency, mutual respect, honesty, and so on.

Plato explored the issue of trust in 380 BC, and thousands have examined this topic over the centuries. The literature is rich and diverse, ranging across disciplines such as philosophy, sociology, psychology, anthropology, management, and more. For those of you interested in beginning a study of trust, we suggest starting with Das and Teng (1999, 2001). Figure 4.4 is a summary of two separate works by them. Their work on alliances links trust and risk and describes risk as the mirror image of trust. In other words, if trust is high, the perception of risk is low.

The propensity to trust leads to subjective trust, which is either goodwill or competence based, and that leads to behavioral trust, or trust displayed by a trustor. The control aspect comes from Das and Teng's work on alliances as well. They suggest that control is a component of trust and can help, or hinder, the perception of risk. Said another way in our definition of trust, "irrespective of the ability to monitor that other party," this is a way to monitor. Goodwill captures most of the descriptors in our model. Competence trust is also covered in the model but is more obvious. In our experience, competence trust is of far less importance on complex international projects. In fact, competence trust alone will not be enough, whereas goodwill trust may be. On short-cycle international information technology projects, competence trust is of much higher importance.

The control part of Figure 4.4 is related to metrics. Behavior control is implemented by creating values and norms, or culture, so that the members of the CPE recognize appropriate behavior. Think back for a moment on the issue of candor and transparency. If the lead project manager honestly and openly practices secrecy, the team will understand that this is appropriate behavior. The CPE members can trust that the lead project manager will not disclose information and that neither should they. Think now about self-sacrifice and imagine that the same lead project manager displays consistently the desire, unrequested, to help others with no anticipation of reward, just to be of assistance. Both are

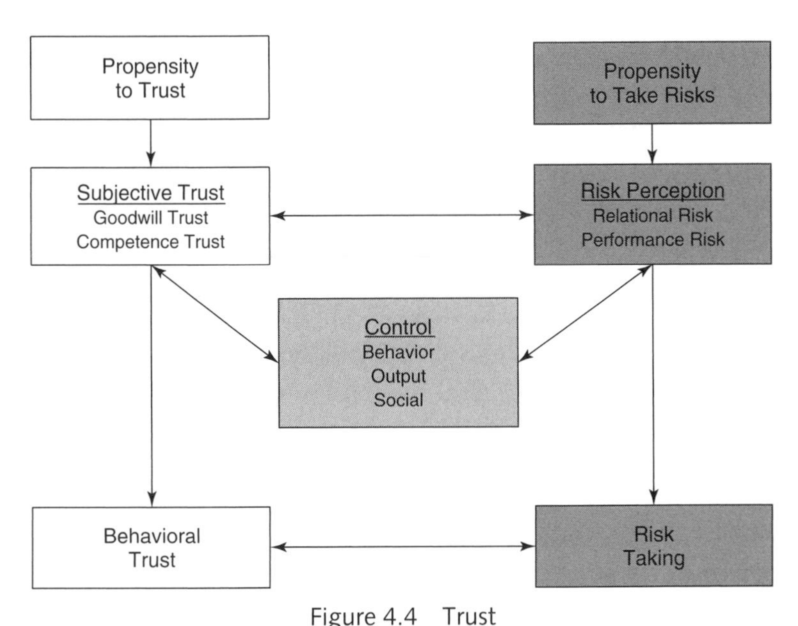

Figure 4.4 Trust
Source: Adapted from Das and Teng (1999, 2001)

examples of goodwill trust and are controlled through the establishment of cultural values and norms.

Output control is performance related and is the primary focus of the *PMBOK*. It is also the primary focus of such things as six sigma, total quality management (TQM), and so forth. It is a measure of output and is related to competence trust. In our model, competence is a descriptor of trust along with judgment. Imagine that you are a renowned small and medium power generation firm. Your books are required reading in many theoretical engineering programs globally. However, you are, alas, an academic, and your reasoning is at best opaque if not unintelligible. If you are the lead project manager, the team may admire your competence but would be quick to intervene if a project problem required judgment and reasoning. There was a film years ago titled the *Absent-Minded Professor,* and you no doubt met such a person. Remember our description of yin and yang, and the need for balance. This is why both competence and judgment are in the model, balance is necessary. For international projects, a lead project manager should have both, and others when necessary. As you have no doubt already concluded, the judgment part of the equation is partially involved with goodwill trust as well.

Social control springs from behavior control. Individuals in the group adopt or adapt the culture of the CPE, but the CPE encourages, rewards, or punishes the behavior displayed. Think of behavior as individual and the measurement of the tribe, CPE, or group as the social control. Imagine that you are the lead for a project with a team in Stockholm and another in Port au Prince. As the lead project manager, you have displayed the culture of the CPE enough times so that people know your ethical attitudes and expectations regarding such things as bribes. Everyone fully understands the behavior expected, but you are not in Haiti so you cannot measure the outcomes. If a local member of the team in Haiti, perhaps an expatriate from the United Kingdom, takes it upon herself to enforce the behavior, this is social control. It is pressure to comply with the values and norms brought by the tribe or CPE.

Imagine you are the lead project manager from the United States for a project in Asia, India, or Africa—a country with a traditional caste system. In the United States, law in the middle of the twentieth century theoretically terminated racial discrimination, but the practice still exists. The same is the case in some in other countries as well. Imagine you have a team in India that includes a Brahmin, generally a teacher, scholar, or priest, and a Dalit, or so-called untouchable. Remember, humaneness and respect are descriptors of trust. If you choose to treat these two people equally, will trust be built or diminished? Think about it this way: The Brahmin's societal culture may be such that he or she holds to the old ways, not overtly but covertly, and assume the same is the case with the Dalit. Both may be uncomfortable with equality but not willing to say so. What norms and values should be established? If equality is simply demanded, will trust result? Likely the opposite will be the case. In such circumstances, another of the leadership attributes, transformation, may be necessary. Equality may be encouraged, patiently, and demonstrated. The U.S. lead can show by action that European Americans and African Americans are entitled to the same respect.

We have selected the United States and India because many know about these issues. But there are thousands in other cultures globally. Knowledge of this diversity is an integral part of XLQ. Trust can take a very long time to grow, but it can be lost in a heartbeat, and it is essential to the success of a CPE. We strongly recommend that trust be top priority for the lead project manager on international projects, and in fact on all projects.

The mirror image of the propensity to trust is the propensity to take risks. It is the natural dichotomy in most all human interactions. If I trust you, I am willing to take risks. If I do not trust you, I am not willing to take risks. So as trust increases, the willingness to take risks increases as well.

Trust is also influenced by how the lead project manager deals with conflict, and that is our next topic.

4.1.3 Conflict Management

One of our key metrics for assessing project managers is how they manage conflict: Are they smooth and comfortable, or nervous and emotional? All human endeavors include conflict in culture, personalities, priorities, or ideas, to name a few. Some conflicts need to be resolved; some need to be managed. Conflict is like change; it should be embraced and guided for it will occur.

Intellectual conflicts are the substance of creativity and the foundation of knowledge transfer; they are far too important to successful business to resolve; they need to be managed. International project leaders need to create what we call intellectual brush fires, where people want to become involved in the debates and discussions. When this process works

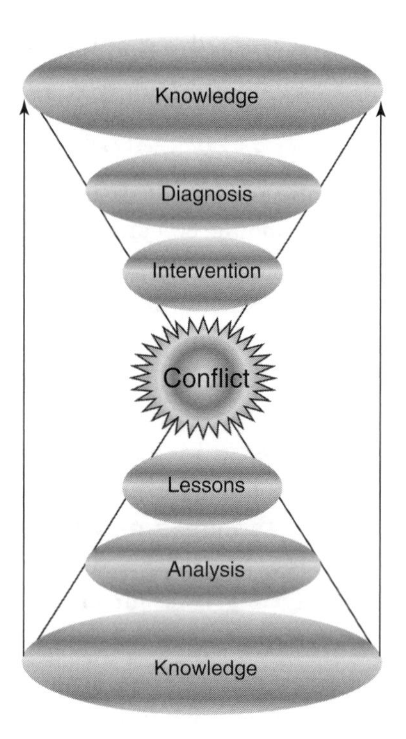

Figure 4.5 Hourglass Model

perfectly, the lead project manager only needs to keep it under control. Figure 4.5 shows that knowledge is the starting point and the key to conflict management.

Knowledge Conflicts in culture are common on international projects, for although thought globalization has brought people closer together, there are still thousands of years of history to consider. An easy example is the political relationship between the United States and Iran—tense, argumentative, with a lot of posturing. If you know that the United States orchestrated the overthrow of the government of Iran in 1953, what followed over the past few decades is not that difficult to understand. The seed of conflict is usually planted in ignorance or misunderstanding.

There are also cultural issues that abound, ranging from hand gestures, to the way people greet one another, to the concept of time. What time does the 9:00 meeting begin? The answer depends on where you are on the planet. There are some more dangerous cultural conflicts though. Think back to Rwanda, the elections recently in Kenya, Darfur, the South and North in the United States, Japan and China: The list is unfortunately a long and bloody one. Humanity still practices intolerance, segregation, discrimination, and much worse, and on international projects the opportunities for such conflicts are almost a given. Lead project managers should view this as a blessing rather than a curse, for it provides a natural way to display their skills under pressure.

Knowledge of culture is essential in dealing with cultural disputes. For example, knowing the underpinnings of the Iranian revolution provides insights into the seeds of interpersonal cultural conflict and thus ideas on how to proceed.

Imagine that a contract dispute arises on an international project over who is obligated to provide a certain portion of the scope. We recommend, first, that the lead project manager make it clear that disputes are to be brought into the open and addressed immediately. Disputes are not like wine and do not improve with age; rather they are like milk and can sour relationships. One key ingredient of the contract terms and conditions, as discussed in Chapter 11, is a dispute resolution procedure. For a CPE, and especially for public-private partnerships (PPPs), we recommend that the two organizations involved in the dispute have a fixed amount of time to attempt to resolve the matter through negotiations. This is always, *always*, the best way to resolve disputes. If the matter cannot be resolved within that time, it needs to be presented to the lead project manager for assistance. If the lead project manager, acting as a facilitator, cannot achieve resolution, then a professional mediation would follow, and binding arbitration if mediation fails. For this discussion of Figure 4.5, assume the CPE has such a process in place. Also, remember disputes include political, personal, cultural, contractual, and goal conflicts, among others.

The lead project manager should set a process in place that requires the parties in conflict to structure their positions in the dispute. In our example, the two parties should begin by setting forth the provisions of the contract that apply to the dispute, their interpretation of these sections, the amount of money involved, the time involved, the risk involved, and any impacts on quality. In addition, the parties would describe when the dispute was recognized and what steps had been taken toward resolution. The lead project manager can now take this explicit information and consider it in the overall framework of the project: past performance, similar disputes, personalities of the negotiators, relationship to the other work of the CPE, and so on. In doing this, the lead project manager can internalize the information and translate it into knowledge.

Diagnosis Diagnosis of the conflict involves structuring the dispute. First, what sort of repercussions will the conflict have on the CPE and the project? Is it symptomatic of a trend developing between the two parties in the conflict, or a larger trend project-wide? What are the common goals of the parties? What are the needs and wants of each party? Are the personalities of the negotiators the impediment to successful negotiations? Does the size of the conflict exceed the authority levels of the project managers who are negotiating? When must the conflict be resolved to avoid project impact? Asking and answering such questions is diagnosis. It includes determining if there appears to be a bargaining zone or if other components must be included to provide a possible range for settlement—this is called "increasing the pie."

Figure 4.6 provides a view of this concept. One component piece of the analysis is to determine the best alternative to a negotiated agreement (BATNA) of each participant. If there is no overlap, or settlement zone, for the parties, then the pie will need to be increased. As shown in the

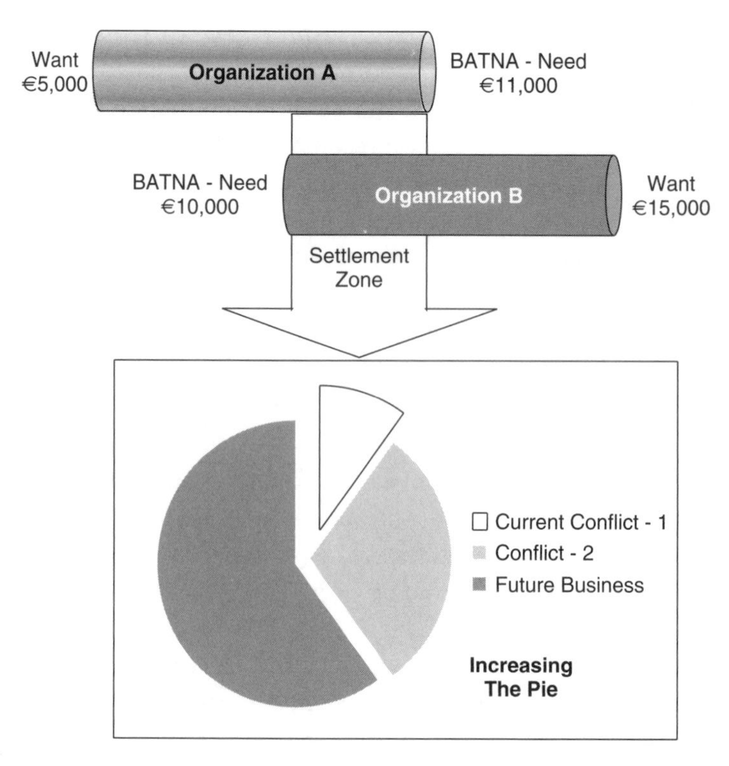

Figure 4.6 Conflict Analysis

figure, perhaps that is accomplished by connecting the current conflict with others. By placing the conflict into a larger perspective, the parties will have more opportunities to find some common ground.

On large, long-term projects, such as PPPs, it is possible to have a rotating panel of conflict facilitators assist the lead project manager. These are called dispute resolution boards (DRBs). These boards provide a wonderful educational tool for teaching people what they need to do to resolve conflicts on their own. They also give the facilitators an external view of the reasons why people cannot resolve their own conflicts. It is said that if you wish to learn about something new, teach it. Similarly, if you want to learn about conflict, help others resolve one. Analysis also instills discipline. It forces the participants to ask and answer the difficult questions that will come up during negotiations.

The diagnosis phase also includes relationship considerations that can improve or deteriorate the long-term health of the CPE. Conflict can contaminate the members of the CPE who are not directly involved in the dispute, for they see how the lead project manager conducts him- or herself and know they could be next. In complex CPEs, often multiple parties are involved in a single conflict, so it is extremely important that the lead project manager create a culture that embraces conflict as part of the project and manages it as work breakdown structure or any other task. The diagnosis must address the short- and long-term aspects of the conflict, for different players will have different needs at any given time. A firm that is publicly traded, for example, will need to meet quarterly metrics, and a governmental agency may have yearly budget restrictions. Figure 4.6 points to this consideration on increasing the pie. Then there are the authority levels and the personalities of the parties involved. To create win-win negotiations and increase the pie, emotional tendencies, personal and cultural, must be planned for and addressed.

Intervention The intervention lens of the model in Figure 4.5 is planning for the negotiations. Once the knowledge and diagnosis of the conflict have been completed, the conduct of the negotiation itself needs to be considered. Think of it this way: If you enjoy playing card games such as bridge, the knowledge is knowing the game; the diagnosis is knowing your cards, opponents, and partner; and the intervention is determining how to play the hand. This includes when and where to have the negotiations, who will be the participants, how many rounds are required, the formality of the process, and the conflict(s) to be addressed are part of the intervention. Sometimes it is useful to segment negotiations and deal with the emotions first; sometimes not. In multiparty conflicts, it might be useful to have the value chain parties come to a consensus with the next level up before bringing in the other organizations. What role the lead project manager intends to play is also a major consideration for the intervention. From our experience, successful win-win negotiations are 90% preparation and 10% engagement.

In Figure 4.5, the next step is engagement in the conflict by the lead project manager. Our best advice here for lead project managers is to help the parties to listen actively to the other party and to see the conflict from their perspective. The parties will gauge the equity and fairness of the lead project manager quickly and will adjust their negotiation accordingly. Look back at the XLQ model for a moment. The parties will gauge the humaneness, integrity and ethics, and truth and justice of the lead project manager, and decide if she or he is trustworthy. The lead project manager needs to assume an attitude like that of a mediator: there to help the parties find their own solution, not to make the decision—at least during the first negotiating session.

Look back at Figure 3.1 for a moment. If the conflict is between service provider A and service provider B, then probity would necessitate the use of an independent facilitator. On large projects, we recommend that the CPE retain the services of an independent mediator or a DRB with experience in the type of work being undertaken on the project. This provides an added opportunity between the intervention of the lead project manager and a formal mediation procedure. If an independent mediator is used, the steps for dispute resolution will include this as one of multiple opportunities to resolve disputes. In chapter 11 we discuss the steps below in more detail.

- *Project managers.* PMs attempt to resolve the conflict in 15 days. In Figure 3.1, an example would be value chain A.1 and value chain A.2.
- *Sponsors.* They attempt to resolve the conflict in 15 days. This also is shown in value chain A.1 and value chain A.2.
- *Lead project manager.* The lead project manager and the two PMs attempt to resolve the conflict. We suggest that there be a session after each monthly progress meeting to attend to conflicts. Of course, critical conflicts may require more immediate attention.

- *Independent mediator.* If the lead project manager negotiations prove unsuccessful, mediation should follow within as short a time a reasonable; we suggest 15 days. Unless the conflict is very large and complex, 1 or 2 days of mediation should be adequate. The parties must bring a decision maker to this session, so the size of the conflict will dictate if the sponsors need to attend. The lead project manager should sit in on only the joint session to provide input for the mediator. Having a dedicated mediator is useful as far less time needs to be spent educating him or her on the project, its status, and the relationship of the parties. Statistics from the American Arbitration Association show that mediation is successful approximately 90% to 95% of the time.
- *Binding arbitration.* As we discuss in Chapter 5, binding arbitration is the best choice for conflicts that simply cannot be resolved through other means. In arbitration, a single arbitrator, or a panel of arbitrators, decides who prevails in the conflict or what portion of the burden is shouldered by each party. It is an expensive process and takes time. For conflicts under €1 million, it could well require a few months, if the negotiations and mediation are admissible as evidence. We recommend that the entire negotiation process explained here be used as the factual basis for the arbitration. This will save a lot of time preparing and reviewing same evidence and will encourage the parties to present their evidence early in the process rather than to hold back in hopes of a better deal later. We recommend that the contract include the specific arbitration rules to be followed, including discovery, that the arbitration be conducted in a country that is a signatory to the 1958 New York Convention, and that the arbitration process be limited to 90 days.

In our experience as an arbitrator, the majority of the time is preparation of evidence, and the desire of attorneys to anchor the process in discovery. Discovery is the acquisition of documents from the opposing party that are then poured over looking for evidence to support a position. Expert testimony may be required in some matters, but we recommend that it be required at the beginning of the process and certainly by the time mediation occurs. In every case we have heard, parties hold back information until forced to reveal it during the hearings to protect themselves. It is precisely this information that holds the promise for quicker, less expensive, win-win negotiations that build rather than undermine relationships.

Lessons At each step in the intervention process, lessons are learned—lessons about personalities, cultures, the firms, the facts, and the process itself. The lead project manager can greatly improve his or her ability to anticipate and avoid not only the conflict du jour but also conflicts in general and patterns of behavior. Some firms have a culture of extreme competition and set stretch goals for their employees, forcing them to compete for resources—think of gladiators and you will have a good idea of what we mean. Such firms imbue a culture of winning at any cost, and that is the ultimate symbol of success inside the firm. We have seen such firms on projects and the behavior it inspires in others. Even the most transparent and fairest of personalities will defend themselves against aggressive attacks.

Imagine that there is a firm, Aggression RUs, on a project. It bought the project, meaning bid far lower than other firms, with the hope of making its profit on claims and conflicts. If the lead project manager finds him- or herself in a dispute on the very first day of the project with an intransigent project manager from this firm, it is clear that more will follow. This is what we mean by patterns of behavior. Generally, such firms encourage people to withhold information and rely more on emotional outbursts rather than facts. That is not to say that facts govern and emotion is bad but rather that both are ineffective at the extremes. If such behavior is mixed with a nonprofit culture, the clashes can be significant; so many PPPs can have intransigent conflicts due to corporate cultural norms.

Interpersonal and cultural conflicts are the most dangerous and the most difficult to resolve. We believe that intellectual conflict should be managed but cultural and interpersonal conflict needs to be resolved. The process just described will work quite well for interpersonal and cultural conflicts. However, the lessons learned from a conflict sometimes means that a person must leave the project. This is a very difficult situation, and the lead project manager must handle it with a great deal of empathy, consideration, and dignity, especially true if the organization and societal culture place a lot of importance on "face."

Analysis A mediator's role is to ask difficult questions, calibrate expectations, and help the parties find a resolution. An arbitrator is not permitted this latitude and should not help a party make its case. The role of the lead project manager should always be that of a mediator. It is difficult sometimes not to take sides as a mediator, but once you slip even slightly, trust is lost and the process likely will fail. For interpersonal and cultural conflicts, the mediator role is the perfect approach. In this role, the lead project manager will also learn where he or she has perhaps gone too far, or not far enough. Analysis also includes how to better anticipate where conflicts are possible in the future and if the current conflict may spread, or is endemic in some way.

Knowing what motivates people, how they communicate (high context and low context cultures), how they are persuaded (logic, or emotion), how well they prepare, and their personalities is a key to successful negotiations. The lead project manager also learns what makes the parties comfortable, how much they trust the lead project manager and each other, and how knowledgeable they are about negotiation skills. This stage presents the possibility to pay it forward, or give something to others with the expectation they will do the same for someone in the future.

Contractual issues are easier to deal with in that they have metrics attached. The analysis of the discussions reveals a great deal about the parties' intentions and boundaries. As we said earlier, sometimes it is best to bifurcate negotiations, as it provides the parties and the mediator with an opportunity to reflect. In addition, information may be discovered during the negotiations that needs to be researched. Not only does this information reveal facts that will help resolve the conflict, but it can also demonstrate the lead project manager's concern for fairness and justice. Unfortunately, it can also point to other parties in the CPE that should be part of the dispute and should have a chair in the negotiations.

Knowledge Figure 4.5 shows knowledge as the bottom of the hourglass. As with an hourglass, this knowledge feeds into the understanding of the parties and the lead project manager when negotiations begin again. By "again" we mean on the same conflict and negotiation or on a future one. Smaller disputes, say those under €50,000, likely can be handled quite nicely in a single negotiation session. Larger ones actually may benefit from two or more sessions, especially if the topic is complex. The session gives everyone the opportunity to convert explicit information into tacit knowledge and to reflect on the first negotiation session.

The knowledge gained through conflict is invaluable. It can be a blessing in disguise if handled in a sincere, equitable, respectful, and just manner. Again, this knowledge is an opportunity for the lead project manager to build trust in him- or herself. Remember the descriptors (and subdescriptors) for trust:

- Care and concern (esteem, face)
- Character (honesty and integrity, duty and loyalty, admiration)
- Competence (technical ability, judgment)
- Dependability (predictability, keeping commitments)
- Fearlessness (confidence, self-sacrifice)
- Humaneness (tolerance, respect)
- Integrator (goal clarity, cohesiveness)
- Integrity and ethics (values, ethics)
- Truth and justice (fairness, candor)

It is easy to see how conflict and trust are intimately linked. If the parties in a conflict do not trust one another and/or the lead project manager, it will be exceedingly difficult to negotiate a win-win solution. In such cases, arbitration will likely be the venue for settlement.

4.1.4 Power

Returning to the XLQ model of Figure 4.1, next let us discuss the attribute of power. The definition we prefer for power is "the capacity to influence behavior and attitudes to achieve intended results." This definition was constructed based on the work of Yukl (1998), who says "the term power is usually used to describe the absolute capacity of an individual to influence the behavior or attitudes of one or more designated target persons at a given time" (p. 142). Bass and Stogdill (1990) warn that power is not synonymous with influence but that all power yields influence. Each attribute of the XLQ model in some way reflects the others, like the facets of a diamond. Think about trust for a moment. If people have trust in a lead project manager, that will give him or her power, power to influence behavior and attitudes. The title alone will not result in power, or if it does, it will be a fleeting sort of power.

Here are the descriptors and subdescriptors of power.

- Knowledge power—also sharing knowledge, mentor
- Position power—also legitimate, political
- Power distance—also locus, communitarianism
- Referent power—also bravery, warmth
- Reward and punishment power—also coercive, reward

Knowledge power has a number of different aspects. On a fundamental level, it is connected to the competence descriptor of trust. For example, it may reflect a lead project manager with 20 years of experience with non-profit health organizations in Africa, in charge of such a project. In such a case, the other members of the CPE will initially tend to trust the lead project manager due to his or her knowledge. Now imagine that this lead project manager decides to withhold knowledge for personal or business reasons. No mentoring will occur, and at best, the team will think the lead project manager does not know enough to share. At worst, the team could believe that the lead project manager intends to take advantage of the others in the CPE by using his or her knowledge—in either case, there is no trust. As we will discuss in Chapter 11, the structure of a CPE can force such behavior by the lead project manager. This happens if an adversarial structure is selected and the lead project manager feels compelled to withhold knowledge to protect the profits of his or her organization.

Now imagine the same knowledgeable lead project manager sharing the knowledge of 20 years with the team and mentoring project managers in the value chain who may never have undertaken such a project before. Think of the trust that could be developed and of the power that flows from that trust and knowledge. The sharing of knowledge is also connected to empathy, communications, transformation, and culture. This is one reason we emphasized the importance of the structure of CPEs. Such open sharing not only engenders trust and increases the power of the lead project manager but also can inspire such behavior in others.

Legitimate position power ranked highly in our study, but political power did not. Legitimate power is more connected with the idea of expert power. It is power that others give to a leader because of their respect for his or her abilities. Like knowledge power, it is also connected to the descriptors of fearlessness, care and concern, and competence. Similarly, as we will see shortly, it is also connected with referent power.

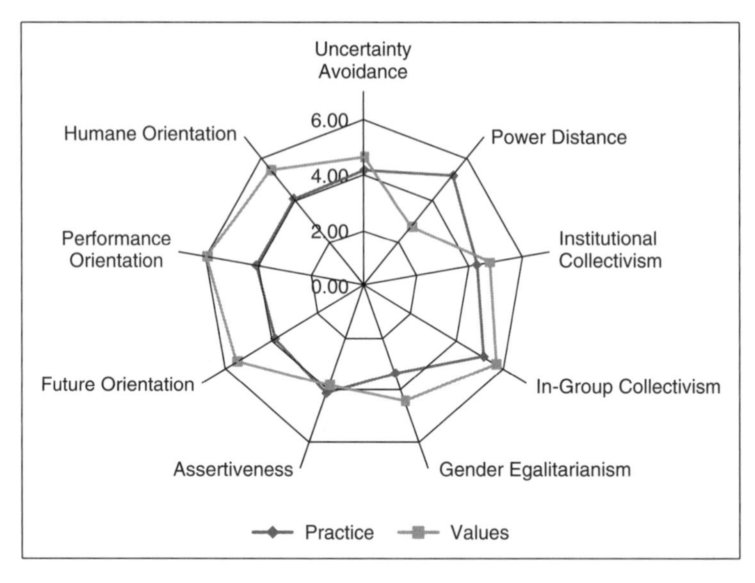

Figure 4.7 GLOBE Survey

The follower bestows, voluntarily, power on the leader and thus defines the leader's legitimate entitlement to it. Political power can be potent, but it is forced, not voluntary. Therefore, it is not sustainable unless combined with serious punishment power. Political power in Myanmar is tangible and carries a very real punishment component. However, that is very different from the political power of a CPE. Political power typically has a short shelf life, and it can breed contempt if misused.

"Power distance" is a term we introduced earlier. It is degree that people expect and agree that power will be stratified and concentrated at high levels of organizations. The GLOBE survey ranked cultures on a scale of 0 to 7, with 7 being the strongest (see Figure 4.7 Notice that for power distance, the average ranking was approximately 4, which is correlated but weakly so.

When asked how things were currently practiced in the country, Denmark scored the lowest with 3.89, and Morocco scored the highest at 5.8. When asked how things should practiced, Denmark dropped to 2.76 and Morocco dropped to 3.11. Think about a lead project manager who considers empowering those in the CPE. If we assume that the numbers are adequate for general tendencies, the organization from Denmark will welcome the confidence and will expect it—the more the better. The organization from Morocco may frown on such a diminution of authority, but likely the people in the organization would welcome it—even more strongly than the Danes. As can be seen in the figure, the values (how things should be) are very weakly correlated at about 3. The globalized view presented is that people everywhere want to be empowered. They just may need support in making taking the risk.

The subdescriptors for power distance, locus and communitarianism, are also useful in thinking about projects. Communitarianism, a sense of community or family or group, is in fact what a lead project manager is striving for in a CPE culture. Building an esprit de corps is of course the goal, but the ancillary benefit is that the community will need a leader and will voluntarily choose to invest power in this person. The locus of power refers to group activities in this context. If a community bestows power on a leader, he or she can then pass along some of this power to a team to achieve a particular goal. Think of power as a currency; it can be acquired, saved, or passed along.

Podsakoff and Schriesheim (1985) found that "the use of referent power by leaders usually contributes to their subordinates' better performance, greater satisfaction, greater role clarity, and fewer excused absences" (p. 235). Referent power is power given voluntarily to the leader by the follower. The reasons for doing so include knowledge, as we indicated earlier, a warm feeling toward or desire to be like the leader, to emulate his or her behavior, and simple respect. Respect ties back to all of the attributes of trust. There is a debate in the academic community about the reality and durability of referent power, for it has not been widely studied.

In our experience, it is the strongest and most durable type of power. It is lent to a leader on a probationary basis. Think of a neighbor who is working on her home and needs to borrow an expensive tool. You know her in passing because your schedules do not permit more, and you are not certain if she will return the tool in good condition and in a timely manner—uncertain trust. You lend it and it comes back completely cleaned, with a package of Chinese Wuji tea (organically grown in the Wuji mountains) and a thank-you note, the following afternoon. Trust is built, and power is given. Imagine the opposite now, and think about trust and power. Power and trust are strongly connected, and we believe it is through referent power. The trouble is that is lent and can be taken back at any time for any reason. The leader has no say in the decision; he or she can only precipitate it.

In all leaders, different attributes of leadership will be more developed, some less developed. Most people around the globe think of Nelson Mandela as a leader. The respect given to him by his people, and the international community, gave him vast amounts of referent power. In his case, it grew stronger and more durable as people came to know him better. Different people respect different things, obviously, but some things are common internationally, and the attribute of trust is one of those. Another is compassion, and honesty.

Lao Tzu, a Chinese philosopher who lived in the 6th century BC and was a central figure in Taoism, said, "When you are content to be simply yourself and don't compare or compete, everybody will respect you" (Hendricks 1989). That means respect is not a single-point metric but a spectrum. Holistically, the maturity and wisdom of the lead project manager will determine if respect is possible. If you have referent power, many things are possible that would otherwise not be for the CPE. In some adversarial projects we have seen, a lead project manager respected by the organizations involved can accomplish what collaborative structures can. We have also seen the reverse: a lead project manager with no respect on a project with a collaborative structure. Such projects also do not end well. Respect is not the only attribute of leadership, but it is a very important one. It would be an interesting hypothesis to test: Projects with leaders who are respected will succeed. XLQ can help leaders increase their sensitivity to how others perceive them and thus learn more about themselves. Lao Tzu also said, "Being deeply loved by someone gives you strength, while loving someone deeply gives you courage"(Hendricks 1989). That is the connection to bravery.

The work of Iacoboni, Molnar-Szakacs et al. (2005) focused on human subjects and tested the relationship among context, action and intention. Their findings suggest that mirror neurons create coding, or the physiological creation of neural pathways, by the intention associated with the actions of others and that these neurons suggest motor acts that are likely to follow in a given context. They also found that intention is ascribed by inference from the action and context. This research suggests that imitation of actions is wired in along with the intention of the action. For a leader then, it is important that the context and intention are made clear when actions are taken. If people have a physiological disposition to mimic, then the leader's behavior is crucial, as it will be mimicked, good or bad. Thus, referent power can be wired into our circuitry; the opposite can be as well.

The last descriptor for power is reward and punishment. These are clear enough and are linked in a fundamental way back to the work of Maslow (1943) and others on motivation. Normally a lead project manager is unable to utilize either of these broadly since she or he are not entitled to have a say in the review of people in other organizations. The lead project manager can certainly dispense rewards, but he or she cannot reach within the organization that employs an individual. Maslow's work is worth studying, for it provides a good framework to think about the different needs of people in developing versus developed countries. Later work by others (Herzberg, Mausner et al. 1959) found that what motivates people is not money but recognition. This is subject to debate, but certainly people who have their families cared for and are satisfied with their ability to enjoy life may be more motivated by recognition than someone who is homeless.

A lead project manager can use the concept of face effectively for motivating behavior in a CPE. Positive face can be gained through recognition, praise, empowerment, and the like. Negative face can be gained by isolation, correction (carefully and respectfully done), demotion, and similar moves. We are not inclined to use punishment, unless the acts are egregious, and unless it is done in private. It is so easy to get punishment wrong and to cause others concern. Recognition is a good simple, usable way to congratulate people on a job well done. We do not favor setting up preestablished bonuses but do believe that good work needs to be acknowledged as it is observed.

4.1.5 Empathy

We like Mullavey-O'Brien's (1997) definition of empathy: "the ability to put oneself in another's place, to know others' experiences from their perspective, and to communicate this understanding to them in a way that is meaningful, while at the same time recognizing that the source of one's experience lies in the other." It means to feel the world through another's heart and to recognize that we are all connected. Here are the descriptors and subdescriptors for empathy.

- Cultural intelligence—also metaphors and customs
- Humaneness—also compassion and consideration
- Servant leadership—self-sacrifice and empowerment

To feel the world through another's heart requires compassion, curiosity, and patience. Just how important is Diwali, the lunar New Year, Hanukah, Ramadan, Vesak, and Christmas, to name a holidays around the world? We cover culture in more detail later, but here we point out that you must have an understanding of culture in order to empathize. As with trust, empathy is connected to the other aspects of the model, such as culture. For now, recognize that part of empathizing is demonstrating knowledge and concern. The depth of the knowledge and concern conveys the depth of the empathy—to show another person that you care enough to gain the knowledge and then demonstrate that you have the ability to communicate it. Metaphors are indispensible for such communications.

Humane treatment is defined culturally, and legal systems define humane treatment differently. Consider the treatment of prisoners, people found guilty of crimes, refugees, those who live in poverty. Humane treatment depends on the cultural attitudes of the society, and changes over time. For example, in the United States, there are by some accounts 50,000 homeless children. Is it humane for a U.S. child to be homeless? There are certainly millions of homeless children in India. Is it humane for an Indian child to be homeless? The United Nations recently celebrated the sixtieth anniversary of the Human Rights Declaration, a living document with 30 articles. Article 1 provides a summary, saying "All human beings are born free and equal in dignity and rights. They are endowed with reason and conscience and should act towards one another in a spirit of brotherhood." This lofty goal is one that we believe should guide those who work internationally. However, there needs to be some latitude for cultures that are at a different place on this path.

Here the lead project manager must deal with a conflict. Justice in Saudi Arabia is dispensed on Friday at the mosque. Those from western cultures may find the punishment techniques harsh and inhumane; those from kingdom may not. The treatment of women in some cultures is considered inappropriate in others. It is incumbent on a lead project manager to recognize the cultural differences; otherwise no empathy, but it may be very difficult to reconcile the differences. In creating a CPE culture, what values and norms do you use? As we indicated earlier, set the bar high, and permit failure to reach it without stigma. We strongly recommend that the lead project manager does not attempt to ignore the cultural conflict. To do so will surely eliminate any trust the CPE has in him or her. The trick is to understand and not judge, encourage and not demean. Be considerate and respectful of the values and norms of other cultures. That is not to say that you must embrace them.

According to His Holiness the Dalai Lama:

> True compassion is not just an emotional response but a firm commitment founded on reason. Therefore, a truly compassionate attitude towards others does not change even if they behave negatively.... So, when a problem first arises, try to remain humble and maintain a sincere attitude and be concerned that the outcome is fair. Of course, others may try to take advantage of you, and if your remaining detached only encourages unjust aggression, adopt a strong stand. This, however, should be done with compassion, and if it is necessary to express your views and take strong countermeasures, do so without anger or ill intent. You should realize that even though your opponents appear to be harming you, in the end, their destructive activity would damage only themselves. In order to check your own selfish impulse to retaliate, you should recall your desire to practice compassion and assume responsibility for helping prevent the other person from suffering the consequences of his or her acts (Gyatso 2009).

This connects back to trust and speaks to an individual's ability to control the self. In the literature, this ability to control emotions is part of emotional intelligence (EQ).

Robert Greenleaf (1997) is generally credited with the concept of servant leadership, although the notion has been around for centuries. In 600 BC, for example, Lao Tzu said, "The greatest leader forgets himself and attends to the development of others" (Hendricks 1989). Self-sacrifice and empowerment capture the essence of this descriptor. Leaders like Gandhi set out to help or serve people. Those who choose to follow define such people later as leaders; service precedes leadership.

Perhaps the starkest version is of a military leader who serves his or her troops by protecting them from the enemy. Empowerment is also a facet of self-sacrifice in that the leader serves by giving a portion of his or her power to another and demonstrates trust in the follower who receives the power. Servant leadership feeds back into trust and, as we will see shortly, into transformation.

Hauser (2006) provides a broad review of the psychological clinical trials that have been performed on how people and animals come to develop moral beliefs and norms. Hauser builds his concepts on a premise similar to the work of the linguist Noam Chomsky (1988), that there may be deep similarities between the development of language and morality. Hauser says that empathy moves as a form of contagion. This idea connects XLQ trust and empathy directly to the values of the leader. Among many references in a broad study of clinical trials, Hauser points to the work of Johnson (2003), who found that 12-month-old children display joint attention (socially important behavior of following the gaze of others). Hauser contends that people have a genetic, as well as social, disposition toward a sort of human moral imperative. A leader must build trust and must establish a benchmark for values. Research again shows that people have an innate proclivity to emulate the physical and emotional actions and deeds of others.

4.1.6 Transformation

Bass (1985) provided the definition of transformation:

> [D]emonstrating charisma (vision, instilling pride, gain respect and trust), inspiration (high expectations, uses symbols storytelling and metaphors to express important principles and purposes), intellectual stimulation (promotes intelligence, knowledge, creative problem solving), consideration (personal attention, coaching, mentoring), and the pursuit of change.

The descriptors and subdescriptors for transformation are:

- Inspiration—also expectations and mentoring
- Charisma—also decisive and uniqueness
- Risk change—also desire to change and security
- Vision—also foresight and goals

Recall our definition of leadership: "the ability to inspire the desire to follow, and to inspire achievement beyond expectations." From Webster's dictionary, inspiration is "to exert an animating, enlivening, or exalting influence" on someone. It also causes people to want to emulate the actions of a leader. If you are inspired, goals look easier, your desire to achieve the goals increases, and people can actually be happy about attacking a difficult goal. Think about an inspirational experience that you have had and the emotions that it stirs. The ability to inspire is a complex commodity. We believe that it requires the lead project manager to the attributes of the XLQ model, especially superb communication skills.

Expectations are multidirectional: those of the follower about self and about the leader; those of the leader about self and about the follower. Imagine a leader who is respected by the project team and who sets high standards for CPE values and norms. Such a leader can inspire CPE members to exceed their own expectations of themselves. Think back on our example of a culture that holds ancient views that are not shared in the western world. It is possible to inspire people to consider another way? Consider a project with near-impossible task durations; is it possible to inspire creativity and the desire to accomplish what is seemingly impossible? We have seen it in practice. We believe the ability to inspire is rooted in trust. More on that in a moment.

One basic tool available to CPE lead project managers is mentoring. A lead project manager can mentor anyone in an organization involved in the CPE. It requires only the desire to do mentor and the desire to be mentored. The desire to be mentored is driven at least in part by respect; the desire to mentor is driven in part by servant leadership. We prefer mentoring to coaching on CPE because the relationship is more bidirectional. Mentors must expose themselves to the person who is seeking help; coaches do not. To be an effective mentor, the person seeking the help needs to know the mentor's personality and experience for context. That is what we mean by bidirectional. It requires more intimacy and thus enables people to form far stronger intellectual and emotional bonds. This can be a great benefit for leaders with high XLQ.

Charisma is a difficult thing to describe, but people seem to know it when they see it. Two subdescriptors that we found in the literature were decisiveness and uniqueness. People who are decisive, or fearless, are sometimes described as having charisma in part because they are confident in their knowledge and abilities and can inspire this feeling of confidence in others. That does not mean that all people who are decisive are charismatic. Charisma must have depth to be effective for a leader. Physically beautiful people may be charismatic until they speak, for example. Generally, charisma takes time to become apparent—it needs to mature like fine wine. It is a very special attribute, completely defined by followers.

Uniqueness is likewise in the eye of the beholder. A person obviously can be horribly unique or admirably unique. A component of charisma is how unusual people consider a leader to be. Mother Teresa was unique and still is a considered a charismatic leader by many. We believe that many of the attributes of XLQ create uniqueness in a leader. To put it on a scale, think of unique as being unlike the norm.

One big issue for all lead project managers is change. As we have said repeatedly, change will happen, for sure. The question is: How do you lead change? It is a complex question with many facets, but there is one bedrock principle that we believe is essential: feeling safe. Imagine a trapeze artist who has been doing a three-revolution flip for the past year without the use of a safety net. Business is down and a new act is needed, so the manager decides that a quad (four revolutions) will do the trick. The performer who does the flip thinks she can easily do a quad but is not certain the catcher (person who catches after the flip) can handle it. If that person fails, serious injury or worse—extreme change—will result. Imagine now that the manager, sensing some trepidation, agrees to put a net under the artist for safety; now let us try five revolutions!

In the business world, people must be given the opportunity to try new things, stretch their abilities, and challenge themselves. This will mean doing things they have not done before, and likely they will make mistakes. The resistance to change is often at this point: the fear of failure. Imagine that a lead project manager empowers a man to try a new task and that he gives it his all but fails. As a result, the project is delayed and money will have to be spent to remedy the problem. The lead project manager now must decide to take responsibility and protect man who made the mistake or to place blame on that person. The former is the way to lead change and is connected directly to trust, conflict management, power, empathy, and of course transformation. The way to minimize the potential downside to such failure is mentoring: help and guide. If the blame option is used, you can forget about leading change, trust, and more. This can be a huge dilemma for some project managers.

The other consideration is that the lead project manager must extend his or her transformation beyond the confines of the organization to the other members of the CPE. For example, look back at Figure 3.1 and consider the relationship between the lead project manager and the project manager for service provider B. Imagine the lead project manager wants to lead change and the wants to empower the project manager for service provider B to take the lead in preparing the monthly report. The project manager for service provider B is experienced, but not internationally and not with virtual teams. What if an honest mistake is made that causes the customer great concern and thus political repercussions for service provider A's sponsor? Think fearless is the right word for the lead project manager? In our experience, international project CPEs are challenging because the lead project manager must blur organizational boundaries. He or she must lead change and must provide a secure place for people to strive and get it wrong from time to time. How a lead project manager meets this challenge helps define the culture of the CPE; it is a big deal.

Vision is another challenge. Organizations that have a Project Management Office (PMO) at a senior executive level have an advantage here. In our experience, the more a lead project manager knows about the direction of the organization he or she works for, the better the person can manage international projects. There is an old saying: "One cannot see the forest for the trees." Vision is the ability to see over the trees. If a lead project manager has a short-term quarterly view of organizational priorities and goals, it will be difficult for him or her to share a long-term vision with the CPE members. Imagine that the international project is a PPP with a projected life cycle of 30 years. Can you imagine the difficulty of trying to project for the CPE what the future holds? You can get better odds in Monaco or Las Vegas.

Unfortunately, many lead project managers do not work in such PMO structures and therefore must undertake to learn as much about international markets and their own organizations as possible. This knowledge actually can mitigate the shortfall of knowledge in the eyes of the participants. Visions can change, but they are not visions if they change every quarter. On large complex international projects or PPPs that have a long life cycle, it is incumbent on management to provide the lead project manager with organizational planning so that he or she can give direction to the CPE. Followers will forgive lead project managers for not having access to such knowledge, and will empathize, but this lack nonetheless diminishes the referent power given to these leaders.

These are the leadership components of the XLQ model. Now let us look at the cross-cultural components of the model.

4.1.7 Communication

Harkin's (1999) book defines communication as:

> [A]n interaction between two or more people that progresses from shared feelings, beliefs, and ideas to an exchange of wants and needs to clear action steps and mutual commitments. Specifically, a Powerful Conversation produces three outputs: an advanced agenda, shared learning, and a strengthened relationship.

The shared learning part is the most challenging, as we will see. As George Bernard Shaw said, "The single biggest problem in communication is the illusion that it has taken place." To that point, let us begin with the most important facet of communication.

Grisham's second law of project management is listen, question, think, and then act. It summarizes theory and practice regarding communications.

There has been a great deal of work on communications, and there are many good sources for those interested in enhancing their understanding. One we suggest as a starting point is Gibson (1997), who discusses the issues surrounding communications across cultures. She begins by separating the process as shown in Figure 4.8. Gibson contends that intercultural differences in communications occur due to differences in cognitive styles (methods) and cultural values (particularly Hofstede's masculinity-femininity, individualism-collectivism, and power distance dimensions). Gibson describes the processes and features from a psychological perspective:

Encoding

- *Message source.* Internal sources are the individual's personality. External sources are the individual's sensitivity to the group. Individualistic cultures tend to rely on internal sources; collectivistic cultures, on external ones. This is also called low differentiation in the academic arena, meaning that individuals make only slight

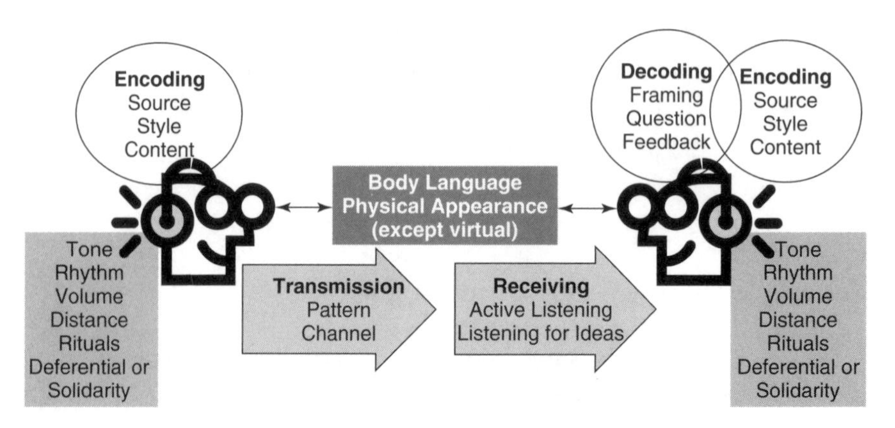

Figure 4.8 Communication

distinctions between themselves and the group. Societies with tendencies toward low differentiation include Japan, China, Indonesia, Mexico, and India.

- *Style.* Collectivistic cultures use implicit meaning that there is more said by what is not said (the use of words "maybe" or "perhaps"). Collectivistic societies include Indonesia, China, Japan, Thailand, Chile, Brazil, Korea, and Pakistan. Individualistic cultures tend to use explicit style, which means what you say is exactly what you mean (the uses of words "exactly" or "precisely"). Some individualistic cultures are Switzerland, the United Kingdom, Germany, the United States, Australia, Finland, and Canada.
- *Content.* Content can be emotional, logical, abstract (metaphors), temporal, nurturing, or many more. Collectivistic cultures generally prefer emotional appeals; individualistic cultures generally prefer rational appeals or logic. Societies that are characterized by a nurturing communal attitude, such as Indonesia, China, Israel, Taiwan, Thailand, or Chile, tend toward a more emotional approach. Content can also be explicit or abstract: "Isn't it a nice day?" versus "Can you explain Plato's *Republic*?"

Transmission

- *Pattern.* Patterns first are recognized linkages between group members. For example, high-context cultures tend to use a large variety of patterns in communications, depending on the subject to be communicated. Indonesians, Chinese, Japanese, and Arabs tend to use this technique, which is required by the circumstances. For example, there are numerous words for "you" in Japanese, and the one selected conveys rank and status. Low-context cultures, including Switzerland, the United Kingdom, Germany, the United States, Australia, Finland, and Canada, tend to use fewer patterns and tend to be more generalized and abstract. In part, the reason is that there are far fewer social nuances required.
- *Channels.* Obviously there are formal and informal channels for communications. People from high power-distance cultures—Indonesia, China, Japan, Thailand, South America, India, and Pakistan—tend to use formal channels of communication. Low power-distance cultures—Switzerland, the United Kingdom, Germany, the United States, Australia, Finland, and Canada—tend to use channels that are more informal.

Receiving

- *Active listening.* "Active listening" means being engaged in the conversation. By "engaged," we mean paying attention: not text messaging, e-mailing, checking voice messages, reading, or watching others. It means simply paying attention to the person communicating: Stop what you are doing, and listen. It also means asking questions for clarification (listen, question, think, and then act). Doing so helps to ensure that the context is the same and helps the receiver of the communication internalize the information. Perhaps most important is the empathy it demonstrates: It shows the person you are truly interested in what he or she has to communicate.
- *Listening for ideas.* Not everyone is direct in communications; remember the collectivistic/individualistic and high-context/low-context cultures. Patience is necessary to permit the person sending the message to get it out, in his or her own cultural or personal way. A few ideas may be riding along on what could be a tidal wave of words. The idea here is like catching fish in a river. A lot of water flows past but only a few fish. Be patient, focused, and wait.

  Listening must be done without filters. In Figure 4.8, the earphones are metaphorical filters. Filters are created by your culture, personality, and XLQ, including those items shown in the shaded boxes; more on those shortly.

Decoding

- *Framing.* Framing is primarily the ability to empathize with the sender. To empathize, you must have the ability to adopt you frame of reference, or context (listen, question, think, and then act). When thinking about the communication, it is often necessary to question the sender for clarification, to adjust the frame, and to make sure both people are in the same context.
- *Question.* In this step, you decide if you understood the message the way it was intended to be received, and if doubt exists, ask.
- *Feedback.* Promptly follow-up. Here you think about the message received and repeat it to see if you understood it correctly (listen, question, think, and then act). This is the action portion of Grisham's second law.

In addition to Gibson's work on communications, we highly recommend the work of Father Michael Oleska (2007) that those interested in a solid practical video watch section 4. Oleska spends the first section of the video describing the process of what Gibson calls encoding a message. In the second segment, he describes the style of transmission, and in section 4, he points to the following considerations. Some call these filters, the shaded boxes in Figure 4.8, meaning they can block the understanding of communications:

- *Tone.* Parents speak to children in tones that vary across cultures. For example, the normal tonal range for Spanish is less extensive than English. English tends to use higher highs and lower lows.
- *Rhythm.* Cultures have different rhythms in speaking. Think of people raised in New York compared to people raised in Madrid. The pace of conversations is quite different.
- *Volume.* Some cultures speak at a higher volume than do others, especially some African cultures. Other cultures, say Japanese, speak at lower volume. In the West, loud volume may mean aggression, for example.
- *Distance.* How physically close can one be to another when communicating, and is touching expected? Research has found that Australians did 25 touches per hour when communicating, English did 0 touches per hour, and Italians did 220 touches per hour (Pease and Pease 2004). Intruding on personal space can send messages.
- *Rituals.* The example used by Oleska is the ritual saying good-bye. Is it permissible to hang up a phone once the message has been received, or are there social rituals that need to be addressed, such as saying something like "Okay, I will talk to you later"?
- *Deferential or solidarity.* The concept here is one of informality or power distance. If an American greets a person from the United Kingdom, the American might refer to the other person by first name immediately: solidarity. The U.K. person might prefer to use honorifics, or the deferential style.

Figure 4.8 connects the concepts from Gibson and Oleska. Before leaving this figure, we want to point out a few items related to physical appearance and body language. Some very good work on the topic is worth a read. First is the work of Paul Ekman (2003) on facial recognition. His work included many cultures and spanned years of clinical investigation. What it shows is that people understand certain facial expressions to be associated with certain emotions. Ekman found that we get some of them wrong rather consistently especially mixing up fear and surprise. Reading one of his books is well worth the time. We have already mentioned the work of Pease. Pease's book is a great reference for body language, from the meaning of handshakes to that of scratching one's nose. Some body language is understood intuitively, and some is quite remarkable. Pease also gathered his work from clinical trials; it is not just anecdotal.

It is easy to judge people superficially. Too often in this fast-paced world, we make snap judgments without having the time or inclination to dig deeper. Honoré (2004) suggests that people could benefit from adopting a more measured pace of life—sage advice for reading people. Be very wary of quick assessments that rely on body language and physical appearance, especially cross-culturally. In fact, virtual teams offer a hidden blessing—you cannot see the other people so you are forced to dig deeper. The goggles in Figure 4.8 are the filters that we can use to make cultural snap judgments before any words come out. Imagine a beggar approaching you on the street as you are rushing to meeting. Have you ever played the tape in your head of what the beggar will say and decided in advance to not help? The thinking here is similar; those strong visceral signals can easily overcome the cognitive ones, particularly if you are in a hurry.

One other point on multitasking: People rely on what psychologists call autoappraisers. These are genetic responses, such as fear, and they happen in milliseconds. Imagine that a woman works all weekend on a priority task for an executive sponsor. There is not enough time to do the work, but she makes a best effort and completes the job on time. She knows that the job is not perfect. Thirty minutes after she turns the work in, she gets a call from the sponsor that begins with "I have a real problem." Fear can set in and can block or eliminate what follows. And while on the topic of emotions, there are two types: etic or universal (Pike 1967) emotions and those that are adopted, in order to fit in with society.

To continue with the discussion of communications, how do you get exactly the same idea that is in your mind into the mind of another person? Oleska uses the word "dog" to illustrate. If two people speak English, both will understand the word, but likely both will have a different visual image of what the dog looks like. Therefore, it is possible to lose the listener if he or she engages the cognitive process to decide if the size, shape, color, disposition, or weight of the dog. For example, consider the next conversation:

> Sender sends: I enjoy pets and have a dog, do you? I first began to enjoy pets when I was living alone....
>
> Receiver hears: I enjoy pets and have a dog, do you? Blah blah blah blah.... (Thinking instead: Hhmm I had a big black dog when I was growing up. Remember when....)

Now imagine that the sender is from northern Europe, and the receiver is from the Middle East. Both understand the word "dog," but the implications of it are quite different. Dogs may be companions in northern Europe but unclean in the Middle East. The conversation might be:

> Sender sends: I enjoy pets and have a dog, do you? I first began to enjoy pets when I was living alone....
>
> Receiver hears: I enjoy pets and have a dog, do you? Blah blah blah blah.... (Thinking: Hmm, why is this person accusing me of having a dog in my house?)

Nonaka (1998) divided knowledge into two components: explicit and tacit. Explicit is information, simply information. So in our example, the word "dog" is explicit information. It must manage to navigate its way through the goggles, listening filters, and encoding and decoding processes. Tacit knowledge means that a listener understands the meaning the sender intended. To do this the seed of explicit knowledge must find a fertile field of context in which to grow. Actually, Nonaka describes knowledge transfer as an organic process.

Imagine listening to the BBC evening news. A report tells of the Maoist insurgency in Nepal and that the leader of the movement has become Nepal's new leader. If you do not know where Nepal is located, what kind of government it had, or what a Maoist is, the news story will just be information. If you are from Nepal, there will be context, and the explicit news story could be transferred immediately into tacit knowledge. The dog example is also related to tacit knowledge. Listen, question, think, and then act—the thinking part is comparing the explicit knowledge with context and attempting to convert it into tacit knowledge, or internalization. Once the comparison is accomplished, the explicit information may not fit with context. If so, you return to the question phase.

Another way to think about the process is to use the simple yet effective phrase "What I understood you to say was. . . ." The idea is to rephrase what you heard and question the sender to see if that was what actually was intended. The question needs to be innocent and polite; otherwise the explicit message (what I understood you to say was . . .) can be tacitly heard as "I cannot believe you could say such a thing." Such messages can be delivered through tone, pitch, content, body language, and definitely e-mail format using all caps. The burden of communications resides with the sender of the message. If you are the sender, it is your responsibility to make certain the listener gets the tacit knowledge you intended to send.

At the CPE level, Szulanski and Jensen (2004) say "a central tenet in viewing transfers of knowledge as the replication of organizational routines is the importance of the template. We hypothesize that a template, i.e. a working example, is essential in replicating knowledge assets effectively" (p. 347). The authors confirmed that templates increase the stickiness (ability of knowledge to be replicated) in the knowledge transfer process. Another way to harvest knowledge is through communities of practice (CoPs). CoPs are made up of groups that have a common sense of purpose and need to know what the others know. For example, a group of people who enjoy cricket or soccer could be a CoP. The concept of a CoP is that when engaging in discussions on topics mutually enjoyed by the members of the group, ideas about work and problems can be shared.

Alan Watts describes how a person learns to fence. The student is given menial jobs to perform at the school, such as taking out the rubbish, washing the floors, and the like. The master then looks for opportunities to engage in surprise attacks when the student least expects it. The lesson is to learn to react to the opponent without thinking. Zen masters teach the same sorts of lessons. The concept of the CoP is that while people are engaged in something they enjoy, they will drop their filters and let slip ideas and knowledge that otherwise might not get out. Enabling CoP's is also a productive way to stimulate creativity in a group (Grisham and Walker 2005).

If you need to send a complex message, one effective way of doing so is the use of storytelling and metaphor (Grisham 2006). Kövecses (2005) contends that linguistic metaphors are expressions of metaphorical concepts in the brain's conceptual system. He showed in his research that a metaphor includes linguistic, conceptual, social-cultural, neural, and bodily components. He claims that abstract concepts are largely metaphorical and that the source and target domains help to explain the universality and particularity of metaphors ("love is warmth"—warmth is the physical or source component, love is the target or abstract component).

Kövecses uses Hungarian, English, and Chinese to espouse a belief that basic metaphors—happiness, anger, time, event structure and self—are in fact near universal. Although the linguistic metaphors themselves are near universal, Kövecses notes that the congruent metaphors are culture specific and are filled in from the near-universal level. One example of congruent metaphors is the concept of *hara* in Japanese ("anger is in the belly"). No other culture uses this adaptation of the near-universal anger metaphor. In this way, the metaphors are congruent with the near-universal level but culturally specific.

We believe that metaphors are a shortcut to convey complex ideas and that storytelling is a perfect vehicle to do so. Here we refer you to the work of Steven Denning (2005), who explored the topic in great detail and provides some practical advice for using storytelling. If you consider the history of human endeavor, storytelling is a central method in all cultures to convey values and norms, educate, and entertain. If you want to encourage active listening, make it interesting—make it a story. Children are told stories from an early age, and each culture has its own repertoire to draw on. The values and norms in children's stories are sometimes similar, but the players in the stories are specific to the culture.

Why do people enjoy literature, the theater, and movies? All are forms of storytelling. We argue that international projects should naturally make use of this basic and universal human form of communication. It is a powerful way to practice leadership skills at all levels, and it can provide a rich and evocative way to communicate complex ideas quickly. We recommend that participants in a project post a biography of themselves that includes a story to tell other people who they are. Based on Denning's work, the story should include the next points and should be no more than about 300 to 500 words:

- *Quick trust.* Must be a true story. Tell where and when the story happened.
- *Relevance.* Must reflect the type of person you are (caring, trustworthy, dedicated, funny, etc.).
- *Clarity.* Must make your values clear.
- *Distinctiveness.* Must reflect your uniqueness.

- *Consistency*. Must reflect your conduct, how you actually act (walk-the-walk idea).
- *Happy ending*. If possible.

To summarize, the key to effective communications is to realize that all communications are only approximations of what is in a person's mind. Communication is at best approximate. You must set aside enough time to communicate; it is vital to success. When communicating, focus on the sender as if the message is the secret to life itself. Give it importance, drop your filters, and listen actively. Then reword the message from your understanding and return it to the sender with respect. Going through the process enough times to synchronize understanding requires compassionate listening, curiosity (to know what the sender has to say), and patience: Listen, question, think, and then act.

With all of this in mind, we turn to a few recommendations on building a communications plan.

Project Communications Plan A communication plan is an essential component of international projects. The lead project manager must address this as a top priority at the start of the project. In general, the plan should address who needs what information, at what level of detail, how often, and in what format. It must consider and make clear whether there is to be a stratification of communications (e.g., whether technical people from one organization are to communicate with their peers in another organization, or if a project manager in one organization is to communicate with a sponsor in another organization). Ad hoc communications will of course occur at functions and in passing, but what we are talking about here are formal lines of communications.

The first focus needs to be the key stakeholders, and what level of detail they require. Some sponsors and executives prefer to know the details of a project and like printed reports. Many today prefer a dashboard graphical version in electronic format. Some key stakeholders need to have reports before board meetings or public meetings. Some key stakeholders do not want to be bothered with frequent reports but prefer quarterly updates. Many key stakeholders do not want their e-mail address made public and will not appreciate a distribution that includes nonkey stakeholders. Spend whatever time is required to ask and get the communication plan right at the beginning of the project.

The same goes for the functional teams, from project manager down. Some projects take advantage of enterprise software to accomplish this communication. Enterprise packages make use of a server platform that is partitioned so that each organization has its own secure area for private information, and there is a common area for shared information. In this way, the levels of access can be determined at the start of the project. As participants come to the project, they are given the access permissions necessary. Philosophically, we prefer to enable access to most information for almost everyone unless it is proprietary or executive level. However, the idea is to determine who needs what information to do their work and then to provide them with the necessary security clearance.

The packages we have tried require a significant amount of bandwidth so they may not be suitable for all locations. However, even if a full-blown enterprise package is not possible, the concept is a solid one: Put the documents on a server so all participants can access them. Using e-mail to send documents is okay, but by having them accessible on demand will greatly improve communications. On some projects, it is possible to convince sponsors that fresher information, on demand, can cut the costs of reports. Equally important, it can help prevent the potential for someone to be missed or the inevitable special need to arise. One of the many benefits is the ability to put up a single schedule on a server that can be updated daily. The entire team has access to fresh information on timing and can report necessary changes daily. This is far better than the weekly or (more often) monthly updates common on projects. It also provides a common platform for e-mail communications, which can save enormous amounts of time. Another great benefit is documentation. Many projects require a library of technical reference material, historical information, or design documentation. It is a big advantage to know that the most current revision is available to everyone. Posting online can save large amounts of money and time.

Imagine a project with 100 organizations in 20 countries and perhaps 1,000 people working on the project at any given time. If the staff turnover rate is 30% per year and the project lasts for five years, by the end of the fifth year, over 3,700 people will have worked on the project. Every time a person leaves and a new person comes to the project, communication links are severed, and there is a learning curve: The CPE must learn who the new person is, and the new person must to learn about the CPE. Assume that each person in the CPE spends two hours becoming acquainted (name, title, duties, e-mail, phone location, etc.). To avoid the waste of time having every single person who participates in the project either using a server or doing it by e-mail, we recommend creating a CPE directory. The lead project manager assigns the task of creating this directory and keeping it current on a weekly basis to an individual. A CPE directory should include these items:

- *Biography.* We suggest a standard format that can be filled in online preferably, or in spreadsheet format if necessary. The biography would include a personal story, photo, names, experience, avocation, and languages spoken. We like to encourage people to put basic information about their family if they wish and to share their religion if they feel comfortable doing so. This is not to intrude on privacy but rather to be able to celebrate the diversity of the team and to recognize important holidays. The lead project manager and sponsor must set the theme here by posting first. If they are not comfortable sharing this information, they should not demand it of others.
- *Contact information.* Include work phone number, cell phone number, emergency contact number, primary e-mail address, emergency e-mail address, fax number, pager number, and Skype address. Also include your time zone, normal work hours, and public holidays recognized by the organization. In addition, supply the name and contact information for the person's supervisor and at least one peer. On virtual teams, it is easy for people to go missing, especially in today's economy. It is important to be able to discover what has happened to the person. Figure 4.9 provides a view of the time map we discussed earlier for determining synchronous time available on a project and of course for scheduling meetings. A similar version is available online at www.timeanddate.com/worldclock.
- *Duties.* List the person's title, function within the organization, and supervisor(s) name.
- *Organizations.* List the name, mailing address(s), phone, fax, Web page, e-mail, and buyer or organization that the organization reports to or has a contract with. Also include a brief statement of the scope of the work being performed on the project. We recommend including the corporate mission statement as well. As we said earlier, ethics and values are vitally important, so organization beliefs should be shared. We also encourage a paragraph or so about the organization.

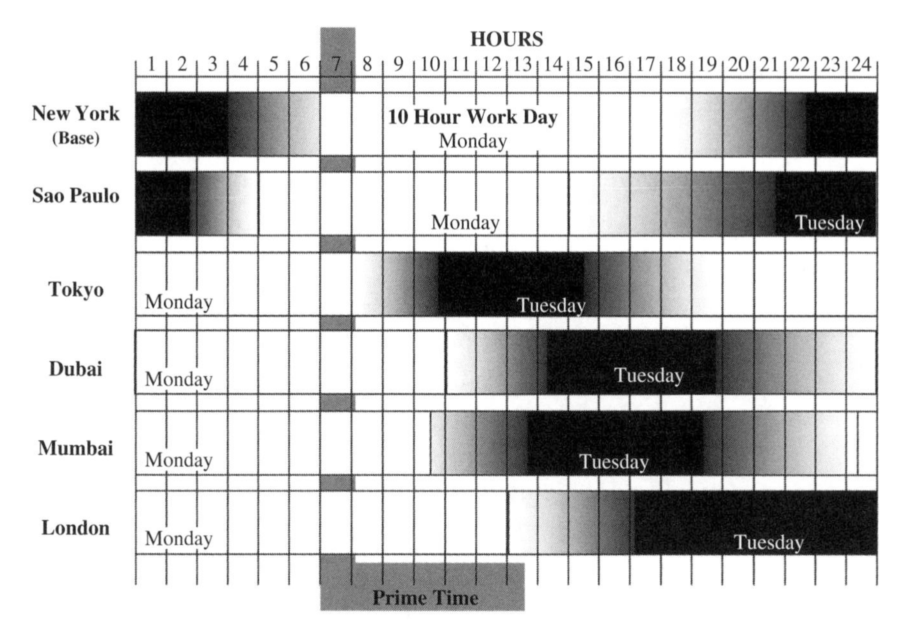

Figure 4.9 Time Map

- *Emergencies.* On many international projects, life-threatening emergencies are an issue. We recommend listing emergency political, security, and health contacts in the same level of detail as discussed for contact information. Consider a project for Médecins sans Frontieres. If the project is located in a remote region with little food or water, no infrastructure, few medical supplies, and a war raging, the lead project manager must provide the best safety nets possible. Said another way, lead project managers should not place others in conditions that they would not feel comfortable in themselves. Therefore, this section of the form should list critical contact information for embassies, consulates, medevac firms, transportation (air, ground, and water), medical supplies, food and water, security forces, United Nations, and the like.
- *Communication map.* Develop a matrix that lists the names (if they are known) or the categories of positions for the project in one column. Use the remaining columns to capture the type of information, frequency, format, time zone, and so on. The map should also include information about a daily communication plan. As will be discussed soon, virtual teams can benefit from regular daily communications, especially if most communications are asynchronous. (The time map

in Figure 4.9 is for synchronous communications.) Daily communications means asynchronous communications, unless there is a need for daily synchronous communications. (These may be necessary in cases of life-threatening emergency, war, social unrest, outages, etc.) For example, an entry might say: Tokyo (originator of communication), 1300 hours, 21 August 2000, and provide contact information and instructions.

Similar information is required if there are weekly communications, on a weekly communication plan. If the project involves regular face-to-face meetings (say monthly), include location information, timing, and participants. Our confirmation rule is *always* confirm what you have said in writing: Say it three times and write it a fourth. Be respectful and considerate of others, particularly those working in a second or third language.

- *Project standards.* State:
  - The project language(s) to be used for letters, reports, meetings, and conference calls.
  - The time and date formats to be used.
  - The protocol expected for external communications.
  - The size of paper to be used. In the United States, $8\frac{1}{2} \times 11$ paper is most common; A4 is used everywhere else. (Try buying an A4 binder in the United States sometime.)
- E-mail. Establish the type and timing for communications. Due to time, cultural, and business differences, it is important to set regular contact intervals for the team. For example, e-mail contact will occur once each day and phone/conference communications will occur once each week. Doing this creates a habit for the team members. It also reduces the uncertainty associated with a lack of response to e-mails and phone calls. Knowing that the team will be reading the e-mail daily gives some comfort and stability to the communications and enables the team to concentrate on the content.

  The type of communications for virtual teams includes various media, including e-mail as the primary mode, phone, fax, CoP software, knowledge software, and conferencing software. The team leader must ensure that everyone understands who must be included on the communications matrix, how often, and the format to be utilized. Setting basic expectations for e-mail communications can prove invaluable. A few rules from experience are (Grisham 1999):

- *Provide an opening and closing.* Think of an e-mail as being a personal card that you are sending. Be polite, starting with the recipient's first or last name and an honorific if suitable (Dear Dr. Smith, Dear Joe, James, etc.). For the closing, think of using the closing for letters, such as sincerely, yours truly, regards, cheers, or ciao. The formality of both opening and closing should be adjusted to fit the respect and consideration you want to convey.
- *Make the subject line descriptive.* Confine the scope of the e-mail to answering a question, posing a question, offering an opinion, requesting an opinion, and so on. Do not try to solve the world's problems with a single e-mail. The subject line should indicate what is in the message. Also, keep to one subject per e-mail.
- *Include contact information.* In the footer, include your phone and fax numbers, location (e.g., country), organization, and title for easy reference.
- *Keep distribution to minimum.* As the author of a first e-mail, try to think about who needs to know the information (from a communications matrix). When responding to e-mails, we suggest reply to all and add people only if necessary. Copying too few people can result in information not reaching those who may need it most, and copying too many can result in hurt feelings and a sense that the originator was attempting to withhold information. If there is a communications matrix, these problems can be avoided or at least mitigated.
- *Keep the e-mail short.* When longer documents are necessary, include them as attachments. We suggest no more than two paragraphs for an e-mail. Do not include photos or graphic files in the text of e-mails; make them an attachment. Help people to respond quickly by making the message easy to read.
- *Keep it simple.* No Booker Prizes are awarded for e-mails. Use simple, easy-to-understand English (or whatever language is the standard for the project). Avoid unusual vocabulary and structure, and use full sentences.
- *Check quality.* Proof names and titles, phone numbers, dates, times, and the like. Run a spell check on the message.
- *Avoid long chain e-mails* (e.g., Re: Re: Re: Re: New Project). Scrolling through multiple responses takes times. Having to do so can make people feel that they are being monitored and scrutinized, or it can create running dialogs with no conclusions.

- *Stop and think.* Experience has shown three major problems on projects that use e-mail heavily: the SEND button, timely responses, and reviews.
  - Recently a colleague sent an internal e-mail, complaining in colorful language about a business partner. One of the people receiving it accidentally forwarded it to the business partner being discussed. One way to avoid this is to keep a personal directory that segregates internal from external contacts. Another possible option is the use of a graphic overlay on the SEND button warning about casual use.
  - A colleague informed me, with a grin, that he had 1,900 unopened e-mails. In the age of quick communications, e-mail must be answered promptly, as you would do with a phone call. A team goal should be to respond to every e-mail each day. Answer priority e-mail within one hour. (The team leader sets the required metrics.)
  - The team leader must review the e-mail traffic for the first two weeks and provide guidance for the team as required. Doing so will help to ensure that any conflicts are worked through immediately and that the guidelines are being followed. After this, periodic checks to see that the responses are timely and appropriate to the guidelines are strongly recommended.

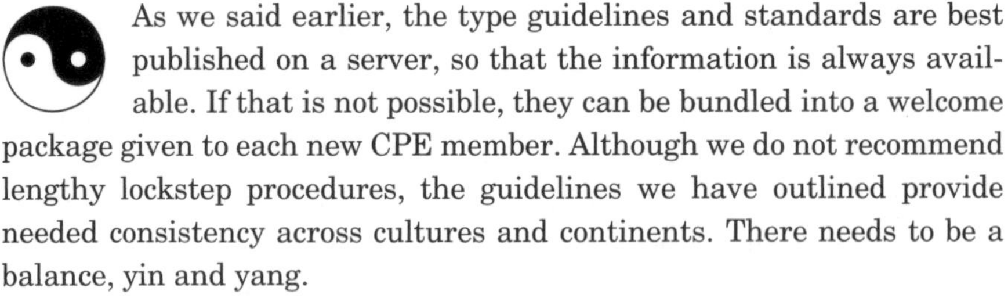

As we said earlier, the type guidelines and standards are best published on a server, so that the information is always available. If that is not possible, they can be bundled into a welcome package given to each new CPE member. Although we do not recommend lengthy lockstep procedures, the guidelines we have outlined provide needed consistency across cultures and continents. There needs to be a balance, yin and yang.

Now let us discuss last component of the XLQ model, culture.

4.1.8 Culture

Back in Chapter 2, we provided the definition of culture by Margaret Mead as "a body of learned behavior, a collection of beliefs, habits and traditions, shared by a group of people and successively learned by people who enter the society." Figure 4.10 provides a view of the different types of cultures we have been discussing in this book. *Societal* culture is

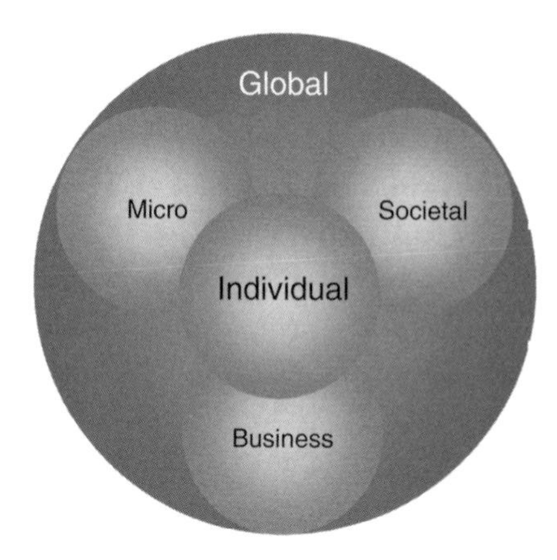

Figure 4.10 Types of Cultures

the most familiar, and is the one that Hofstede, the GLOBE survey, and most academics address. We have been discussing the "business" organizational culture as part of each organization that makes up a CPE. Consider the culture of Nokia compared to that of General Electric, for example. The "individual" culture is the personality of the person and the values and norms absorbed from family and friends. The "micro" culture is the culture of the CPE. The term "micro" is used because it is temporary, and not formally a business per se, because not all of the organizations are bound by a single contract.

We mentioned that the GLOBE survey asked people their opinions about norms and values in different countries. The survey was seeking information on the way things were in each of the 60 countries studied and the way people thought they should be. In our experience and research, we think that this was a measure of globalization, or global culture. What we have begun to witness is a normalization of values and norms internationally. As the Japanese do so well, people seem to be adopting parts of other cultures that they admire or think valuable. With increasing mobility of labor, we often run into the children of expatriate parents; they have a truly global perspective on culture. The paradox is that such people, while globalized, are complex and far more difficult to understand than less-mobile people who spend their lives in a single country. You may be surprised if, in a business meeting, when you are expecting tea and small talk, you instead get direct and to the point conversation.

In brief, in becoming more normalized, the world has become more complex. Therefore, the need for cultural intelligence is greater than ever. Our concept of CQ is a model based on some six sigma terminology: green belt, black belt, and master black belt levels. At the green belt level, people should understand some basics of other cultures so as not to offend catastrophically. The book by Morrison, Conaway et al. (1994) provides this level of understanding. A Web page for the book (www.kissboworshakehands.com) offers tests of CQ. Another Web page and test that may prove useful is the Cultural Orientation Framework Questionnaire (www.cof-online.com). Hofstede also has an interesting interactive Web page that enables readers to compare cultural dimensions on two cultures simultaneously (www.geert-hofstede.com/hofstede.com). Also Gannon's Web page (www.csusm.edu/mgannon) provides education tools for metaphors, as does the Peace Corps (www.peacecorps.com). Use any of these resources to increase cultural intelligence and to make people aware of the different values, customs, and norms in other cultures.

At the green belt level of training, people should know how to read business cards (to determine, e.g., which is the family name in China or Spain), how to exchange business cards (e.g., in Japan compared to Russia), and how to greet people (honorifics, protocol, familiarity). People at this level should know about gifts, entertaining, and how other cultures process information in negotiations (logic versus emotion). At this level, people could do rudimentary business on small projects with few cultures, and little planned interaction, and should have some knowledge of following aspects of culture:

- Geography
- History
- Type of government
- Language
- Religion
- Demographics
- Time zone
- Cognitive styles
- Negotiation strategies
- Value systems
- Appointments
- Negotiating
- Business entertaining
- Greetings
- Titles

- Gestures
- Dress
- Gifts

Some organizations consider this adequate cultural training for expatriate assignments; we do not. We have seen many project managers at this level or training or less have made big messes of things and need someone to bail them out. We have seen this occur on single-meeting assignments, where a project manager or SME is sent to meet with a customer. Our rule of thumb is simple: If it is okay to offend, no training is required. To use an educational example, green belt would be similar to a grade 12 level of CQ.

At the black belt level, a person has a much richer understanding of a culture including:

- Religion
- Early socialization
- Family structure
- Small-group behavior
- Public behavior
- Leisure pursuits
- Total lifestyle (work/leisure)
- Body language
- Sports (cultural values)
- Political structure
- Educational system
- Traditions (established order)
- History of the society
- Food and eating behavior
- Social class structure
- Rate of technological change
- Rate of cultural change
- Organization and work ethics
- Aural space
- Roles and status
- Holidays and ceremonies
- Greeting behavior
- Humor
- Languages

A quick look at this list indicates that more than just training is required. Here people will need to have a curiosity about others, the desire to study on their own, and some experience traveling. For example, you would need to understand, and appreciate, cultural metaphors. As we have said earlier, metaphors offer a way to communicate complex concepts efficiently. They can convey, in a short time, rich, complex information; people who understand the context of the metaphors can quickly be transform their information into knowledge. Look at the next list of items from Gannon's (2001) work and see if you could explain the metaphors to someone else.

- *Authority-ranking cultures.* In these cultures, there is a high degree of collectivism and a high degree of power distance (GLOBE and Hofstede). The metaphors are:
    - Thai kingdom
    - The Japanese garden
    - India: dance of Shiva
    - Saudi Arabia: Bedouin jewelry
    - Turkish coffeehouse
    - Brazilian samba
    - Polish Village
    - Korea: kimchi
- *Equality-matching cultures.* In these cultures, there is a high degree of individualism and a low degree of power distance. The metaphors are:
    - German symphony
    - Swedish stuga
    - Irish conversations
    - Canadian backpack and flag
    - Danish Christmas luncheon
    - French wine
- *Market-pricing cultures.* These cultures assume it is possible to compare individuals using statistical scaling. The metaphors are:
    - American football
    - Traditional British house
- *Cleft national cultures.* These cultures have major ethnic groups that are clearly separate from one another. The metaphors are:
    - Malaysian Balik Kampung
    - Nigerian marketplace

    - Israeli kibbutzim and moshavim
    - Italian opera
    - Belgian lace.
- *Torn national cultures.* These cultures have had major assaults on their values and have been torn from the roots that nurtured their societies. The metaphors are:
    - Mexican fiesta
    - Russian ballet
- *Same metaphor, different meaning.* These cultures have the same metaphor, but it has radically different interpretation. The metaphors are:
    - Spanish bullfight
    - Portuguese bullfight
- *Base cultures and their diffusion across borders.* As with the influence of the British on America, Australia, New Zealand, and Canada, the influence of the Chinese outside the borders of China is strong. The metaphors are:
    - China's Great Wall
    - Chinese family affair
    - Singapore hawker centers

The metaphors convey a blend of the items we listed above for black belt and green belt. For example, consider the difference in cultural attitudes between Chinese in the People's Republic of China and Chinese in Hong Kong or Singapore. Alternatively, consider the huge changes in Russia and the impact they have had on the Russian people and culture. Metaphors are macroviews of a culture and have numerous subtle facets. We believe that people must have traveled to a country, received training, and have done their own research to develop a black belt level of knowledge. By research, we include reading the literature and history, listening to the music, understanding the social customs, and the rest on the list. Black belt would be the educational equivalent of undergraduate or bachelor-level CQ.

We must one warning give at this point. Cultural studies and metaphors are useful in understanding *general* tendencies in cultures. They give an idea of social attitudes and momentum. That is it, period. If metaphors are used to define differences they become stereotypes. We are adamantly opposed to the use of stereotypes and want to emphasize that they are the opposite of our beliefs about cultural diversity. Each person is unique, and as we have indicated, layers of culture must be understood to develop CQ.

Master black belt introduces another dimension: the ability to synthesize cultural knowledge. People at this level have worked in other countries and have learned how to learn about another culture. What better way is there to empathize with other cultures than to have enjoyed, or suffered, the evening commute in Moscow on a rainy evening, the joy of crossing Bangkok in a taxi, buying food on the street in Kolkata, traveling across Sao Paulo and not stopping for traffic lights?

The most radical experiences are those where people go from a developing economy to a developed economy, or vice versa. Imagine a large complex project in Vietnam, which has a CPE with organizations located in Taiwan, the United Arab Emirates, Poland, and Mexico, and the lead project manager works for an organization located in Tokyo. Imagine that you are the sponsor for the Japanese organization and need to select a lead project manager. For the exercise, assume all candidates have similar personalities. Choice 1 is Suzuki-san: He is young, extremely bright, and a project management professional (PMP) who has never traveled out of Japan. Choice 2 is Siew-san: He is a young, extremely bright PMP from Singapore. Choice 3 is Ester-san: She is 40, a PMP, with expatriate assignments in Africa and the Middle East, and she is from South Africa. Which one do you assign?

Figure 4.11 shows our answer. From experience, international projects are consistently complex and generally hold high risk for the organizations involved. Rarely do today's projects include teams from just a few countries; the global value chain has almost assured that. The size of the circles in the figure indicates this graphically. Some simple low-risk projects do exist, and some need a higher level of XLQ, but most need a master black belt level of XLQ. Therefore, we would pick Ester.

For organizations that offer international project management services, or for those that are often involved in international projects, we *strongly* recommend XLQ training. We also recommend that organizations assign potential lead project managers to expat assignments as part of this training. Some organizations require people to work outside of their home country before becoming a partner. In our experience, there is simply no other way to help people achieve the level of knowledge that they will need except through practice or life experience. No level or duration of classroom training will do it.

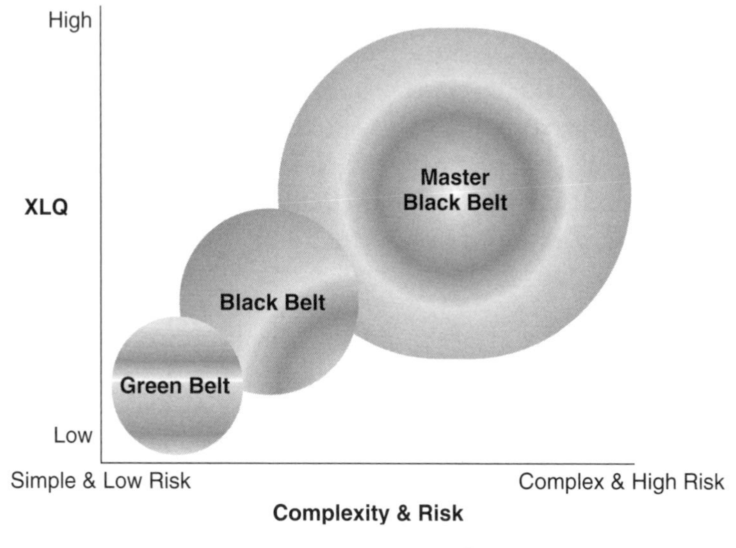

Figure 4.11 XLQ Ranking

We also strongly recommend that trainees be assigned a mentor to guide them with the process and that they not be assigned front-line duty immediately. As we have said, people need time absorb information and internalize it, time to think. The best training then is when people have time to observe and think. Complex international projects sometimes feel like a tennis match where experts on the other side of the net are hitting multiple balls simultaneously. Although a seasoned lead project manager with high XLQ may enjoy the adrenaline rush and the challenge, a person in training will not benefit from such an experience. Some exposure to the speed of project implementation certainly is necessary, but not a consistent diet of it.

Think back for a moment to transformation. If a trainee is partnered with a lead project manager as a mentor, the lead will know how much empowerment is appropriate and when a safety net is needed. The key to development is a foundation of theory, the ability to observe in practice, and the opportunity apply the knowledge gained. In addition, of course, the comfort to strive and fail is required. Unfortunately, all of this mentoring represents an investment in the person by the organization, and that is becoming a risk for organizations. This is especially the case in places like India where people can be encouraged to transfer to another organization for small monetary incentives.

This section on leading diversity is a critical skill set for international project managers. The ability to lead and motivate people toward a common goal is invaluable to the success of a project—in fact, to the success of a business and to the success of a CPE or PPP. In our judgment, this chapter alone contributes about 70% to the success of complex international projects. For most international projects, some team members must receive XLQ virtually.

4.1.9 Virtual Teams

On international projects, there will likely be a combination of face-to-face interactions and virtual ones. Some CPE members will never see the other members of the team and will participate only virtually. This section focuses on those in the second category. Jarvenpaa and Leidner (1998) explored the relationship between communication and trust in virtual teams. As the authors said, "Trust is pivotal in a global virtual team to reduce the high levels of uncertainty endemic to the global and technologically based environment" (p. 1). Their study, of MBA students from various countries who had practical work experience, looked at initial or swift trust (Meyerson, Weick et al. 1996) and trust that developed over the course of the assignments. We summarize their findings next.

Communication Behaviors Facilitating Trust Early On

- *Social communication.* Social messages seemed to facilitate trust early on in the team's existence. Teams with low initial trust exchanged few social messages in the first two weeks, whereas teams beginning with high trust were largely social. However, the high-trust teams appeared careful not to use social dialog as a substitute for progress. The concept of the biography and short story in the preceding section are targeted toward this finding.
- *Communication conveying enthusiasm.* In teams with low initial trust, the message content showed little enthusiasm or optimism. In high-trust teams, there was a great deal of excitement about the project. Measured exuberance and curiosity are key.

Member Actions Facilitating Trust Early On

- *Coping with technical and task uncertainty.* Teams with low initial trust were unable to develop a system of coping with technical and task uncertainty, whereas teams with high initial trust developed schemes to deal with these challenges. The high-trust teams used numbering systems so that all members would be aware if they had missed an e-mail message. These teams also informed other members of when they would be working and would be unavailable. The section on communications suggested both of these actions.

- *Individual initiative.* Teams with low initial trust had members who did not take initiative. By contrast, high-trust teams were characterized by initiative. Even though a leader emerged on high-trust teams, most of the members took initiative at different times.

Communication Behaviors Maintaining Trust Later on

- *Predictable communication.* Inequitable, irregular, and unpredictable communication hindered trust, not so much the level of communication. As we indicated earlier, knowing that daily communications are required helps to bind the team together and remove uncertainty.
- *Substantive and timely response.* A key difference between low-trust teams and high-trust teams was that in the latter teams, members received explicit and prompt responses that their messages were thoroughly read and evaluated. Responses to e-mails are important. Just a simple acknowledgment is a big step, and verifying by the quality of the content of the response that one read the e-mail. As we suggested, keep messages short, so that they can be read, understood, and responded to in a brief amount of time.

Member Actions Facilitating Trust Later On

- *Leadership.* A problem common to high-trust and low-trust teams was ineffective and/or negative leadership. Team leaders were chosen not based on their level of experience but apparently because they were the first to communicate or sent the largest number of initial messages. Read this point again. The leaders of the high-trust teams emerged after they had produced something or exhibited skills, ability, or interest critical for the role: in other words, referent power. The leadership role on high-trust teams was not static but rather rotated among members, depending on the task to be accomplished. This ability to lead then follow is critical, in our experience. Think back to referent power, empathy, transformation, and trust from the XLQ model.

- *Transition from procedural to task focus.* All of the teams seemed to make the transition from discussion of procedures, such as e-mail, to tasks. Thus, the project manager should explain the tasks early on in as much detail as possible. Once the team is functioning, less detail is required.
- *Phlegmatic reaction to crises.* The study found that high-trust teams remained phlegmatic during crises. Think back to our discussion on leading change. The team's trust in the other members to help and the project manager's commitment to protect are vital in cool response to crises.

Jarvenpaa and Leidner (1998) summarize the importance for practitioners: "For the manager of a virtual team, one of the factors that might contribute to smooth coordination early in the existence of the team is a clear definition of responsibilities, as a lack of clarity may lead to confusion, frustration, and disincentive." Particularly if the work is only part of the team members' organizational responsibilities, which is likely to be the case, providing guidelines on how often to communicate and, more important, inculcating a regular pattern of communication will increase the predictability, and reduce the uncertainty, of the team's coordination. This is where the communications matrix and e-mail guidelines recommend daily communications. The authors continue:

> [E]nsuring that the team members have a sense of complementary objectives and share in the overall aim of the team will help prevent the occurrence of desultory participation. Another critical factor will be the effective handling of conflict. One strategy is to address perceived discontent as early as noticed: emotions left unchecked in the virtual environment might erupt into sequences of negative comments, which will be difficult to resolve asynchronously. Another strategy in handling conflict will be to address as much as possible only the concerned individual and to avoid copying the entire team to messages that might be best to address to a single individual where a potentially problem has arisen.

As we indicated in the section on conflict, the lead project manager and the other project managers must pay close attention and intervene early on.

Jarvenpaa and Leidner (1998) add one last warning:

> Finally, not all individuals may be equally adept at handling the uncertainty and responsibilities inherent in virtual work. Managers should carefully choose individuals for virtual teamwork; such qualities as

responsibility, dependability, independence, and self-sufficiency, while desirable even in face-to-face settings, are crucial to the viability of virtual teamwork.

Think back on the issue of transformation, especially change and empowerment. Seldom can a project manager or lead project manager for a CPE select the team members within all organizations. More often, individuals chosen who do not have the necessary skills will need help. This help must come early.

Research has shown that virtual teams may in fact have advantages over collocated teams, and we have seen this in practice. Walther (1996) and others have found evidence that people can do social networking without interfering with tasks. From personal experience, people will find time after normal task hours to communicate socially, especially if there is a social bond. While Ekman (2003) has shown that people misread body language, especially when they are from different cultures, this problem does not exist in an asynchronous environment. People compensate for the loss of face-to-face interaction in other ways: social discussions, more reflection on the way the message is encoded, more time to reflect on the way the message is written, focus on sending the message at hours of one's choosing (e.g., after work hours). Walther's work points to numerous empirical studies that show there are opportunities in asynchronous or computer-mediated communication, opportunities that do not exist in face-to-face communications.

Virtual teams are an opportunity to increase the diversity in a CPE and benefit from the knowledge of that resides in that diversity. They offer time to reflect and consider messages without the pressure of immediate interaction. They present an opportunity to read and write English (the business language) instead of being forced to speak it as a second or third language. They are opportunities to leverage the expertise and lower cost of the global economy and to lead a 24/7 international project and shorten the cycle time. In our experience, these are a few of the benefits of virtual teams.

The disadvantages of virtual teams are that they require high XLQ. Harvesting the fruits of the diversity takes compassion, curiosity, and patience. It is like riding a wild horse; if you have done it before, you know the dangers, you know your limitations, and you are prepared to meet the dangers with a sense of humility and humor. Doing so requires high XLQ.

Chapter 5

Integration Management

The day of the storm is not the time for thatching.

—*Irish proverb*

Force, no matter how concealed, begets resistance.

—*Lakota proverb*

In this chapter, we discuss the binding together of the project management processes into a holistic integration plan. The Project Management Institute and International Project Management Association integration process describes this effort from the perspective of one organization. For international projects, the view must be extended to encompass all organizations in the collaborative project enterprise (CPE).

5.1 CPE STRUCTURE

We recommend that international project managers think of the project as a CPE when designing the structure. Projects are truly temporary organizations formed to perform the project, then they are acculturated, and finally they are dissolved (Mintzberg 1983; Brown and Duguid 1996; Toffler 1997; Winter, Smith et al. 2006). International projects that make use of alliances, joint ventures, consortiums, and value chains must function in a way as to promote the successful completion of the project. Doing this requires each organization to change and adapt to the CPE, but it also requires each organization to change and adapt internally to the external CPE (Turner and Mueller 2003).

Grabher (2004) summed it up by saying that "the formation and operation of projects essentially relies on a societal infrastructure" (p. 211). Our experience confirms this idea. It is useful to think of starting an international project the way you would think about starting up a business.

All of the communication patterns and protocols, power structures, financial and time responsibilities, quality assurance and control systems, risk management, human resource management, leadership, corporate social responsibility (CSR) and triple bottom line plus one (TBL+1), ethics, and cultures need to be considered and addressed.

If the project has a large global organization with headquarters in Bern partnered with a 20-person local firm in Yemen, partnered with a midsize Korean manufacturer, the corporate and cultural differences will be quite different. The job of the international lead project manager is to create a CPE culture that celebrates the differences in each culture while creating and maintaining CPE culture with its own unique values and norms. If the lead project manager does not create a CPE culture, each of the organizations and each individual participant will adapt their own. This situation engenders me-them thinking and precipitates a culture of separation rather than of commonality. We discuss CPE culture throughout the book.

Most international projects require different organizations to come together to provide services and expertise to a customer. Figure 3.4 provided a map of contracting approaches for CPE structures. In the next section, we discuss the relationships and structure shown in the lower left quadrant of this figure.

5.1.1 Partnerships, Joint Ventures, and Alliance Agreements

As we have discussed, the design of the CPE is critical to the success of a project. The culture, experience, skills, and personalities of the individuals will of course make a big difference in how a project performs, but the structure of the environment will dictate the extent to which they can positively affect the project. For example, an adversarial environment may be just the thing for the one-time purchase and installation of a server; the construction of a new e-business hub might be better served by an alliance. The terminology for describing these agreements can be confusing, so we begin by describing the terms we use.

In Figure 5.1, a transactional relationship is a one-time relationship; the current project is a one-time-only engagement between the parties. In such a relationship, each organization attends to its own goals and objectives. The sharing of knowledge is on a need-to-know basis, and only when required. Often this is the case on projects that are competitively bid using the adversarial environment shown in the figure.

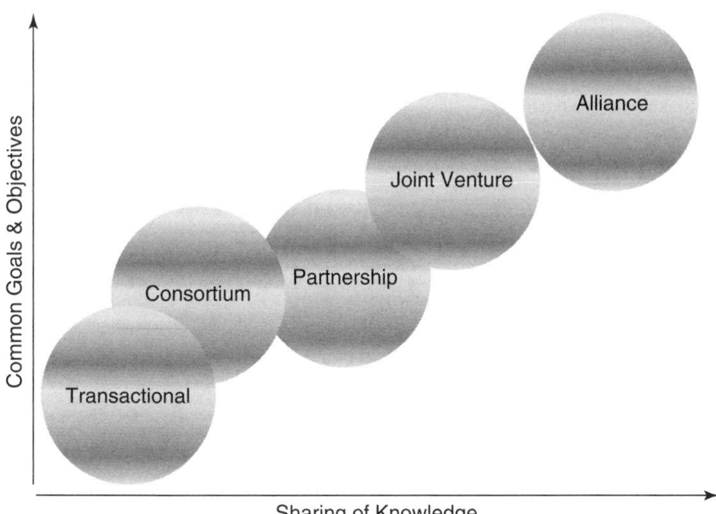

Figure 5.1 Agreements

A consortium and a partnership can be transactional or long-term agreement undertaken to provide a project (or projects) that is beyond the abilities of the individual organizations. The firms recognize a partial set of common goals and objectives, and agree voluntarily to share knowledge to advance their own internal agendas. There is no quantitative difference between a consortium and a partnership; their location on the chart is a relative one that depends on how the agreements are constructed. The agreements tend to be loose ones that can go partway toward creating a faux long-term relationship, but not always. To use a family metaphor, a consortium and a partnership might describe the relationship between distant relatives of the same culture.

Joint ventures are generally thought of as equity-sharing agreements, but again, they are not always. In these agreements, organizations fund and own an enterprise established to meet a set of goals and objectives that tend to be medium term or long term. These ventures require a closer relationship in culture, management, communications, and financial information. The appropriate family metaphor here might be that of siblings—more intimate, more sharing, more common needs.

Finally, there is the alliance. The family metaphor here is like that of a husband and wife, that is to say, intimate. The parties recognize that they have long-term goals and objectives that are aligned, that these objectives require a close relationship where trust is the foundation, and they share knowledge and information freely and openly. One organization does not let the other fail; the organizations consider the ones they are allied with as part of themselves. The relationship is close, enduring, open, and trusting.

The key differences in these agreements are the degree of relative goal congruence, the degree of knowledge sharing, and the amount of trust present. In fact, much of the business literature looks at all of these agreements as degrees of an alliance. We need to address one other aspect of collaborative structures: whether the structure is an inclusive or exclusive one. The collaborative environment in Figure 5.1 will help to illustrate this point. Imagine any of the collaborative agreement structures (alliance, joint venture, or consortium) was utilized to perform a previous project, and that agreement is now to be used as a model for a current project. The agreement could include or exclude the customer. In our experience, the difference in trust, and in the level of information sharing, is with inclusive being the preferred approach.

The Construction Industry Institute Australia reviewed 32 partnering projects in Australia to look for the benefits of partnering (alliance, joint venture, consortium). The study found that there are three clusters of benefits (Walker and Hampson 2003), confirming what we have indicated:

1. Good communications and high levels of trust
2. Improved profits and safety, and reduced disputes and delays
3. Knowledge sharing as a norm

Larson (1995) did a study of 280 U.S. Army Corps of Engineers projects with an average cost over $10 million and a duration of over 22 months. Approximately 140 of the projects were awarded based on low bid (adversarial); the remaining were not based on a low bid. (In other words, they were collaborative, quasi-collaborative, or quasi-adversarial.) In the study project managers, executives, resource managers, and consultants were interviewed and asked to rate project performance based on the dimensions of cost, time, technical performance, customer needs, avoidance of litigation, and participant satisfaction. The responses were then categorized into relationship descriptions rated on a scale of 1 to 5, with 5 being a successful project. The results were:

- Adversarial—2.34
- Guarded adversarial (quasi-adversarial)—3.15

- Informal partners (quasi-collaborative)—3.89
- Project partners (collaborative)—4.51

Larson concluded that project partnering works, regardless of whether the contract relationship is a low-bid one or a non–bid one. The difference Larson found was that collaborative projects are more successful in the eyes of the participants.

Yeung, Chan et al. (2007) explored alliances and their requirements. According to their study, two contractual elements are required: a formal contract and a real gain-share/pain-share agreement. In other words, contracts should be equitable and balanced. The authors also suggest three important relationship gains from an alliance: trust, long-term commitment, and cooperation and communication—all of which are aspects of cross-cultural leadership intelligence (XLQ). Based on results of more than 20 international projects, our experience mirrors their findings. There must be a formal agreement that binds the parties; the agreement must be portable, or moveable from project to project, and from organization to organization, and understandable globally; and it must address explicitly the sharing of risks. In our experience, the benefits are the relationship-based issues, especially trust. When there is no trust, there is ineffective communication and cooperation, and there is a transactional feel to the entire project. Transactional feel underscores the need for high XLQ for international project managers, particularly international lead project managers.

Once the structure for the relationship is established and the CPE is designed, the international project manager should turn attention to the CPE project charter.

5.2 CPE PROJECT CHARTER

A project charter can be defined in several ways.

- According to the Association of Project Managers (APM) a project charter is "a document that sets out the working relationships and agreed behaviors within a project team" (p. 130).
- According to the International Project Management Association (IPMA), "once a project has been approved for investment, the project owner should produce a project charter that defines the scope of the project, its objectives and deliverables, budget, timeframe, review points and team membership" (p. 44).

- According to the *Project Management Body of Knowledge* (*PMBOK*), a project charter is "a document issued by the project initiator or sponsor that formally authorizes the existence of a project, and provides the project manager with the authority to apply organizational resources to project activities" (p. 368).

The definition of a project charter that we use in this book combines parts of each standard:

> Project charter—A document prepared by the lead project manager and signed by the sponsors that formally authorizes the start of the project. It provides the lead project manager with the authority to lead the project on behalf of the CPE; defines the goals and objectives, the scope of the project, the deliverables, the budget, and the time frame; and establishes the ethical standards for the CPE.

The standards look at projects and project charters from the perspective of each participant organization. On international projects, each organization must prepare its own internal project charter and manage its own portion of the project. We strongly recommend that there be one project charter for the CPE, especially on collaborative projects; we therefore prefer our broader definition. Each firm prepares its own "internal project information" section; it can contain significantly more detail than is shown, depending on the needs of each firm. If the collaboration agreement is an alliance, often the internal information of each participant is shared. Imagine sharing your internal profit margins with other organizations, and you can get a sense of the difference in trust between an alliance and an adversarial agreement. Now let us look at the individual participants in the project charter.

5.2.1 Lead Project Manager

In Chapter 3, we began a generic discussion of the project charter in regard to its position and importance in the overall project life cycle. This section focuses on the substance and management of the project charter, from the perspective of a normal international CPE. Consider the example of a project that has been won by an equipment manufacturing organization, which now must find a local partner and an implementer. The equipment manufacturer will create an internal project charter to formally recognize the project internally, as will the local partner and the constructor, once their individual agreements with the equipment manufacturer have formalized. These project charters are the essential starting point for creating the CPE project charter.

The Welsh have a saying: "*Pan fo llawer yn llywio fe sudda'r llong*"—When the steersmen are many, the ship will sink. Military commanders, business leaders, and project managers know this from experience. Project teams need to have clear, consistent set of goals, objectives, ethics, values, and norms for the CPE. To facilitate this, there must be one lead project manager. When committees of project managers attempt this role, consistently poor performance and lingering disputes result. It is not impossible to run a project by committee, but it has a very low probability of success. In such cases, the leadership of the CPE are open to competition: the strongest, most aggressive, least passive, most informed, and so on.

Some customers and users in a collaborative environment, shown in Figure 3.4, will require that parties to a contract be jointly and severally liable for performance. In such agreements, often one party is authorized to make decisions and to be responsible for the partnership. Other customers and users prefer to have a contract with only one of the partners so to avoid confusion. When users and customers are actually a part of an alliance that undertakes a project, they will of course prefer that one of their own is the lead project manager. We recommend that customers consider the level of involvement that they wish to have in a project and their level of expertise in leading projects.

On one project in Thailand and another in the United States, the customer actually knew more about the project that did the consortium partners. In these cases, the customers actually were the designers and the users. The customers dictated the means and methods (techniques, sequences, methods, and resources) to the consortiums and insisted on keeping all of the authority while giving the consortiums all of the responsibility. These were not Machiavellian approaches to project management. Indeed, they were cases where customers knew exactly what they wanted, were the subject matter experts, and really just wanted a consortium to provide resources for the execution. The customers also wanted to transfer the risk associated with the execution and wanted to manage the work without taking the risk. It was not a successful strategy, for there were two project managers (customer and consortium) always vying for the lead. It served to confuse and divide the CPE.

CPE PROJECT CHARTER

Project Name and Location

Customer Project Information

- User names and locations
- Customer names and locations
- Sponsor names and locations
- Project manager names, locations, and level of authority
- Customer goals and objectives, and the reason for the project
- Project deliverables—the products
- Budget and duration
- Major risks
- Ethic standards of the CPE

Partner Project Information (repeat if multiple)

- Sponsor names and locations
- Lead project manager name, location, and level of authority
- Partner goals and objectives
- Project deliverables—the products

Internal Project Information (each firm prepares its own)

- Sponsor names and locations
- Project manager names, locations, and level of authority
- Resource manager names and locations
- Internal goals and objectives
- Project deliverables—the products

Approval

The project managers designated above are hereby authorized to begin work on this project and are responsible to ensure customer satisfaction with the product and services, including quality, timeliness, budget, and scope. The lead project manager is the single point of contact for the customer and is responsible for the work of all of the organizations participating in the project.

Selecting the lead organization and lead project manager for the CPE should be based on four considerations:

1. Experience and expertise on similar projects
2. Ability to lead, and manage, the project
3. Depth of financial and technical expertise
4. Stake in the project

If projects are competitively bid, customers should qualitatively assess the potential service providers based on their judgment and on the experiences of other customers who have used the service providers. Customers should also seek the advice from their consultants (often the design or process management firms) about their experience with the service providers. We recommend that customers actively decide what weight to assign to previous experience and the other dimensions, and clearly state these priorities in the tender documents. If there are bid protests, the critical components are transparency and equity. Preparing bids is expensive, and many service providers have a success rate of only about 10%. Be considerate if you are the customer, for doing so likely will reduce the cost of your future tenders.

On negotiated agreements, all of the organizations should be consulted and screened before agreements are signed. This step enables the customer to make more informed decisions and to find organizations that are well suited to undertake the work. In the case of a collaborative contract, customers must actively decide what authority and participation they want to have. We have seen customers who believe that collaborative means no involvement; often they find that they must insert themselves into the project—normally at the worst possible time—to ensure it produces the results they anticipated.

As we mentioned, on international projects, we strongly recommend selecting the lead organization and lead project manager before the initiation and planning of the project, and that this be done before CPE project charter is prepared. There is another equally important piece to the project charter, however, and that is stakeholder analysis.

5.2.2 Stakeholder Analysis

All the international standards—*PMBOK*, IPMA, APM, and projects in controlled environments, second major revision (PRINCE2), describe the necessity of doing a stakeholder analysis on each project, and all project managers should know the importance of this task. One major issue on large international projects is the number of participants and of course their remote locations. To illustrate the challenge, look back at the CPE structure shown Figure 3.1. Assume that Service Provider A and B have offices in four countries and that each is participating in the project. Certainly the chief executives, vice presidents, regional managers, sponsors, functional managers, project managers, and project teams are stakeholders for these organizations. Likewise, the customer and user will have the same types of stakeholders, as will the value chain organizations.

On most international projects, we recommend that active consideration should be given to the public groups that have interests in the environment, human rights, transparency, governance, and corporate citizenship associated with the project. Assuming that there is only one person in each category for our example, there could be well over 100 stakeholders. On large international projects, the number can easily be in the thousands, and it is simply not possible to analyze all the stakeholders in detail. Therefore, we recommend that stakeholders are managed in two categories: key stakeholders and stakeholders. Key stakeholders—those people or organizations that can significantly help or hinder the project—must be identified and managed actively. The remaining stakeholders then should be managed passively. By "passively" we do not imply that they should be ignored; rather, those who have a greater potential impact on the project receive most of the attention.

On international projects, we recommend starting with a list of all of the participating people and organizations—we call a project directory. Keep the list current as the project progresses. This list can serve as the starting point for performing a stakeholder analysis to determine who the key stakeholders are. Projects without project directories waste inordinate amounts of time, and there is less of a sense of camaraderie. On one five-year project in India, there were 1,000 people on the team on any given day in at least 20 countries, and the turnover rate was approximately 30% per year. Imagine the cost if each person had to spend just one hour making a project directory, and imagine the mistakes that would ensue.

We recommend that the project directory list the people involved in or affected by the project (including the value chain) and include the listed information for each person. (We presented more details in Chapter 4.)

- Name, title, and firm name
- Curriculum vitae
- Location and time zone
- Phone and fax numbers, and cell phone number
- Skype address if used
- E-mail address
- Photographs

We also suggest that each person share a bit about him or herself, such as their culture, religion (so the CPE can celebrate holidays), and interests.

The project directory helps the team to commemorate each new participant and say farewell to each person leaving the project; do not remove participants from the directory when they depart. The directory can also be used for responsibility assignment matrices, communication matrices, process mapping, knowledge management, and of course stakeholder analysis. The old adage is if a person picks up a calf every day from birth, then, when fully grown, the person will be able to lift the cow. Unlikely, but the idea is to start early and be diligent. Both aspects can be accomplished by having a lead project manager lead.

On a detailed level, Lester (2007) suggests that stakeholders be organized into positive and negative categories, then direct (participating in the project) or indirect (not participating) stakeholders, then internal (to the firm) and external stakeholders. He suggests that the stakeholder analysis then be performed by considering the interest, impact, probability of the stakeholder exerting influence, and actions to enhance or mitigate. Our view is similar and is as shown in Table 5.1 In this template, we have used the same hypothetical project firms as shown in Figure 3.1 and have listed some typical entries to convey the concept.

When completing the template, first make sure to establish the correct name and title of each stakeholder to be analyzed, paying particular attention to the culture. For example, most Hispanics have two surnames, one from the mother, which is listed last, and one from the father, which is listed first. The father's surname is used in addresses. So Señor Juan Carlos Martínez García would be addressed as Señor Martínez. A married woman named Señora Ana María Garcia Hernando de Martinez (de Martinez being the husband's surname) would be addressed as Señora Ana María de Martinez (Morrison, Conaway et al. 1994).

Table 5.1 Stakeholder Analysis Template

Stakeholder Name and Title	Organization	Power (0–10)	Involvement (0–10)	P x I[1] Rank	Key Stakeholder	Management	Communication Needs	Action Plan
CEO	User	8.0	1.0	8.0		Passive		
Sponsor	User	6.0	1.0	6.0		Passive		
Project Manager	User	5.0	5.0	25.0		Passive		
Division Managers	User	2.0	1.0	2.0		Passive		
Resource Managers	User	1.0	1.0	1.0		Passive		
Project Team	User	2.0	1.0	2.0		Passive		
CEO	Customer	10.0	4.0	40.0	Yes	Active	Quarterly reports	
Sponsor	Customer	8.0	5.0	40.0	Yes	Active	Monthly reports	ASAP: major problems
Project Manager	Customer	6.0	8.0	48.0	Yes	Active	Weekly reports	Weekly meeting, daily e-mail
Division Managers	Customer	4.0	3.0	12.0		Passive		
Resource Managers	Customer	4.0	3.0	12.0		Passive		
Project Team	Customer	4.0	8.0	32.0	Yes	Active		
CEO	Service Provider A	8.0	4.0	32.0	Yes	Active	Quarterly reports	
Sponsor	Service Provider A	7.0	5.0	35.0	Yes	Active	Monthly reports	ASAP: major problems

Project Manager	Service Provider A	6.0	9.0	54.0	Yes	Active	Weekly reports	Weekly meeting, daily e-mail
Division Managers	Service Provider A	4.0	4.0	16.0		Passive		
Resource Managers	Service Provider A	4.0	4.0	16.0		Passive		
Project Team	Service Provider A	4.0	9.0	36.0	Yes	Active		
CEO	Service Provider B	7.0	4.0	28.0		Passive	Quarterly reports	
Sponsor	Service Provider B	6.0	5.0	30.0	Yes	Passive	Monthly reports	ASAP: major problems
Project Manager	Service Provider B	5.0	9.0	45.0	Yes	Active	Weekly reports	Weekly meeting, daily e-mail
Division Managers	Service Provider B	3.0	4.0	12.0		Passive		
Resource Managers	Service Provider B	3.0	4.0	12.0		Passive		
Project Team	Service Provider B	4.0	9.0	36.0	Yes	Active		
CEO	Supply chain A.1	7.0	4.0	28.0		Passive		

(continued)

Table 5.1 (*Continued*)

Stakeholder Name and Title	Organization	Power (0–10)	Involvement (0–10)	P x I[1] Rank	Key Stakeholder	Management	Communication Needs	Action Plan
Sponsor	Supply chain A.1	6.0	5.0	30.0	Yes	Passive		
Project Manager	Supply chain A.1	5.0	9.0	45.0	Yes	Active		
Division Managers	Supply chain A.1	3.0	4.0	12.0		Passive		
Resource Managers	Supply chain A.1	3.0	4.0	12.0		Passive		
Project Team	Supply chain A.1	4.0	9.0	36.0	Yes	Active		
Politicians	Local	9.0	4.0	36.0	Yes	Active		
Interest Groups	Local	6.0	1.0	6.0		Passive		

[1]Probability x impact

In completing the organization column, make sure to get the correct name of the organization (some global firms operate under different corporate names) and the correct location and contact information. The power and involvement of the stakeholders is the numeric analysis portion of the matrix. Both of these assessments are purely subjective, and should be. Numerical ranking helps to increase consistency and provides a basis for comparing across projects and organizations. For example, on some projects where the customer is involved on a daily basis, it may be necessary to increase the ranking for resource managers. There are no heuristics for the rankings of the power and participation, as each project is different. The product of these two dimensions provides a convenient way of ranking the stakeholders to determine which are key stakeholders and which are not.

Table 5.1 provides a column to specify the action plan for managing the key stakeholders, and some of the passive ones if deemed necessary. It is a good place to record personal information, preferences, friends, colleagues, and other important characteristics of individuals. If the key stakeholders, especially the sponsors and project managers—are encouraged to provide their project directory information, all team members can get to know a little bit about those who make the decisions on the project. This information also can encourage storytelling, which we discussed in Chapter 4. The template includes a column for the communication preferences and needs of the stakeholders. Use this column to establish the type, method, and timeliness of the information desired by stakeholders.

As with all aspects of international project management, stakeholder analysis is a dynamic and ongoing task. Stakeholder analysis is *essential* to the success of international projects, and the customer is always the number-one key stakeholder. The return on investment in understanding your stakeholders is extremely high.

5.2.3 Customer Goals and Objectives

Most scope documents provided by customers describe the work to be performed in ways that vary from performance specifications to detailed specifications. It is unusual, however, to see a customer describe the reason why the project is being undertaken and his or her goals and objectives. Everyone has this information by the end of the project. From our experience in forensic work, we know that often this information could have made a difference in the conduct and outcome of the project if it had been known at the beginning.

In the United States, we were involved in forensic work on a private residence that was nearing completion. The original budget for the residence, about US$15 million, had already been exceeded by a factor of 3, and there was no end in sight. There were of course issues with time, cost, scope, and quality—there always are—but what struck us was the fact that none of the parties understood what the customer really wanted. Contract documents described the work, but not the goals and objectives, nor the reason for the project. We found that the customer wanted to make a statement about status, wanted to have personal tastes validated, and wanted to interact with others in creating a work of art. Despite three years of conversations, meetings, and correspondence, the true reason for the project was never articulated. This created a lack of trust, ill feelings, and alienation of the participants simply because what the customer needed was not said, and no one asked.

Customers are strongly encouraged to articulate the goals and objectives for the project, and the reasons for why the project is being undertaken. Seeing the project from the customers' perspective helps the participants to focus their efforts on the primary dimensions of customer satisfaction, or what is critical to quality (CTQ). (More on quality in Chapter 10.) If a customer chooses not to provide this information, then the lead project manager should seek to define it at the outset. In creating a CPE project charter, we believe that knowing why the project is being undertaken is critical to its success. We have found that customers actually enjoy being asked and explaining their thinking; it shows respect and consideration as well. Frequently, customers will offer a detailed accounting of the occurrences that led to the initiation of the project. Knowledge of the history that led to the project, from the outset, provides a wealth of knowledge about the customer. It also provides an excellent opportunity to structure the project planning to synchronize the goals and objectives of each participant to those of the customer.

By synchronizing, we do not mean that the goals and objectives of all participants will be the same; far from it. What we mean is that the participants will all know where they are in conflict from the start and how they can best manage these differences. The CPE Project Charter previously discussed provides the vehicle to capture this information in a portable and consistent way. Think of a project that you have undertaken. Consider what you knew at the beginning and at completion about the customer's needs and why the project really was undertaken. Think about when you knew, and how you learned, the customer's implied needs.

Although the primary focus is understanding the customer, it also is important to understand the other organizations. Apply the same considerations that you use with the customer to the primary organizations on the project. In the course of constructing a CPE project charter, a lead project manager can discover much about the participants, far more than what is provided on the charter. The lead project manager must take the lead in gaining a deep understanding of project users, customers, and participants. Look past the superficial; try to know the people and organizations as they know themselves; empathize with them. This is one way to build trust, establish a CPE culture, build the team, and demonstrate leadership.

5.2.4 Initial Project Budget and Duration

The budgetary and time constraints on a project are more tangible than the goals and objectives. International projects have are added twists that we need to discuss. First, consider the structure of the CPE. If the customer has a separate agreement with a designer in an adversarial environment shown in Figure 3.1, the design professional's contract is not normally divulged. The designer, however, would know the value and timing for the implementer. In practice, this imbalance is not helpful, for it creates inequity, an air of secrecy, and an exclusive (us versus them) culture. Why? Both parties are tied to the completion date of the same project and to the actions of one another, like it or not. Throughout history, conflict finds fertile ground when people permit differentiation. If there is no legal, competitive, or privacy issue associated with such a decision, it says: "You are different, and will be treated accordingly."

If the structure for the project is the collaborative one shown in Figure 3.1, the importance of sharing information is even more critical for the reasons described earlier. However, in a collaborative environment, information privacy is more complex within the partnership entity, because the relationship is more intimate, and therefore dealing with privacy is more difficult. If the structure is an alliance, there is really no issue, for financial information is expected. However, if the entity is a partnership or consortium, some information probably should be kept private. On one international project, the equipment manufacturer had a profit margin of 25% and the contractor had a margin of 3%. The relationship was a consortium, with the contractor viewing the relationship with the customer as being transactional, whereas the equipment supplier saw it as long term. The customer had a reputation of not being forthright and of demanding added scope at no cost. As you can well imagine, the equipment supplier had no incentive to share financial information; doing so would cause the contractor to look to the equipment supplier for "free" scope changes, or changes at no cost. The differential in profit margins also provided leverage for the customer in negotiations, where the customer was able to play the parties against one another. Yes, the entire CPE had an idea of the imbalance, but no one knew exactly what it was. Although project cost amounts were shared among the parties, no one knew the overhead or margin amounts. Effectively, all of the parties had to keep two sets of books; more on this in Chapter 7. Worse yet, it can—and in this case did—create a divisive culture of secrecy and competition.

Consider the practical implications of this secrecy. There were 4 people dedicated to preparing monthly reports on this project, and then there were the reviews, copying, mailing, and so on. It cost at least US$320,000 per year to produce the monthly reports, and we estimate that at least 25% of the time was spent trying to get the two sets financials to balance. Then there is the cost for keeping two sets of books. We know that there were at least 10 people on the site associated with quantity surveying, and certain that there were at least as many in home offices. Being conservative, the yearly cost for accounting and quantity surveying was in the range of at least €2 million. The effort expended was at least 50% focused on getting the books to balance. It was costly in simple financial terms, but imagine the frustration of those trying to get the numbers "right." As someone from Enron once told us, "Do you know the definition of a good accountant? A person who will ask what you want the numbers to be." Couple the extra costs with the damage done to relationships and the lack of trust, and it is easy to see why transparency is a better option.

In addition, each party will create a contingency fund for its individual work, so from a project viewpoint, the contract amount in the project charter will include the sum of the individual organizational contingencies. If this information is not shared, the estimate at completion will be significantly inflated. It will likely cause a reevaluation of the project and, if accepted, will skew the progress calculations, the S curve, and more. Using the sailing metaphor, it is like having a nautical chart that is

+/-60% accurate when you are striving for +/-90%. We have seen projects fail too many times because the plan and baseline have this hidden dimension. We find it infuriating that organizations undertake detailed project management processes built on a foundation of concealment.

Consider a project that you yourself have undertaken. How might it have been conducted if everyone from the outset knew the contingencies of all the organizations? We are not suggesting that all projects must do this; depending on the structure of the CPE, doing so may be inappropriate. We are recommending, however, that transparency is the best approach in collaborative environments. It is far less costly and far more effective in many ways. The only caveat is, again, in cases where sharing the information is illegal, exposes proprietary information, or would disclose competitive knowledge.

The structure of the CPE also has an impact on how the parties envision the time required to complete the project. Most international projects contain penalty clauses and liquidated damages (LDs). We discuss LDs in more detail in Chapter 11, but for now, recognize that they are penalties for completing the project late or for failure to meet contract performance requirements. On a power project in the Caribbean, for example, there were daily penalties of US$100,000 for completing the work late and a complex penalty for performance shortfalls. On this project, the CPE was a consortium; the customer was a member of the consortium. To achieve completion by the contractor, performance had to be demonstrated by the equipment manufacturer; to demonstrate performance, the facility had to be operated by the customer with the customer's fuel. Completion, and acceptance, required a team effort. The structure of CPE did not facilitate a cooperative culture, and the parties looked to their own interests first and to the project second. As a result, when the performance fell short, the arguments began and the LDs began to pile up. The completion date in the contract was clear enough, but in the real world, knowing when one is complete often is not easy.

Parties in contracts assume that they have the entire time of the contract within which to perform their work unless it is divided up during project initiation and planning. For example, if the contract duration is 700 days, with no further breakdown, all of the parties in an adversarial environment will assume they can use all of the time to perform their work and will price their work accordingly. Of course, once the detailed scheduling begins, the need to have a natural sequence of work shortens everyone's available work duration. In an adversarial environment, the parties must make some assumptions about the work of others when they prepare their own bid. If, however, the project is in a collaborative environment, there is less of a problem because the costs and sequence can be developed in a coordinated and cooperative manner, if our recommendations are followed.

Customers frequently do not possess the expertise internally to do the estimating and scheduling required to set forth a detailed plan for a contract duration or a professional estimate. They turn to external professionals to provide these services, professionals who themselves may be only partially qualified. For example, customers will hire an architect to design their project and ask the architect to do the estimate and schedule. Depending on the organization, a design professional may not have the experience and knowledge required and may not have the necessary resources to draw on. In such cases, the estimates and durations are not appropriate to the market conditions.

On international projects, the market conditions are truly global and require people with experience at both a global and a local level. Locally, a short-term shortage of raw material in China can easily ripple through a project that uses goods from there. Globally, to leverage economies of scale, centralized procurement can alter prices significantly. The same applies for labor and labor costs. In our experience, it is far better for customers to use a collaborative environment, where the experts who will perform the work estimate the time and costs for the project transparently. Again, we recommend the collaborative environment whenever possible, and particularly on large complex projects.

More difficult than budget and duration, however, are the governance and ethics of the CPE.

5.3 GOVERNANCE AND ETHICS

In Chapter 2, we discussed governance and ethics from the global viewpoint. Now let us look at the same issues from a CPE perspective. Ghoshal (2005) proposed that "by propagating ideologically inspired amoral theories, business schools have actively freed their students from any sense of moral responsibility" (p. 76). Her paper explores some of the reasons that debacles such as the Enron scandal have occurred. Ghoshal puts forward a compelling argument that research over the past 30 years has moved education, management literature, and corporate ethics away from a focus on moral responsibility. The focus has been on the bottom line or increasing shareholder value. Let us consider the implications for a CPE.

The lead project manager will seek to optimize her or his firm's shareholder value first, and the project managers for the other organizations probably will do the same. Regardless of the contract environment, the first allegiance likely will be to one's employer. The natural tendency is to think of other organizations as "they" and our organization as "us." This sows the seeds of competition and encourages the growth of a win-lose CPE. On a short-term or transactional basis, the individual organizations will seek to maximize their profits and reduce their costs, for it is what we are all trained to do.

Imagine an internal project charter where the stated goal is to maximize profit. Use the metaphor of a boxer who is trained to knock out the opponent and extend it to a ring full of boxers concurrently seeking to do what they do. It is really quite remarkable that 50% of projects actually succeed. Some of the successes can be attributed to luck, but some—many, we think—can be attributed to the character of the people who lead projects. If a project fails, most of the parties will suffer in some way. That suffering can be a transactional loss on the project, but it can also be a long-term loss of reputation.

At another level, in psychological studies of group behavior, it has been found that people who are completely ethical can as a group do unethical things. The reason is that the culture of the organization condones the behavior as an accepted norm or value—some so-called whistleblowers are ostracized by the organizations that they are attempting to protect. Taking such a stand requires XLQ and a fearless nature. Since there will likely never be a project charter that gives a lead international project manager the level of authority required to change internal organizational values and norms, in practice, it comes down to this personal courage. This courage comes from knowing yourself and being willing to risk the repercussions of standing up for what is right.

It is an imperfect world, with many views of what constitutes ethical behavior. Ethical behavior varies across cultures, across organizations, and across personalities. Although there are international standards, the application of those standards varies by person and organization. We hope someday all organizations will adopt the CSR approach and heed. Ghoshal's warnings. In the interim, lead project managers are left with their own personal values and ethical standards. If a person is to lead an international project, it is critical that his or her beliefs are made know to the CPE and that they remain unwavering. People likely will not agree with the standards displayed but most often will respect the lead project manager for taking a position. As we said earlier, it is important to set the ethical bar high, recognize that not everyone will reach it, and show compassion when people fail.

If firms feel that this is a risk, the best than can be done is for all of the project managers to look for no-cost ways to support the CPE and avoid increasing estimated profits at the expense of the other participants. This is perhaps a more palatable compromise for firms in today's business environment. It will encourage true partnership, and research shows that it improves the success rate on projects, regardless of their structure.

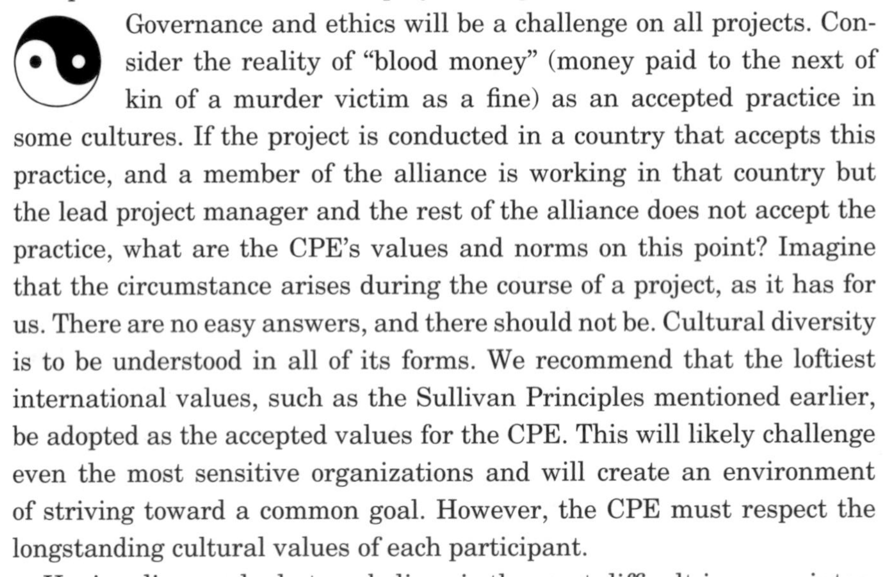

Governance and ethics will be a challenge on all projects. Consider the reality of "blood money" (money paid to the next of kin of a murder victim as a fine) as an accepted practice in some cultures. If the project is conducted in a country that accepts this practice, and a member of the alliance is working in that country but the lead project manager and the rest of the alliance does not accept the practice, what are the CPE's values and norms on this point? Imagine that the circumstance arises during the course of a project, as it has for us. There are no easy answers, and there should not be. Cultural diversity is to be understood in all of its forms. We recommend that the loftiest international values, such as the Sullivan Principles mentioned earlier, be adopted as the accepted values for the CPE. This will likely challenge even the most sensitive organizations and will create an environment of striving toward a common goal. However, the CPE must respect the longstanding cultural values of each participant.

Having discussed what we believe is the most difficult issue on international projects, let us look at the overview for the international project management plan.

5.4 CPE PROJECT MANAGEMENT PLAN

According to the *PMBOK*, a project management plan includes what it calls subsidiary management plans. These plans reflect the *PMBOK* knowledge areas, and we have used this outline for the basic organization of this book. We selected this format so that practitioners who are familiar with the *PMBOK* or are project management professionals will recognize the structure. However, we do not recommend the sequence set forth in the *PMBOK*. In developing a project management plan, we recommend that a particular sequence be followed that will save time:

- Develop a diversity plan (human relations and communications in the *PMBOK*)
- Develop a scope management plan
- Develop a cost management plan
- Develop a risk management plan
- Develop a schedule management plan
- Develop a quality management plan
- Develop a procurement management plan

Each project manager is responsible for preparing a project management plan for his or her organization. The lead project manager then incorporates these individual plans into a CPE project management plan. In many projects, particularly those that fail, the CPE project management plan—if there is one—is not shared with all of the participants. This sharing of information on a need-to-know basis is understandable, as we have discussed. However, the risk in withholding information is that the participants cannot utilize their expertise (which is often why firm in the value chain are in the project to begin with) to its fullest potential; they cannot fully coordinate their work; and they cannot anticipate. In forensic work, when we question firms on why they withheld information, we often are told indirectly that they had to protect their competitive advantage and their profits. In fact, on most of these projects, the opposite effect is achieved by withholding. On failed projects, the transparency is often low or nonexistent, and the projects fail due to a lack of trust, cooperation, and communication. The project fails, the company loses large amounts of money, and participants end up in a dispute to try to recover their losses.

In building a project plan for the CPE, even in a collaborative environment, some aspects of the project will not be fully known, or detailed, at the beginning of the planning. It is not always feasible to get 100% of the organizational project plans before building the CPE project plan. For example, at the start of a project, it could assumed that existing water distribution piping system is serviceable. Later it is found that new lines are necessary, so the portion of the project plan associated with this change could obviously not be built into the original project plan. Generally, however, most, and certainly the major, organizational project plans can be in hand. We prefer to have these submitted prior to the initial CPE meetings so that the lead project manager can study the ideas presented, look for the gaps, and assemble a draft CPE project plan to serve as the basis for discussions. Doing this will save huge amounts of time on international projects.

We discuss this in detail as we progress through the book. For now, the main idea is that a CPE is required and that it needs to be shared transparently with the organizations that participate. That does not mean that the organization that manufactures the doorknobs for the building needs to have the full plan; judgment is necessary. However, there must be a culture to share rather than withhold information. The doorknob manufacturer simply would not benefit from having the details of how the foundations were planned, but it would appreciate having the information on the types of doors and the criticality of the delivery date—the true criticality, not a manufactured one.

In the event that the project fails or there are disagreements, it is essential the lead project manager helps to resolve disputes.

5.4.1 Disputes

In Chapter 2, we reviewed the different legal systems that come into play on international projects. Each organization must first attend to the laws and professional standards of its home country or country of incorporation, then to the other standards. Home-country laws should be well known and understood. On most international projects, the customer determines the legal standards (i.e., "In keeping with the laws of Malaysia . . . ") for conducting business and for dealing with disputes. Organizations always attempt to negotiate the use of their own laws as a standard for the contract, but usually the customer wins out. If the negotiation is multilateral (i.e., an alliance), the dilemma is compounded.

Organizations are subject to the laws of the countries where they conduct business, and so the dilemma focuses mostly on the legal rights of the parties and the options for resolving disputes. A full description of the legal dilemmas is beyond the scope of this book and must be addressed by competent attorneys with international legal knowledge and experience. However, we would like to make some specific recommendations regarding disputes. Resolving disputes effectively and efficiently is an attribute of successful international projects, and it needs to be addressed in three ways: legal framework, project process, and project attitude.

Despite all good efforts and intentions, the project team cannot resolve some disputes. The legal framework for a contract sets forth the manner in which intransigent disputes are resolved. With the problem of multiple legal systems, the method of choice has become a meditation-arbitration process, which is also part of what is called alternative dispute resolution (ADR).

The International Center for Dispute Resolution (ICDR 2007) offers this ADR procedure:

> In the event of any controversy or claim arising out of or relating to this contract, or a breach thereof, the parties hereto agree first to try and settle the dispute by mediation, administered by the International Centre for Dispute Resolution under its Mediation Rules. If settlement is not reached within 60 days after service of a written demand for mediation, any unresolved controversy or claim arising out of or relating to this contract shall be settled by arbitration in accordance with the International Arbitration Rules of the International Centre for Dispute Resolution (www.adr.org/sp.asp?id=32745).

There are also guidelines promulgated by the United Nations Commission on International Trade Law (UNCITRAL; www.uncitral.org/uncitral/en/uncitral_texts/arbitration.html), most of which deal with the ways that nation-states can draft laws to take advantage of arbitration and of course the rules governing arbitration. There are also rules promulgated by the International Chamber of Commerce (ICC; www.iccwbo.org/drs/english/adr/word_documents/adr_clauses.txt). The ADR clause suggested by the ICC is very similar to that of American Arbitration Association (AAA).

Alan Redfern (2003), a barrister in London, suggests that in writing an arbitration clause, the parties should consider the arbitrators, the location, and the New York Convention. From our experience, this is sound advice. The parties can agree to mutually select a single arbitrator, they can agree to a panel of three arbitrators where each party selects one arbitrator and then a neutral arbitrator is selected, or they can agree to two arbitrators. A single arbitrator reduces the cost, but it may be difficult to find one who has technical knowledge, legal knowledge, and arbitration process knowledge.

A location should be selected that is neutral in the sense that it is not the home country of any of the parties and backup services readily available for accommodations, access, communication, support, and the like. The location should be a place where the local courts recognize the validity of international arbitral awards. Location is a guiding principle of the UNCITRAL Model Law on International Commercial Arbitration. Last, and most important, the place chosen should be a country that is a party to the 1958 New York Convention (a list can be found at www.uncitral.org/uncitral/en/uncitral_texts/arbitration/NYConvention_status.html) on the Recognition and Enforcement of Foreign Arbitral Awards. We often use London, Singapore, or New York, depending on the location of the project.

Organizations are *always* best served by negotiating their differences between themselves. If they are unable or unwilling to do this, the next best solution is mediation, which is just a facilitated negotiation, and the parties still make their own decisions. If both of these fail, arbitration is the least repulsive alternative, but it has become expensive in the last 10 years, and if attorneys are involved can transmute into what we call l'arbitration (a faux litigation).

Building on Redfern's views and the AAA wording, we suggest the next wording for an ADR clause:

> In the event of any controversy or claim arising out of or relating to this contract, or a breach thereof, the parties hereto agree first to try to settle the dispute between the project managers. If the project managers cannot reach settlement within 30 days, then the matter shall be submitted to the project sponsors.
>
> If the project sponsors cannot reach settlement within 30 days, then the matter shall be submitted to mediation, administered by the International Centre for Dispute Resolution under its Mediation Rules.
>
> If settlement is not reached within 60 days after service of a written demand for mediation, any unresolved controversy or claim arising out of or relating to this contract shall be settled by arbitration in accordance with the International Arbitration Rules of the International Centre for Dispute Resolution. The parties agree to select a tribunal of three arbitrators. The parties also agree that the arbitration shall be conducted in London (or New York or Singapore).

5.4.2 Cultural Norms for the Project

In Chapter 2, we introduced the concept of a CPE. Experience shows that the lead project manager must establish a set of cultural norms for the CPE, based on a compassionate, empathetic, and patient understanding of the diversity on the team. If this is not done, people will construct their own from self-interest or protection—an "I" mentality. The lead project manager can "contaminate" the CPE with inclusive behavior through his or her actions by building a "we" mentality. This can preclude the natural exclusiveness that differences in culture and business goals engender. Even on a project that has a contract structure that naturally pits the parties against one another, the lead project manager can inculcate a modest sense of "we."

There is infinite diversity among and across cultures, but some generally accepted basics can be used to anchor the culture of a CPE: trust, empathy, transformation, power, and communications. These are not quick fixes but require compassion, time, consistency, patience, and repetition. They also demand that the lead project manager articulate and demonstrate them. "We will treat one another with respect" is one example of a CPE cultural norm. Respect for a junior partner, respect for a social culture, respect for a disabled individual, respect for the age of an individual, respect for other ethnicities, respect for other genders, respect for the laws in a country are some examples to illustrate the point. The level of respect will determine the level of effectiveness. It does not need to be a contractual ironclad definition; in fact, it is better if it is not.

Since the so-called black ships first landed, the Japanese have masterfully adopted select portions of other cultures and technologies. Likewise, with globalization, some emulation naturally occurs, and is easy to see it in the global markets today. The challenge for leaders is that emulation can be accurate, or can be an adapted version of another culture. People enjoy learning about other people (people watching, literature, operas, soap operas, etc.), and the chance to test one's knowledge and curiosity is the opportunity to apply the honest communications. We strongly believe that storytelling is an excellent way of starting this process (Grisham 2006).

For example, each person can share a story from his or her culture—a children's story, a poem, a fairy tale, or a folk tale. Doing this is safe (third person), educational, and enjoyable; most people enjoy talking about where they come from. Asking each party to write a story using other characters representing their perceptions of the positions of the parties in the dispute is an effective way of opening up barriers between the parties. However, a deep and respectful understanding of personalities and cultures is required when using this technique for disputes. The efficacy of the XLQ displayed by the lead project manager will have a great impact on the degree to which the participants adopt the norms of the CPE. They must feel that they are part of the tribe and must decide voluntarily to follow the lead project manager.

One other aspect of international project integration flows naturally from the leadership of the lead project manager, and that is survivability. What we mean by this is that corporate priorities change and customer attitudes change, which can easily result in the redeployment of the lead project manager and/or individual project managers from one of the organizations in the CPE. The culture of the CPE survives changes and in fact can become even stronger when change occurs.

5.5 JOINT PROJECT PLANNING MEETINGS

As introduced in Chapter 3, once a project is selected and a CPE project charter executed, the team can begin initiating and planning the project. If the structure of the CPE is similar to that shown in Figure 3.1, we recommend that the next four steps be used for the joint planning meetings:

1. *Create an organizational plan.* Each organization should conduct an internal joint planning session based on the CPE project charter issued by the lead project manager. This provides the context for the scope of work undertaken by each organization. By conducting an internal session, each organization can prepare a preliminary project plan and get ready to address potential project challenges, such as scope questions. It also shows respect for the organizations involved in the project, for ultimately all of the organizations must come together in a holistic joint project planning session.
2. *Initiate a value chain plan.* Conduct external joint planning sessions with the value chain. Therefore, it would be best for service provider A in Figure 3.1 to arrange for joint planning sessions with the value chain A.1 and A.2 individually, after each organization has completed its own internal session. This will enable service provider A to determine where gaps in scope exist, how interface points with the other organizations will be managed, how risks will be allocated, how communications will be maximized, and much more. These meeting begin the process of forming the CPE from the grassroots level. They also begin the process of building trust, empathy, transformation, power, and communications.

3. *Plan with partners.* Here service provider A in Figure 3.1 would conduct an external joint planning session with the partner (service provider B). By this step, both organizations should see clearly where challenges exist and where leadership must focus. The partnership can now prepare a holistic scope split document, project schedule, project budget, project risk register, project communication plan, project quality plan, and project human resources plan. As with the external value chain sessions, this step begins the process of trust and transparency within the partnership. From our experience as a customer, we found that the more intimate the relationship between partners, the more comfortable we were with the service providers. It is easy to tell how close and trusting the relationship is in a partnership and where, from the customer's view, challenges will emerge.
4. *Create a CPE plan.* This step includes an external meeting between service provider A, service provider B, the customer, and possibly the user, depending on its role in the project. If the user will be directly engaged in project acceptance, we strongly recommend that it should be included. The importance of this session cannot be overemphasized. By showing up well and fully prepared, the partnership can demonstrate the importance of the project to the customer and user. It builds trust.

The ideal approach is to have all of the key participants in the project present for the scope split discussions. Here the lead project manager must decide which organizations fall into this category. Figure 5.2 provides one way to evaluate the participants. We do not recommend only looking at an organization's financial stake in the project; that may be dangerous.

Each organization should perform its own project plan for what it believes is its work and should carefully review of the contract documents. Doing this will facilitate a productive and trusting dialogue between the parties. The sponsor, project manager, and salesperson, if there are any, for each organization should attend the scope split discussions, along with any support people required. It is recommended that the lead project manager assemble the project plan of each organization into a single document and share it with all participants at least a week in advance of the planned meeting.

The partnership will have designed the preliminary plans that we discussed earlier and in the process will have developed a list of challenges

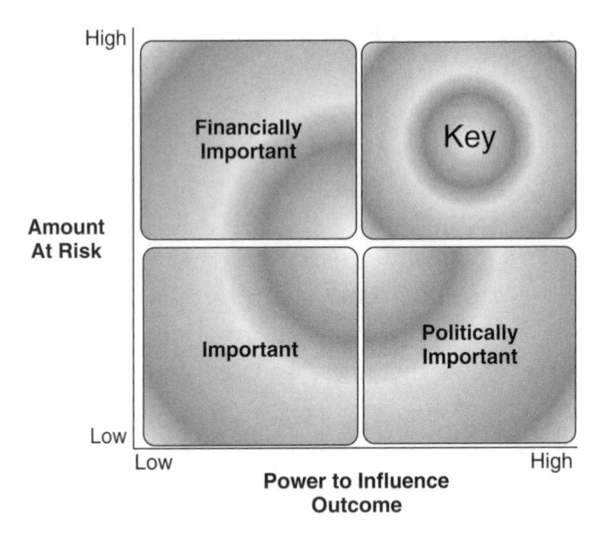

Figure 5.2 Key Participants

that need to be addressed. The partnership will have learned about the customer and user, and should have a solid understanding of their organizational culture and structure. The partnership should also have developed an understanding of the cultures and personalities of the key stakeholders. The lead project manager is on trial at this meeting. The participants will observe how this manager conducts him- or herself and whether words and deeds match. The meeting objectives for the lead project manager should include:

- *Building the team.* International projects rely on virtual teams, and this is a golden opportunity to have everyone meet face to face. Time must be reserved for people to interact and perhaps for team-building activities, if appropriate. The lead project manager establishes the CPE value of respect for the team. She or he builds the sense of "we" and encourages commitment, cooperation, and coordination.
- *Learning the cultures and personalities.* Each person has a different cultural background with different values and norms. All will adapt the corporate culture in a different way. Depending on their experience base, they will have different views of ethnicities, genders, age groups, and so on. This is the first opportunity to make lasting impressions, good and bad, and to let people get to know with whom

they will be working with for the rest of the project. The lead project manager establishes the CPE value of empathy for others.

- *Opening the communications.* Depending on the structure of the contract, communications may be limited from the start. This is the first chance to show how transparent, capable, and willing people are going to be when it comes to sharing information. The lead project manager establishes the CPE value of honest and open communications.
- *Building a culture of change.* This involves demonstrating that change is natural and encouraged and that acceptance of change is expected and rewarded. There will be preliminary plans at the start of the meeting, and most definitely changes will be necessitated during the course of the meeting. This step provides the lead project manager with the opportunity to demonstrate that change is a healthy thing and to be welcomed.
- *Building a culture of negotiation.* There will be conflicts on all projects, but the key is how they are addressed and how the participants are affected. Most contracts will have dispute resolution processes (as discussed), but people must understand that conflicts are to be managed just like every other aspect of the project.
- *Seeing the big picture.* One attribute of leadership is vision. The joint planning sessions enable the lead project manager to learn the personalities and cultures of the participants, look for potential points of friction, and build trust and relationships. In more practical terms, though, the sessions enable the lead project manager to look at such things as material and labor demands for the overall CPE.

For example, on a major international airport project, a hurricane struck. Plans were implemented to protect and cleanup afterward. A spike in costs for materials and labor occurred, in some cases as much as 100%. This placed the members of the CPE in competition for precious resources and left critical path activities understaffed because the organization could not get the people they needed. The craftsmen had already been offered higher wages to work for another participant organization whose work was not critical but whose profit margins would be eroded by staying on the project longer. Knowing the manpower and material requirements for the CPE permits the lead project manager to anticipate problems.

The above listing is a lot of work, and we have not yet talked about scheduling, estimating, quality, risk, and scope. However, the tasks are the foundation for a successful international project. If they are not present, they will greatly diminish the chances of success. One added warning: On a project in Pakistan, we had a Chinese design firm, a Japanese supplier, and a U.S. global firm as service provider A, an international U.S. firm as service provider B, a Pakistani firm as the customer, and a U.K. firm as the user. Due to the high costs of getting everyone to one location, we decided to do a single joint planning session. It proved to be a bad idea, creating more problems than we solved. The reason was that each organization had not followed the steps we recommend in this book. Each organization must create its own project plan by creating a virtual version of its product, including its value chain. The partners do the same, then the customer and the user become involved. The reason for our failure was that the individual organizations came to the meeting without a plan, just many questions. This caused the customer and user unnecessary concern and destroyed trust in the CPE.

If a lead project manager focuses at this stage, the other aspects that must be addressed can serve as opportunities to foment these attitudes. From the practical side, we suggest that the lead project manager *lead* the meetings but not manage them. The lead project manager should have an assistant manage the meeting so that the lead can focus on the items listed, learn the personalities, and look for the areas where conflict will arise. A person cannot do this while busy with managing the meeting. These initial joint planning meetings and final acceptance are the two most important days of the project.

The next chapters describe the details that need to be planned, starting with scope.

Chapter 6

Scope Management

A courtyard common to all will be swept by none.

—*Chinese proverb*

Never doubt that a small group of thoughtful, committed people can change the world. Indeed, it is the only thing that ever has.

—*Margaret Mead*

In this chapter, we define some of the primary issues surrounding scope. From the initial development to the validation with the customer, the scope of the project determines all of the other aspects. The scope establishes the boundaries, the resource requirements, the quality, and the other aspects of a collaborative project enterprise (CPE) plan. On international projects, this is even more critical due to the diversity of the project organizations and the cultures.

6.1 DEVELOPING A PROJECT SCOPE

Normally, the customer establishes the scope of a project, unless the project has a public-private partnership (PPP) structure. Developing a scope description for a new project is a challenge, as anyone who has attempted it knows. Imagine writing a specification for a new international information technology (IT) platform for a financial organization. It is an internal project, and the customer is also internal. The project must communicate with the existing back-end systems, must seamlessly integrate with the other financial platforms operating in each country, must function in 10 different languages, and must be easy to use—sound familiar? How much detail is needed to describe the scope? How much money does the organization want to spend writing the scope document? Should we use agile project management and just get on with it?

Normally, three general approaches are used to describe the scope on international projects:

1. *Technical specification.* A specification that describes every component of the facility in detail. It would include a complete description of the generator, the boiler, turbines, and all of the associated equipment. It would take many months to assemble and would require a detailed knowledge of the components and processes. Most customers would need to hire a consultant who is expert in the field. This is the most expensive approach.
2. *Equal to specification.* A specification that references components available in the industry. It would include the same listing of components as in number 1 but would provide a list of model numbers from acceptable existing manufacturers.
3. *Performance specification.* A specification that defines the ultimate performance of a facility. Here the scope would specify the power output (megawatts), the energy efficiency ratio (coal in megawatts out), the characteristics of the effluent pollutants, and perhaps some availability guarantee. The details are then left to the discretion of the designer/builder. This is the least expensive approach.

"Perfect information" is an economic concept that considers 100% certainty to be too costly. Perfect knowledge is not required; accurate knowledge is. The difference is how much of a standard deviation is acceptable. Projects undertaken by organizations like Nielsen Ratings are founded on this idea. To poll people about attitudes toward a product or politician, these organizations take a statistically significant sample that is an accurate, but not perfect, representation of attitudes. If there are 100 million people in the population, you could sample all of them and have perfect information. Or you could sample, say, 5,000 of them who are generally representative of the entire population and be accurate within perhaps +/-5%. If it costs €1 to poll each person, the question is, does the benefit of having perfect information at a cost of €100 million provide more utility that an accuracy of +/-5% at a cost of €1,000? Probably not. Later we discuss how this connection between perfect information and cost fits into control charts and deviations. For now, apply the same concept to writing a scope for the IT project using the three options listed.

Imagine a technical scope for the IT project described down to a thorough technical definition of every line of code required. The cost of perfect information could be astronomical, and the cost/benefit ratio would be unacceptable. At the other extreme, imagine the customer uses a performance specification, such as the one described, that leaves everything to the imagination of the project manager. Likely few issues will arise until the customer and user can see the product. The customer user likely understands the business need and of course the financials, whereas the project manager might understand only the technical side. The product is ready to test drive, and the customer and user see it for the first time. International project managers must protect against such acceptance surprises.

Most customers and users therefore tend toward the equal-to specification approach since it is more moderate. Internal projects within multinationals, particularly IT projects, tend to be performance specifications. The reason is obvious: The projects are funded internally, so the need to be specific is not as pressing. It should be, however. We prefer to apply the same discipline to internal projects as are applied to external projects.

There is also a spectrum of contracting structures that can be utilized to procure external services, and we discuss the details in Chapter 11. In general, there are two primary genre of contracting structures:

1. *Fixed price.* Contracts that have a "guaranteed" price for the scope required. Such contracts often are competitively bid, and governmental agencies frequently use this approach. These may have bonus or penalty incentives to encourage certain performance. This type of contract has limited communication about the scope until after award of the contract. Generally, such structures are better suited to a technical approach or to an equal-to approach where the level of detail on scope is high.
2. *Cost reimbursable.* Incurred costs are repaid by the customer. Costs are paid contracts that are often negotiated. Many times these contracts have bonus or penalty incentives as well. This type of contract has open communication about the scope prior to award of the contract. Generally, such structures are better suited to a performance approach or to an equal-to approach where the level of detail is moderate or low.

On international projects, customers can be national organizations or consortiums, and the CPE is made up of international organizations that provide the services. Generally, the customer retains the services of an organization with subject matter expertise to prepare, or aid in the preparation, of the scope documents necessary for tendering. It is at this stage that the structure of the CPE is determined by the general and special conditions of the contract. We discuss this in more detail in Chapter 11. The customer, and perhaps the user, with the small and medium enterprise often define project scope. That is not the case with PPPs, where the service provider plays a central role throughout the selection, implementation, and planning processes.

Imagine an IT project to create an online human resources platform for a government entity in Sierra Leone, with the government procurement office being the customer and the government human resources office being the user. The customer might contract with a service provider that has expertise in designing such platforms but might not have technical expertise in the existing software, hardware, or the country. This service provider would have limited abilities. Thus customer would need expertise from other service providers, such as hardware, to fully describe the scope. If the service provider doing the design hires a team of experts in these other areas to prepare the scope, clearly the scope will be more complete. However, the cost will be higher, and the design service provider may select sole-source suppliers, thus eliminating competition. In public tendering, this creates problems, both political and legal.

If the design expert prepares the scope, the prospective bidders obviously will have to complete the scope details to complete the project. If the project is bid competitively, completion of the scope will have to be done postaward. If the project is negotiated, it would be done prior to contract finalization. Consider the expert in hardware. The scope would likely provide an equal-to specification for the components themselves and include some sort of training, spare parts, warranty, and maintenance requirements. If the scope document is written without specific current knowledge of the market in Sierra Leone, it may not be possible to provide the equal-to equipment specified in the document. Assume that the equipment can be imported at a higher cost but that there is no infrastructure in place to provide for maintenance. People would have to be flown in from Europe, and this would eliminate any availability requirements in the contract. There would be a similar issue on the software side, but that issue may be able to be corrected virtually. Who takes the risk for overall system availability regardless of whether the issue is software or hardware? The expert writing the scope would of course pass this risk on to the service provider purchasing and installing the components.

Let us assume that the customer hires an organization from the United Kingdom to develop the scope and contract. Then assume that an IT project management organization from Latvia wins a competitive bid contract and assembles a consortium of organizations to provide the service. In the consortium, 33% of the members consist of the IT project management firm, 33% are from a software organization from China, and 33% are from a hardware organization from Japan. All three firms sign a competitively bid contract with the customer and must determine, in detail, who will provide which portions of the work. In reality, responses to tenders often are competitively bid on a to-each-his-own basis, which means that each organization prices the risk associated with the failure of its scope of supply at a high level. After the award, the consortium must come together to perform a scope split and develop a detailed scope assessment.

Figure 6.1 is a graphic representation of the basic options available to a customer when developing the scope for a project. The vertical dimension is risk, with a competitively bid technical specification and a fixed price contract representing the lowest risk to a customer. The reason is that the risk of cost increases and performance is borne by the service provider.

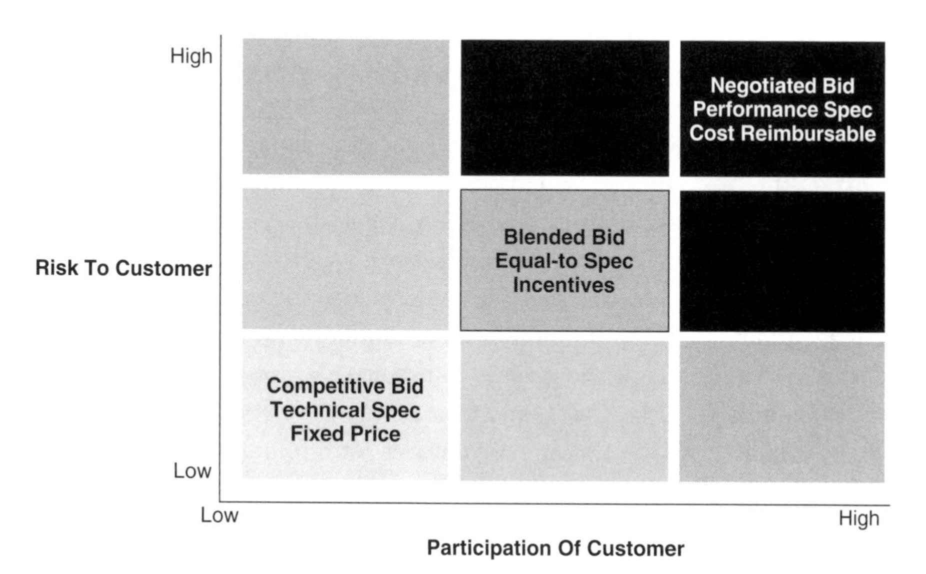

Figure 6.1 Developing Scope

At the other vertical extreme, a fully negotiated performance specification and a cost-reimbursable contract yields the greatest risk to the customer. If a customer does not have a clear and defined scope, the risks of changes in scope are better managed by the organization that can best control the risk—in this case the customer.

In our experience, we have seen customers use a competitively bid format for a project that has a partially defined scope for transferring the risk to the service provider. It results in many disputes and an increase in legal participation. On one large industrial project in the United States, we were responsible for the start-up of the facility. Start-up requires that the process equipment, instruments, and IT systems function properly. Therefore, a start-up team will include people who are specialists in different disciplines, such as IT, electrical, mechanical, and so on. On this project, we had one extra specialist, an attorney. Imagine trying to untangle an IT bug while an attorney with tape recorder is asking why the problem arose and who caused it.

The other side of the coin is that experienced service providers will see the gaps in the scope and increase their bids accordingly. We have seen customers who are repeat users of services ratchet up costs in the market because of their reputation to leave out scope. Unfortunately, some service providers will not increase their bids but rather submit artificially low bids with the intention of flooding the project with change requests to recoup the lost costs. This is one of the many reasons the collaborative approach, and transparency, is recommended.

The horizontal scale in Figure 6.1 shows the participation of the customer. Competitively bid technical specifications with a fixed price contract tend to require less participation of the customer, for the scope and the risks should be well defined. A negotiated performance specification with a cost-reimbursable contract will require ongoing interaction between the customer and the service provider to refine the scope and manage the risks. If the contract is cost reimbursable, the customer must attend to the details of the project to protect its interests in containing and managing the scope, cost, and time.

The majority of international structures tend to be toward the center of the chart in Figure 6.1. In the equal-to specification, there is a sharing of risk, moderate participation of the customer in the project, and a contract structure that includes incentives and penalties. Our experience has shown that a moderate approach, such as this, more often produces successful projects. We discuss much more about contracting approaches when we examine procurement in Chapter 11.

6.2 DEVELOPING A WORK BREAKDOWN STRUCTURE

Developing a work breakdown structure (WBS) for the entire project requires a sequence of steps that were introduced in Chapter 3. The first step in the sequence is the individual organization.

6.2.1 Step 1: Organizational WBS

Each organization must first understand their scope for the project. In fact, this is exactly the perspective the *Project Management Body of Knowledge* (*PMBOK* 2004) takes when discussing the scope knowledge area, and the IPMA would consider this Level C competence. The CPE WBS would be Level A competence according to the International Project Management Association (IPMA), and possibly a program under the *PMBOK*. In this book, we use the CPE WBS to describe a detailed list of activities required for the complete project, regardless of the performing organization. Figure 6.2 illustrates that the term project is confined to an organization, or service provider A in the figure. When we talk about CPE WBS, we mean for the entire scope of the work, including any portions provided by the user and customer—financing, for example.

To do scope planning, the *PMBOK* indicates that the inputs are organizational process assets, enterprise environmental factors, project charter, preliminary project scope statement, and a project management plan for the organization. Organizations value chain A.1, service provider A, the customer, and the user would each prepare a WBS for its individual organization according to the *PMBOK*. All of these are internal organizational efforts. We do not imply that the *PMBOK* ignores the other levels; just that it is designed for this level. To get to the higher levels of coordination requires broadening the perspective taken in the *PMBOK* and by the IPMA. We emphasize this point for people who have become certified and who want to use the standards on real-world projects.

If a project is competitively bid on by value chain A.1 to service provider A in Figure 6.2, it is highly unlikely in practice that value chain A.1 would be given the scope definition that the customer gave service provider A. Most likely, the scope provided for the tender would provide a very high-level description of the project—project charter level perhaps—and a

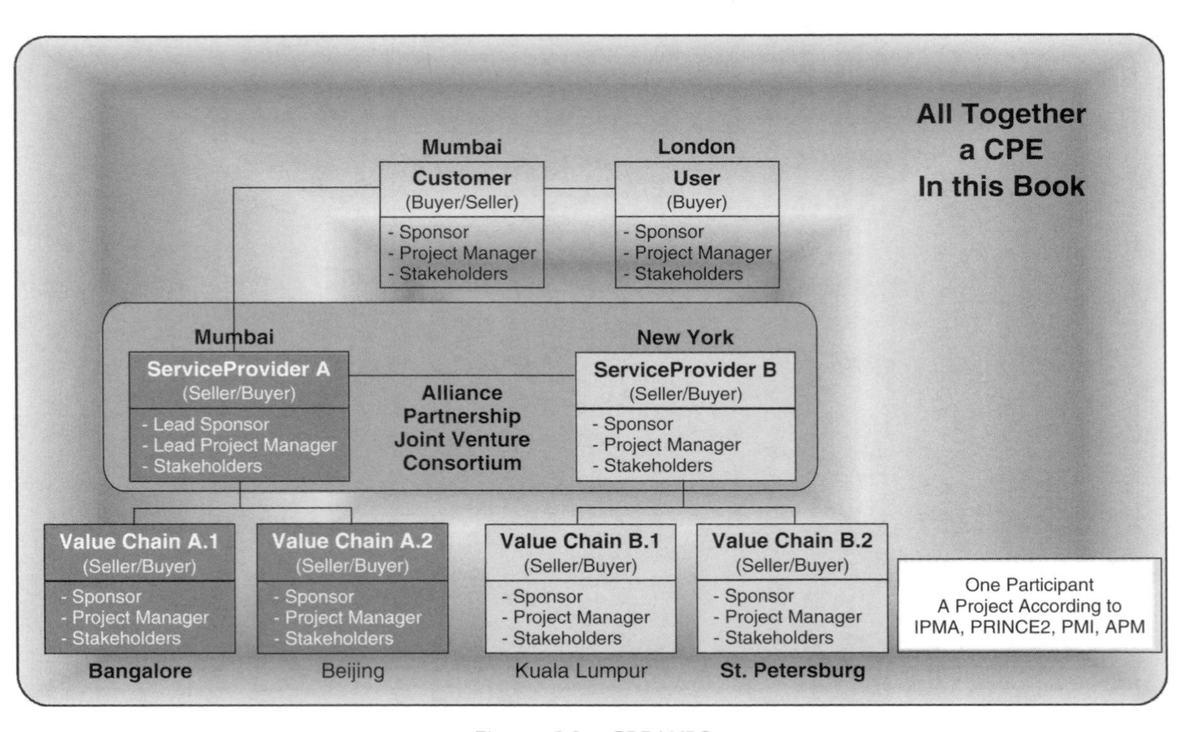

Figure 6.2 CPE WBS

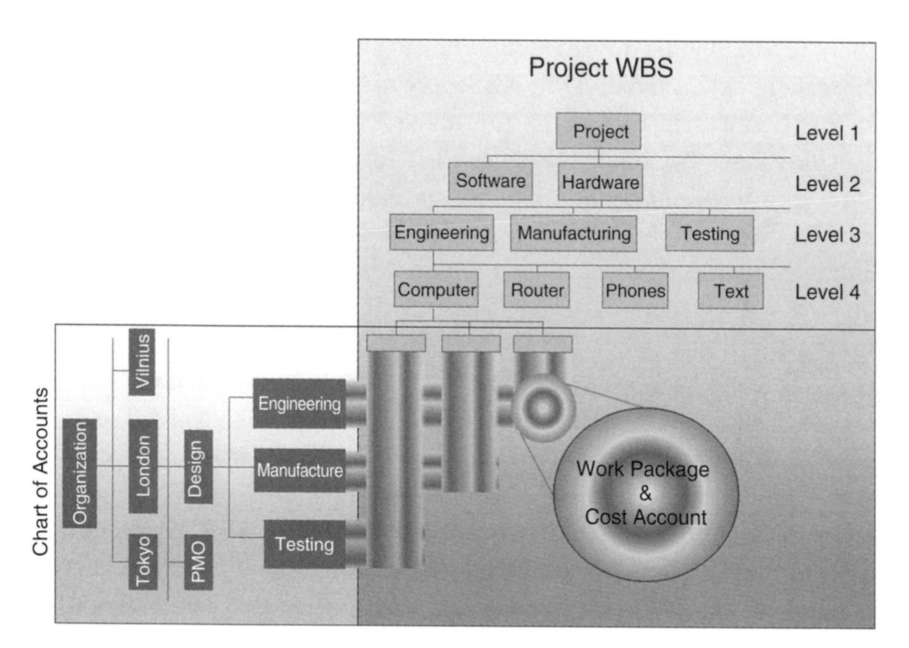

Figure 6.3 WBS and Chart of Accounts

detailed description of the only portion of the scope to be provided by value chain A.1. As we noted earlier, value chain A.1 likely would not consider assembling the bid to be a project, but if the organization were successful in its tender, then an internal project would be initiated. Of course, in this case, the organizational WBS would have been done in order to tender for the work, but normally a tender is not detailed enough to enable the work to be executed.

Figure 6.3 provides a graphical view of the project from the perspective of one individual organization and the intersection between the WBS and the corporate accounting system (chart of accounts). The WBS in our example could have been at Level 3 for the tender, but executing the work might require a lower level of detail. In neither event would value chain A.1 know what the WBS for value chain B.2 was, in practice.

If value chain A.1 is invited to participate in a negotiated tender for a performance specification, it is more likely that it would know about the WBS for value chain B.2, and vice versa. Value chain A.1 would likely assemble a more detailed—perhaps Level 5 or 6—WBS, for its work ultimately would be reimbursed. In many organizations, this approach to work would constitute the beginning of a project and would have a project charter attached to it. Some organizations refuse to open a project until there is a signed legal instrument, but in either case, the WBS would be more detailed. By making the WBS more detailed, more questions emerge as the technical professionals—in this case value chain A.1—looks at the specifications. The other benefit is that at this level, the questions from value chain A.1 may need input from value chain B.2. Therefore, the coordination begins before the actual work begins. The organizations not only know what they are to do at a low level of detail, but why as well. Moreover, this exchange builds a culture of cooperation in the CPE from the start.

A WBS for each organization must be at a level of detail that will enable the project manager to manage the work effectively. Too much detail and the project manager will spend all of his or her time chasing and evaluating the numbers; too little detail and the project manager will continually be speculating. According to the *PMBOK* (2004), this level of detail is called a work package: "a deliverable or project work component at the lowest level of each branch of the work breakdown structure. The work package includes the schedule activities and schedule milestones required to complete the work package deliverable or project work component" (p. 380). In Figure 6.3, this might be below Level 4, or it might be more detailed. The other point to bear in mind is that a Level 5 for the overall project may be an organizational Level 1 for value chain A.1.

Why does this matter? It is critically important that the lead project manager understand the level of detail being supplied by each organization and the timing of that information relative to the tendering process. On a project in Tajikistan, the lead project manager was negotiating a contract with a customer and needed to get pricing from potential partners for some of the scope that the lead organization could not provide. It was needed very quickly and was provided, but with caveats. The lead project manager could not be certain how much detail was considered and thus if all of the intended scope had been factored in. As the partner's work represented about 40% of the value of the total project, it was critically important. The project value was approximately $100 million. As we said earlier, if the accuracy for a project is +/-20%, the control limits would be +/-$8 million. Say the profit for the lead firm was 8%. If the potential partner is within the control limits, it makes no profit, and if because of the rush the accuracy is more like +/-30%, it loses money.

From experience, we have a few heuristics for where to stop in detailing a WBS. The WBS dictionary (to use the *PMBOK* term) includes estimates of cost, which of course require an estimate of duration, resource utilization, and labor rates. A project manager and lead project manager should spend a majority of their time working with the people on a CPE. To create space, or time, a project manager, and certainly a lead project manager, must reduce the burden of fooling around with details. We do not mean ignoring details but rather spending time on those things that bring the greatest return on investment. Therefore, build processes, and use tricks, to let the technology do the detailed work, not you. With that in mind, here are our suggestions:

- *Shortest duration.* Work packages, the lowest level of detail, generally should not have durations that are shorter than 1% of the total project length. If your project is 120 total workdays in duration, work packages should not be shorter than 1 day. We also suggest that organizations look to their cost accuracy. For example, if the organization requires that employee time be recorded on a monthly basis, trying to manage resources for a 1-week activity will not be a useful endeavor. The reason is that at each progress update, the project manager will have to estimate the actual costs. Of course, if the project involves an outage of a short period, say 48 hours, or for a long period, say 10 years, then the heuristic needs to be adjusted.
- *Longest duration.* Work packages should not have a duration longer than the reporting period. If service provider A in Figure 6.2 is to provide monthly progress reports to the customer, then no work package should have a duration that exceeds 1 month. In practice, activities will have either finished or not started in about 66% of the cases. This leaves a smaller percentage of activities that must be measured for physical progress, or earned value.
- *One Work Package = One Chart of Account = One Schedule Activity.* The term historically used to describe an organization's internal accounting system is "chart of account"; these are systems like SAP or Oracle.

In Figure 6.4, assume that all of the groups represent internal organizational branches participating in the project for the customer. The project manager in Bangkok will need to know the corporate chart of accounts charges assigned to his or her project. The project manager will also need to know what work package these resources performed during the period in question. If one work package uses resources from portions of four charts of account numbers, allocating the costs will take

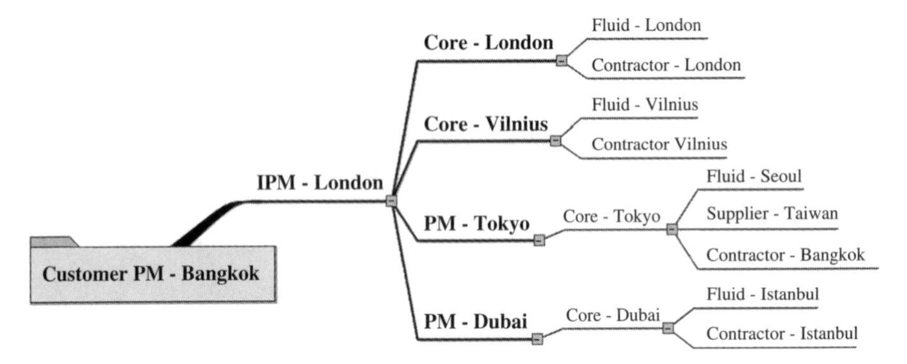

Figure 6.4 International Team

time. Similarly, if multiple work packages use resources from one chart of account numbers, the project manager will also need to spend time to allocate them.

It is not always possible to do this, but we recommend it as a goal. In Figure 6.3, the work package, again, is the intersection between the corporate chart of accounts and the project WBS. Remember this is for one organization at this point.

Having laid the foundation, now let us look at the internal organizational WBS itself. According to the *PMBOK*, a WBS is a list of tasks, whereas a WBS dictionary includes estimates, assumptions, risks, resources, and more. The WBS dictionary template we use to explain the concept is too wide to display fully, so it has been broken into sections. Each figure shows a portion, a few columns, of the Excel workbook. We describe the columns starting at the left (See Figure 6.5.).

- The first two columns provide the work package connection to the project WBS and the corporate chart of accounts. The third column is the description of the work package itself.
- "# of People" refers to the number of internal core resources—employees of the organization—who will be needed to accomplish the task.
- "Hours per Day" are the hours each employee will devote to the work. Each person may be fully dedicated to the project or devote only a portion of his or her workday to the activity. For example, the project manager may only spend 2 hours of her day reviewing the work of others. Breaking this down requires linking resource spreadsheets and is beyond the scope of this book. In this example, assume everyone works the same number of hours on the task each day.

from corporate	from PMO	Analagous or Brainstorm Bottom-up or Top/down	# of Resources		PERT				calc'd	Amount people are paid	Office, management, support, etc.	calc'd
Corporate Chart of Accounts	WBS #	Work Packages	# of People	Hours per Day	Best Case Days	Most Likely Days	Worst Case Days	PERT Work Days	Person Hours	Base Hourly Rate	Overhead & Profit Factor	Burdened Personnel Cost
										$100	3.10	
07.01.03.01	01.01.10	Joint Planning Meeting (Service Provider A)	8.0	8.0	2.0	3.0	5.0	3.2	203	$100	3.10	$ 62,827
	01.01.20	Medical Mentoring	10.0	4.0	20.0	35.0	60.0	36.7	1,467	$100	3.10	$ 454,667
	01.30.10	SC.A.1 - Medical Training Facilities & Documents						-	-	$100	3.10	$ -
	01.30.11	SC.A.1 - Prepare Medical Training Documents						-	-	$100	3.10	$ -
	02.01.10	Deliver Drugs to Hospitals						-	-	$100	3.10	$ -
								-	-	$100	3.10	$ -
								-	-	$100	3.10	$ -
07.01.03.02	01.01.90	Acceleration (Risk of Liquidated Damages)						-	-	$100	3.10	$ -
								-	-	$100	3.10	$ -
		Subtotals							1,669			$ 517,493
		Project BAC										
		Contingency Reserves										

Figure 6.5 Organizational WBS (first section)

- The next three columns are a PERT (Program Evaluation and Review Technique) analysis of the number of workdays required to perform the task. For the Joint Planning Meeting, the example says that the best case will require 2 days, the worst case 5 days, and the most likely case 3 days to do the meeting. The formula is:

$$\frac{\text{Best Case} + (4 \times \text{Most Likely Case}) + \text{Worst C}}{6}$$

- "Base Hourly Rate" is the amount the organization actually pays the person; some call it the person's salary. Each person's rate will be different, and hourly people may be entitled to overtime rates. Breaking this down requires linking resource spreadsheets and is beyond the scope of this book. In this example, assume everyone has the same rate of pay. It is critical to understand that the salary of individuals is not public knowledge and that divulging it is against the law in certain countries and organizations. We have displayed it here only to illustrate the point. We recommend that this column be hidden in all internal and external reports. The reason is that some organizations and cultures consider this to be personal information. Others do not.

  One other point that needs to be mentioned is internal billing rates. Many firms have internal billing rates that one profit center charges another for the use of resources. For example, in Figure 6.3, the London profit center may charge US$350 per hour for internal personnel assigned to work for the Vilnius profit center. It is not unusual for an internal billing rate to be greater than an external billing rate in some companies.
- The "Overhead & Profit Factor" and "Burdened Personnel Cost" columns capture the personal benefits, corporate overhead, and profit. The personnel benefits would be such things as medical insurance, 401K, vacation, sick leave, and so on. The corporate overhead would be such things as the chief executive's pay package, management pay packages, the buildings, the IT systems, and the like. Many firms allocate their overhead to projects; one way is through their contribution ratio to the corporate revenues or costs. Overhead figures for organizations are proprietary information, and divulging it is against the law in certain countries and organizations. We have displayed it here only to illustrate the point. We recommend

that this column be hidden in all internal and external reports for the same reasons noted earlier.

Then there is the profit on the project to consider. Clearly most organizations, except nonprofit ones, do not wish for their profit margin to be known internally and certainly not externally. Therefore, the project manager must make a choice of how to capture profit in the budget at completion. Our suggestion is to allocate it proportionally across all of the work packages. In the example, we used a factor of 3.1 to include benefits, corporate overhead, and profit.

The "Base Hourly Rate" and the "Overhead & Profit Factor" columns are highlighted to indicate that they should be hidden and that access is restricted to the sponsor and the project manager for the organization.

- The "Burdened Personnel Cost" column totals the hourly rate and the overhead and profit factor. It may need to be divulged internally, particularly if a work package is assigned to an individual to lead. It should not be divulged externally except in rare circumstance, such as an alliance relationship.
- You will notice that the last work package is titled "Acceleration," and it is for extra work or a scope change. Here the organization has another choice to make regarding disclosure of information. Assume an organization has assessed the extra work and believes it will require 10 added days of time. Assume that there are liquidated damages (LDs) of US$30,000 per day so the total exposure is US$300,000. The organization assesses the risk and determines that it can be mitigated by working overtime at a cost of US$40,000. The question is: Should it be divulged to the CPE or hidden in the amounts? We suggest you do divulge. The reason is that it is information that may well directly bear on the other participant's ability to perform their work on time, and it builds trust.

Now let us look at the next section of the organizational WBS dictionary shown in Figure 6.6.

- "Fixed Cost" refer to those items that are purchased externally to the firm. Contract labor, materials, supplies, travel, value chain vendors, and the like would be examples.
- The "Contingency Amount" column captures the contingency. As shown, we recommend that the contingency be assigned uniformly

WBS Dictionary

from corporate	*from PMO*	*Analagous or Brainstorm Bottom-up or Top/down*	*Cost for material, equipment, subcontractors, vendors, travel*	*Amount to cover estimate being less than 100% accurate*	*from risk register*	*calc'd*	*from Chart of Acounts*
Corporate Chart of Accounts	**WBS #**	**Work Packages**	**Fixed Cost**	**Contingency Amount**	**Risk Fund**	**Total Budgeted Cost**	**Actual Cost**
				80%			
07.01.03.01	01.01.10	Joint Planning Meeting (Service Provider A)	$ 4,750	$ 13,515	$ -	$ 81,092	$ 84,000
				$ -		$ -	
				$ -		$ -	
				$ -		$ -	
				$ -		$ -	
				$ -	$ -	$ -	
				$ -	$ -	$ -	
07.01.03.02	01.01.90	Acceleration (Risk of Liquidated Damages)		$ -	$ 40,000	$ 40,000	
				$ -	$ -	$ -	
		Subtotals	$ 4,750	$ 13,515	$ 40,000	$ 121,092	$ 84,000
		Project BAC				$ 121,092	
		Contingency Reserves				$ 53,515	

Figure 6.6 Organizational WBS (second section)

and then adjusted as necessary for special items. One reason for this is consistency across projects. In Chapter 7, we provide a detailed discussion regarding contingency.

- The next column, "Risk Fund," comes from the risk register and captures the cost of the risk management efforts. We discuss more about risk in Chapter 8. Here is a conundrum. Should an organization divulge the contingency and risk amounts? In some organizations, the risk and contingency amounts are restricted to the sponsor and the project manager. In some organizations contingency is verboten. When such policies are instituted, project managers are forced to conceal it in the work packages or risk overruns.

  If contingencies are divulged, everyone knows the amount of money set aside to cover that +/-20% accuracy range we discussed earlier. Some people and organizations will attempt to use this knowledge in a predatory way, and others will applaud the transparency.

- The "Total Budgeted Cost" column presents the total cost for each work package including contingency and risk. The "Actual Cost" column presents the total cost for each work package. Both of these columns could be divulged to the internal and external teams and to other organizations participating in the project. We say "could" because it depends on the type of CPE. If the contract environment is adversarial, competitively bid fixed price, the organization may not wish to make this level of information available. If the CPE is structured as shown in Figure 6.2, service provider A likely would need to share this information with service provider B. Our measure is that if withholding of the information could conceivably affect the ability of service provider B to fulfill its obligations, it *must* be divulged. The key is to divulge the information in a way that maintains privacy but enables performance.

Now let us look at the final section of the organizational WBS dictionary shown in Figure 6.7:

- The "Delta" column takes the difference between the total budget and the actual cost. The "Variance" column measures the variance outside the established control limits. For example, we used +/-10%, as you can see in the figure. That means that any variation in the

WBS Dictionary								
from corporate	*from PMO*	*Analagous or Brainstorm Bottom-up or Top/down*	*from Chart of Acounts*	*total cost minus actual costs*	*checks if actual cost is outside control limits*	*Software tech, programmers, designers, iron workers, etc.*	*Assumed no resourse constraints, or assume resource availability, etc.*	*from schedule*
Corporate Chart of Accounts	WBS #	Work Packages	**Actual Cost**	**Delta**	**Variance**	**Resource Category**	**Assumptions**	**Schedule Activity #**
					10%			
07.01.03.01	01.01.10	Joint Planning Meeting (Service Provider A)	$ 84,000	$ (2,908)	$ -	Sponsor, PM, Planner, Estimator, Technical (4)	Fixed cost based on food for 30 people	
				$ -	$ -			
				$ -	$ -			
				$ -	$ -			
				$ -	$ -			
				$ -	$ -			
				$ -	$ -			
07.01.03.02	01.01.90	Acceleration (Risk of Liquidated Damages)		$ -	$ -			
				$ -	$ -			
		Subtotals	$ 84,000	$ (2,908)	$ -			
		Project BAC		$ 121,092				
		Contingency Reserves		$ 118,184				

Figure 6.7 Organizational WBS (third section)

delta column that is either 10% above or below the estimated total costs requires review and explanation. Anything within the control limits is simply a lack of perfection. These two columns are for internal measurements and lessons learned and are not needed by external organizations.

- The "Resource Category" and "Assumptions" columns record the resources that will be needed to perform the work package task and the assumptions made in preparing the estimates. (This will be covered in Chapter 7.) For obvious reasons, the resource column is best populated by generic position titles rather than the names of individuals. This column links the project managers' needs to the corporate resource pool and enables resource managers to understand how best to budget their resources.

  The "Assumptions" column captures the thinking when the WBS is being developed. In practice, project managers may work on multiple projects at a given time, and remembering the particulars of an individual work package on one project may be difficult. If the particular work package exceeds the control limits, it will be necessary to understand why. Having recorded assumptions can help in the investigations. In addition, if a project is delayed after the WBS has been completed, these assumptions will help the project manager to retrace his or her steps. External organizations may need the assumptions column and resource column information.

  For example, when building the WBS, if a work package requires work by both the internal organization and an external organization, then both organizations must provide their input. This is where the make-or-buy decisions are made, and from it flows a list of services and goods that need to be procured. Similarly, if the work package is actually a risk, and the description and assumptions of the risk are needed to perform the work, then both parties must first agree to recognize it as a risk.
- The last column, "Schedule Activity #," presents the number so that the work package can be related back to the schedule.

The organizational WBS provides a listing of work packages at a level that enables the project manager to manage the work without becoming a slave to detail. It represents a unilateral view of the project from one perspective. On international projects, there will be numerous interface points with the other organizations, some of which are obvious, some subtle. The time spent the project manager for the organization spends in considering these intersections will pay dividends. The next step is to incorporate the value chain work.

6.2.2 Step 2: Value Chain WBS

Once each organization has completed a WBS, work on the value chain WBS can begin. The first step is to have an internal joint planning session that creates the WBS for the organization. The second step is to conduct an external joint planning session to create the value chain WBS. In Figure 6.2, this would be a meeting between service provider A and value chain A.1. We will use the CPE structure that you will see in Figure 6.11 and the example of a medical project in Cambodia to illustrate.

In Figure 6.8, we show one example to demonstrate the process. Medical mentoring and training are part of the scope for the consortium, and specifically for service provider A. This next series of charts are from the viewpoint of service provider A, and are services outsourced by service provider A to value chain A.1. As you will note, the "Medical Mentoring" row shows internal resources being used to perform the work package.

For the row "SC.A.1—Medical Training Facilities & Documents," no internal resources are shown, as this work is outsourced from the perspective of service provider A. In Figure 6.9, fixed costs of $20,000 are shown since this is outsourced to value chain A.1 at some fixed cost. In this case, service provider A has added $4,000 in contingency to the fixed cost. Ultimately, when the WBS for the service provider A is published, value chain A.1 will see $24,000 for this work package, not the $20,000 it estimated, so it is extremely important for service provider A to discuss and share this information with value chain A.1. If it is not shared, trust will be lost.

Figure 6.9 also shows an activity called "SC.A.1—Prepare Medical Training Documents." Imagine that value chain A.1 included these costs in the previous activity, but service provider A needs to manage this particular activity closely. Assume the documentation needs to be in Khmer, French, and English and must be reviewed for content by the doctors employed by service provider A and by the customer and the user. In this case, we would recommend that value chain A.1 be instructed to break the estimated cost out for this activity and possibly break down the activity into more detailed work packages that capture these steps. In addition, imagine that value chain A.1 has bid the work based on a

WBS Dictionary

from corporate	*from PMO*	*Analagous or Brainstorm Bottom-up or Top/down*	*# of Resources*		*PERT*				*calc'd*	*Amount people are paid*	*office, management, support, etc.*	*calc'd*
Corporate Chart of Accounts	WBS #	Work Packages	# of People	Hours per Day	Best Case Days	Most Likely Days	Worst Case Days	PERT Work Days	Person Hours	Base Hourly Rate	Overhead & Profit Factor	Burdened Personnel Cost
										$ 100	3.10	
07.01.03.01	01.01.10	Joint Planning Meeting (Service Provider A)	8.0	8.0	2.0	3.0	5.0	3.2	203	$100	3.10	$ 62,827
	01.01.20	Medical Mentoring	10.0	4.0	20.0	35.0	60.0	36.7	1,467	$100	3.10	$ 454,667
	01.30.10	SC.A.1 - Medical Training Facilities & Documents						-	-	$100	3.10	$ -
	01.30.11	SC.A.1 - Prepare Medical Training Documents						-	-	$100	3.10	$ -
	02.01.10	Deliver Drugs to Hospitals						-	-	$100	3.10	$ -
								-	-	$100	3.10	$ -
								-	-	$100	3.10	$ -
07.01.03.02	01.01.90	Acceleration (Risk of Liquidated Damages)						-	-	$100	3.10	$ -
								-	-	$100	3.10	$ -
		Subtotals							1,669			$ 517,493
		Project BAC										
		Contingency Reserves										

Figure 6.8 Value chain WBS (first section)

WBS Dictionary

from corporate	from PMO	Analagous or Brainstorm Bottom-up or Top/down	Cost for material, equipment, subcontractors, vendors, travel	Amount to cover estimate being less than 100% accurate	from risk register	calc'd	from Chart of Accounts
Corporate Chart of Accounts	WBS #	Work Packages	Fixed Cost	Contingency Amount	Risk Fund	Total Budgeted Cost	Actual Cost
				80%			
07.01.03.01	01.01.10	Joint Planning Meeting (ServiceProvider A)	$ 4,750	$ 13,515	$ -	$ 81,092	$ 84,000
	01.01.20	Medical Mentoring		$ 90,933	$ -	$ 545,600	$ 550,000
	01.30.10	SC.A.1 - Medical Training Facilities & Documents	$ 20,000	$ 4,000	$ -	$ 24,000	$ 40,000
	01.30.11	SC.A.1 - Prepare Medical Training Documents		$ -	$ -	$ -	
				$ -	$ -	$ -	
				$ -	$ -	$ -	
07.01.03.02	01.01.90	Acceleration (Risk of Liquidated Damages)		$ -	$ 40,000	$ 40,000	
				$ -	$ -	$ -	
		Subtotals	$ 24,750	$ 108,449	$ 40,000	$ 690,692	$ 674,000
		Project BAC				$ 690,692	
		Contingency Reserves				$ 148,449	

Figure 6.9 Value chain WBS (second section)

previous project that used different drugs. It could have assumed that it could simply reuse training material from that project. By exploring this level of detail, service provider A will have a better chance of discovering such misunderstandings before they occur and the project is affected.

Figure 6.10 shows the final columns of the spreadsheet. Under the "Resource Category" column, it shows value chain A.1. The "Delta" and "Variance" columns would be particularly important if this contract has a cost-reimbursable structure.

Once service provider A has completed joint planning sessions with the value chain organizations and has consolidated the knowledge, it is ready to move to the joint planning session with its partner.

6.2.3 Step 3: Partner WBS

With the value chain details in place, the two partners can effectively undertake the partner WBS, which will serve as the basis for the partner scope split matrix. A scope split document provides a breakdown of which organization has responsibility for performing each of the project tasks. For example, in Figure 6.11, value chain B.2 might be responsible for the drug shipment until the drugs are at the port prior to customs, where the local delivery firm value chain B.1 takes control after customs. Who has responsibility to see that the shipment clears customs can become an issue. If service provider B is responsible for logistics and service provider A is the medical provider, again using the example shown in Figure 6.11, it is possible that service provider A will suffer added costs for the personnel on the ground ready to treat the patients but having no medicines. If there is a penalty for failure to provide the treatment by a specific date, which member of the consortium is responsible? One common way to deal with LDs and penalties is that the owner of the scope that created the penalty is responsible to pay it. This assumes, of course, that you know to whom the scope belonged. That is why assembling the WBS for each of the value chain organizations along with the organizational WBS for the service providers is so critical.

Figure 6.11 shows how the level numbering can be different by adopting different perspectives of the project. The Level 7 work package is illustrative only, a not a preestablished requirement for projects.

It may be necessary to go to Level 12, or to stop at Level 5, depending on the task and the control needed to manage the project. We recommend that the CPE use level numbering to ensure consistency across the

WBS Dictionary

from corporate	*from PMO*	*Analagous or Brainstorm Bottom-up or Top/down*	*total cost minus actual costs*	*checks if actual cost is outside control limits*	*Software tech, programmers, designers, iron workers, etc.*	*Assumed no resourse constraints, or assume resource availability, etc.*	*from schedule*
Corporate Chart of Accounts	WBS #	Work Packages	Delta	Variance	Resource Category	Assumptions	Schedule Activity #
				10%			
07.01.03.01	01.01.10	Joint Planning Meeting (Service Provider A)	$ (2,908)	$ -	Sponsor, PM, Planner, Estimator, Technical (4)	Fixed cost based on food for 30 people	
	01.01.20	Medical Mentoring	$ (4,400)	$ (4,400)	Doctors, Nurses	Equipment & facilities by Supply Chain A.1	
	01.30.10	SC.A.1 - Medical Training Facilities & Documents	$ (16,000)	$ (16,000)	Supply Chain A.1		
	01.30.11	SC.A.1-Prepare Medical Training Documents	$ -	$ -	Supply Chain A.1		
			$ -	$ -			
			$ -	$ -			
07.01.03.02	01.01.90	Acceleration (Risk of Liquidated Damages)		$ -			
			$ -	$ -			
		Subtotals	$ (23,308)	$ (20,400)			
		Project BAC	$ 690,692				
		Contingency Reserves	$ 125,141				

Figure 6.10 Value chain WBS (third section)

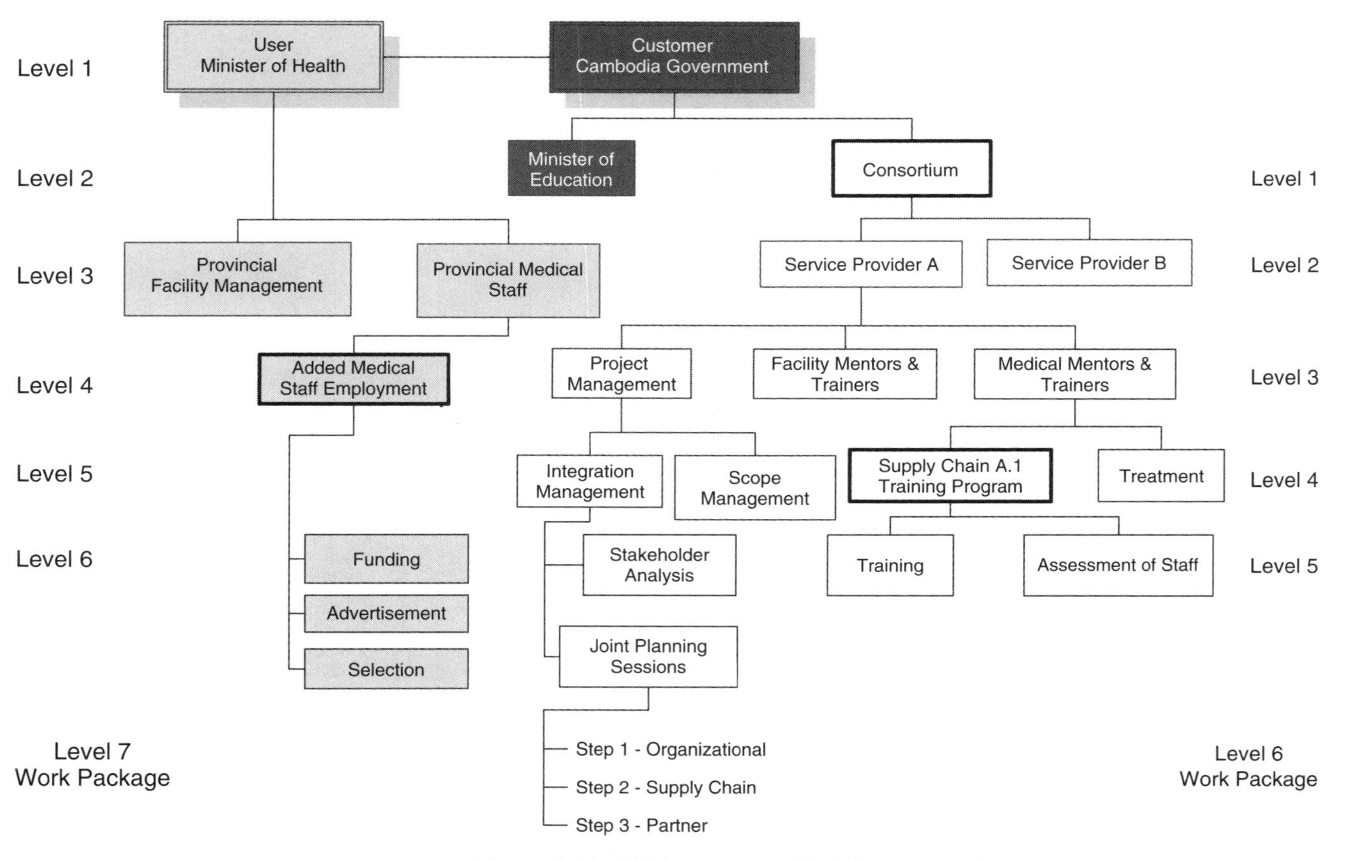

Figure 6.11 CPE Structure of WBS

organizations. As the figure shows, different levels may occur on each side of the structure. The numbers themselves are unimportant; the relative relationship is key.

Development of the partner WBS follows the same pattern as the value chain WBS. The two partners come together with a complete listing of work packages from their previous individual efforts to create a picture of the project. Each would have anticipated the customer and user interfaces and the interfaces with the other partner. Now is the opportunity to compare thinking about how the project could be executed most effectively.

Using Figure 6.11, imagine you are attending the joint planning session of the consortium and the question of training is discussed. Service provider A is to train the user's provincial staff. This will certainly be a function of the user having hired, or selected, the new staff. Service provider B would be interested in knowing the same information because the drugs being delivered will need to be stored properly; that is a function of the customer having the people and service provider A having made sure that they are trained in how to do it. If service provider B is responsible only to get the drugs to the hospitals, it is easy to see the mutual dependencies built into the project for this item. Let us look at how this might be displayed on a WBS that has been codified by combining the efforts of service provider A and B in Figure 6.12.

First, notice that the "WBS#" for the "Deliver Drugs to Hospital" row begins with 02, designating service provider B. "Medical Mentoring," "SC.A.1—Medical Training Facilities & Documents," and "SC.A.1—Medical Training Facilities & Documents" are work packages associated with readiness for delivery. "Deliver Drugs to Hospital" is a work package that service provider A needs before treatment can occur. At this point, neither service provider knows when the new doctors will be available from the customer, but that will be confirmed later. What each party sees, assuming the highlighted columns are not provided, is the budget and actual cost. Actual cost would not be known at this point so the numbers in that column are illustrative only. Now imagine that both partners in the consortium saw the highlighted columns, and consider how much richer the information. Here again is the challenge for the lead project manager: transparency or not.

Assume that both of the consortium partners believe that they will have to accelerate their work to overcome a delay in the customer providing new doctors. Each partner has political intelligence that says the funding

WBS Dictionary										
from corporate	*from PMO*	*Analagous or Brainstorm Bottom-up or Top/down*	*from risk register*	*calc'd*	*from Chart of Accounts*	*total cost minus actual costs*	*checks if actual cost is outside control limits*	*Software tech, programmers, designers, iron workers, etc.*	*Assumed no resourse constraints, or assume resource availability, etc.*	*from schedule*
Corporate Chart of Accounts	**WBS #**	**Work Packages**	**Risk Fund**	**Total Budgeted Cost**	**Actual Cost**	**Delta**	**Variance**	**Resource Category**	**Assumptions**	**Schedule Activity #**
							10%			
07.01.03.01	01.01.10	Joint Planning Meeting (Service Provider A)	$ -	$ 81,092	$ 84,000	$ (2,908)	$ -	Sponsor, PM, Planner, Estimator, Technical (4)	Fixed cost based on food for 30 people	
	01.01.20	Medical Mentoring	$ -	$ 545,600	$ 550,000	$ (4,400)	$ (4,400)	Doctors, Nurses	Equipment & facilities by Supply Chain A.1	
	01.30.10	SC.A.1 - Medical Training Facilities & Documents	$ -	$ 24,000	$ 40,000	$ (16,000)	$ (16,000)	Supply Chain A.1		
	01.30.11	SC.A.1 - Prepare Medical Training Documents	$ -	$ -		$ -	$ -	Supply Chain A.1		
	02.01.10	Deliver Drugs to Hospitals	$ 40,000	$ 52,000	$ 60,000	$ (8,000)	$ (8,000)	Supply Chain B.1		
			$ -	$ -		$ -	$ -			
			$ -	$ -		$ -	$ -			
07.01.03.02	01.01.90	Acceleration (Risk of Liquidated Damages)	$ 40,000	$ 40,000		$ 40,000	$ -			
			$ -	$ -		$ -	$ -			
		Subtotals	$ 80,000	$ 742,692	$ 734,000		$ (28,400)			
		Project BAC		$ 742,692			$ 742,692			
		Contingency Reserves		$ 190,449			$ 162,049			

Figure 6.12 Partner WBS

will not be there in time to hire the new doctors. Each partner has included $40,000 to accelerate his or her work, but it is not obvious in the work package "Deliver Drugs to Hospital." This is one simple example of the importance of sharing knowledge early and often. If the consortium partners recognize that the two US$40,000 additions are for the same risk, they can provide the customer with a total risk budget for this item. They could also propose to increase their services to include funding and hiring, particularly if this is critical to the success of the project.

Also required for this meeting is a risk register that describes the risks facing the project. The risk register is discussed in Chapter 8. The reason for combining risk with this scope discussion is that it is part of the scope. Imagine an IT project, and consider the question who has responsibility for consequential damages. Consequential damages result from action, or inaction, by a party to a contract. In our example, this would be the loss of reputation or profits to the customer's ongoing operations if the project causes a loss of functionality in the existing systems. One of the organizations must assess this risk, price it, and manage it. Risk is a problematic issue on most projects because the structure of the CPE often does not encourage open discussion.

As with the WBS, development of the risk register and the scope split document may need to progress concurrently, depending on the type of contract and the status of the negotiations. On a project in the Philippines, we needed to do an initial risk register and assessment in order to continue with the negotiations, which in turn enabled us to finalize the scope document.

Clearly if both partners show up for the joint planning meeting with this level of detail, it will demonstrate commitment, concern for the other participants, an open approach to communications, and willingness to trust the other organizations. The lead project manager has a wonderful opportunity to demonstrate leadership here. The starting point is answering the question: How many of the highlighted columns should our internal organization divulge? Part of the answer is based on the type of relationship, as shown in Figure 5.1. From our experience, the more transparency, the greater the chances for success.

To recap, a partner WBS should be developed in the Step 3 joint project planning session. Knowing the expectations of this meeting in advance will encourage each party to:

- Do its own internal project planning before attending the meeting.
- Use its expertise to question the completeness of the scope.

- Use its expertise to question the management regarding the risks.
- Learn about the members of the CPE and the key stakeholders.
- Interact with counterparts. (With virtual teams, this may be the only occasion.)
- Build a teamwork atmosphere.
- Create a common enemy—us against the project.
- Learn about the leadership style of the lead project manager.

6.2.4 Step 4: CPE WBS

The customer is the center of a project, and all of the work involved in Steps 1 through 3 is in preparation for the joint planning session with the customer. A satisfied customer is success. From our experience, as a customer and a service provider, customers want to feel that they made a wise decision in selecting the organizations they did to perform the work. Here "wise" means the organizations show respect, leadership skills, and preparation. Respect the fact that the customer has an agenda, may not know the technical aspects, and is placing its faith and trust in the service providers. The lead project manager, and indeed all of the other project managers, can increase their chances of success by acquiring or improving cross-cultural leadership intelligence. Considering only trust, the amount of information that the lead project manager offers to the customer will definitely affect the level of confidence and trust. In our experience as a customer, we quickly became cautious when we detect that a service provider is withholding information or has not prepared for our meeting.

The last metric of preparation goes further than just the development of the WBS. Customers, rightly or wrongly, expect that the service provider has taken the time to learn about their organization, including its structure, politics, long-term goals, power structure, challenges, and the like. The lead project manager can build trust by demonstrating that she or he has done research about the customer, by showing the connection between the work packages and the customer's situation. The beauty is that even if the research is inaccurate, the customer has the opportunity to add value, and talking about its organization is worthwhile. There will always be scope that is not clear, excluded, or misplaced—always. If the lead project manager presents these problems along with a list of solutions, imagine the positive impact this will have on the project and on the customer.

The Step 4 joint planning meeting with the customer is the final step in building a WBS for the CPE. In planning for this meeting, we recommend that the lead project manager address these items:

- *Attendees*. The upper-echelon sponsors of the user, customer, service provider A, and service provider B should at a minimum make an appearance. Depending on the complexity of the project, logistics, and timing, this may not be possible, but it is strongly recommended. Value chain representatives may also be needed at this meeting, depending on the technical and political challenges and expertise that they will supply, especially if they are ranked as key.

  The teams from each organization should be selected based on their involvement in and responsibilities for the work. This is an opportunity to build the CPE culture for the lead project manager. It is also an opportunity to inculcate attendees so that they can contaminate the organizations that are not present. As a practical matter, we recommend that the lead project manager have staff available to record the discussions and information contemporaneously and provide copies to each participant at the end of each session. If the CPE is technologically advanced, this can be put on a project server, and the other organizations in the CPE can comment overnight.
- *Leadership*. In Figure 6.2, a lead project manager is shown with a direct link to the customer. Frequently, customers prefer to structure a contractual relationship in this manner. Sometimes, however, they prefer a joint-and-severable contract in which both service providers sign an independent contract with the customer. In either case, it is necessary for someone to take the lead in the joint planning meeting, and we recommend it be the lead project manager. This meeting is critical, and the lead project manager can utilize the time to build relationships and look for potential conflict, not be a clerk. As a mentor once told us, "You cannot listen when your mouth is open." Better to be observant rather than consumed with the management. From the leadership side, what better way to demonstrate your intention to empower?

  We recommend that the lead project manager open and guide the meeting and then rely on someone else to manage and run it. One great choice would be the project manager for service provider B. This choice shows that the consortium is an integrated entity and that the project manager for service provider B is an important member of the CPE. Make certain the first organization to have its say is the customer and user.

- *Duration*. For complex international project with a budget of $200 million and above, this meeting it may last three to five days. It is important to keep it short enough so that everyone stays stimulated and engaged but not so short that discussion is squashed. Here again is an opportunity for the lead project manager to demonstrate leadership: to show people that their opinions are valued and that time needs to be spent on communications.

  It is also important to budget time for people to mingle, talk, and become acquainted. We recommend that at least 20% of the time be devoted to this stage. Make sure breaks are scheduled at least every two hours, with two hours for lunch and dinner. Also, we suggest a morning or afternoon activity—sightseeing, for example—where people can interact in a nonbusiness environment. Consideration should also be given to team-building exercises that can be interwoven into the fabric of the meetings. To be successful, the interactions must be genuine, not contrived.
- *Venue*. The ideal location is the country where the project will be conducted, unless security, accommodations, or infrastructure prohibits it. It is ideal if the venue has living accommodations in or near the facilities to minimize travel time, and of course food services. Living accommodations provide innumerable opportunities for participants to engage in impromptu meetings and conversations, particularly if the venue offers a range of activities.

  The meeting area should be large comfortable, and well equipped with computers, white boards (we have seen a number of facilities that actually have white walls), smart boards, projectors and screens, flip charts, digital cameras, and of course printers, copiers, and computers. Communications must be dependable so that participants can pull e-mail and make their contacts in the evening after the sessions are completed. Multimedia, computer equipment, and materials need to be readily available.

  As we indicated, the ideal situation is to have information from Step 1 through 3 already on a project server. Then the meeting can utilize the server information and the draft project plan, which is manipulated and recorded during the meeting. The comments, suggestions, notes, photographs, impromptu meeting, ideas, brainstorming, and such can be collected each day by the staff and made available to all of the CPE, whether present or not. We also recommend that at least one synchronous time is set aside each day for virtual participants to engage with the people at the venue in real time. If virtual organizations cannot attend the synchronous sessions, the staff for the lead project manager should monitor the project server for messages and questions that need to be included in the discussions.

- *Deliverables*. The outcome of the meeting is a CPE project management plan that includes individual plans for managing scope, time, cost, procurement, communications, risk, quality, and diversity. We recommend that each participant, as well as those who did not attend but are part of the CPE, receive a compact disc (CD) that captures the CPE plans if a project server is not used. We prefer to distribute these CDs at the time of the closing ceremony. For those not attending, we suggest that the opening and closing statements, at least, are captured on video and included on the CD.

During the meeting, the customer has an opportunity to be directly and completely involved in the discussions as a valued member of the CPE. Some customers prefer a more distant relationship, and that attitude must be respected. However, we believe that it is in the best interest of the project to create a standing and consistent invitation for customers to be involved. They then can make the choice to be involved or not. Most projects require customers to participate in some way, and it is better to engage them early and often, but in a respectful, patient, and understanding way. The user is another issue.

Imagine a project where the customer is a capital investment firm building a cement plant in Yemen. The user of the plant will be a cement organization that has a contract with the customer to take over the facility, run it, and share the profits over time. The service providers have a customer between them and the user that is knowledgeable about money, but not about process. The process knowledge resides with the user. The maintenance of the new plant will be the responsibility of the user, and it certainly will have preferences about such things as access for maintenance. In our experience on such projects, the user may not become involved until start-up and of course the punch list at close-out. This represents a very large risk, for the knowledge about process and preferences resides with the user in our example. We have seen some truly amazing disconnects between the customer and user at start-up time. Some colleagues like to chuckle about the lack of communication, but the problems always end up back in the laps of the service providers. Perhaps this is reasonable, because the service providers they know the processes and should have anticipated the problems that any user would confront. Therefore, it is recommended that lead project managers encourage customers to have the user present for the joint planning session and for the progress meetings.

The joint planning meeting with the customer—and user, it is hoped—the service providers should strive to validate their understanding of the scope at a level of detail that will avoid most if not all of the ambiguities in the scope documents. This is the first of many progress meetings and conversations where we recommend that the lead project manager make it a habit of confirming the scope for the ongoing work. The joint planning session should simply be the first in many scope understanding conversations and meetings. A metric we suggest is having the lead project manager describe the scope for the customer and user to an independent party; if the customer and user simply nods their heads in agreement, things are working as planned.

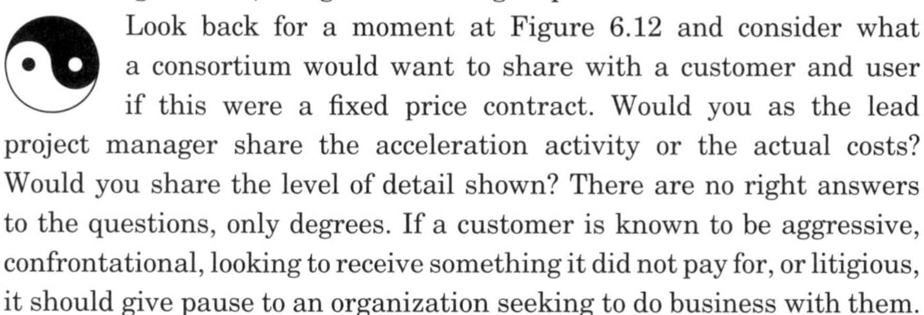

Look back for a moment at Figure 6.12 and consider what a consortium would want to share with a customer and user if this were a fixed price contract. Would you as the lead project manager share the acceleration activity or the actual costs? Would you share the level of detail shown? There are no right answers to the questions, only degrees. If a customer is known to be aggressive, confrontational, looking to receive something it did not pay for, or litigious, it should give pause to an organization seeking to do business with them.

In our experience, such customers are not common, but they do exist. Under such contracts, we have had legal representatives assigned day 1. You must protect your organization and those in the CPE in these circumstances. Except for such customers, we favor transparency beginning on the first day. We prefer to divulge everything that we reasonably can—we would answer yes to both questions. The details will emerge as the project moves forward anyway, so why not make some use of them? What is purchased with such transparency is trust and open communications, and in our experience the return on investment is large.

If the project is conducted on a cost-reimbursable basis, the detailed information will almost certainly be required. Think about the difference between a cost-reimbursable and fixed price contract, why people withhold information on one, and whether withholding the information actually help get the project completed successfully. In our experience, from tendering to claims, most secrets do not stay secrets. When they are exposed, there is consequential damage to relationships.

Now we need to explore how to manage scope change, for it will begin during the joint planning meeting.

6.3 SCOPE CHANGE CONTROL

Imagine you are the lead project manager on a £100 million electronic medical records (EMR) project that will be executed in Brazil, Panama, Chile, and Ecuador for a customer in the United Kingdom. You have the U.K. customer and a U.S. user, and a joint venture partner. During the first day of the joint planning meeting, customer's sponsor is taking her espresso and begins a conversation with you. She tells you how pleased she is with the meeting and asks for your help with one small issue. The user needs to have a nonfunctional Web page for advertisement purposes so that its customers know of the new EMR product, but she tells you that her team neglected to include the requirement in the specification. She is clearly embarrassed to be asking and is implying she wants you to do it, for free. You are sure that it is about a day's work for a Web designer, £100, and do not want to appear cheap. What do you do?

Our suggestion is you deliver something like this: "My job is to make sure that we complete this project on time, under budget, and to make certain you are satisfied with the service and the product. Changes will be required, and we must adjust the project plans accordingly each time a change is required. Our change control system requires us to assess every request, regardless of how large or small, and determine if it has any impact on the cost, time or quality of the project. We will provide you with an assessment on each change and then ask you if you want to proceed with it knowing the effect on the project."

On large projects, we prefer to this additional bit, "For small changes, it will be impossible to assess them efficiently. Therefore, we have set up small account for items that are under £500 and have included it into the budget for the project. We will keep track of these changes and if the total does not exceed £30,000 by the end of the project, there will be no added costs to the project. Are you amenable to such an approach?" Scope documents are not perfect, and we have found this approach to be an effective way to deal with the myriad of details that got missed. The approach enables customers to request small changes without having to suffer through signing a £100 change order that costs £1,000 to process. The work can be recognized and the costs bundled. It also serves to put a number on the "goodwill" of the service providers. Perhaps most important, it records the small requests; it is not uncommon to have numerous sponsors and project managers during the course of an international project. Each of them will make requests, and this technique will commemorate them. It also demonstrates leadership by building the confidence that you are transparent in your transactions. Transparency can be troublesome in some countries where bribery is prevalent. Giving warning at the beginning is far more considerate of the customer than waiting until the pile of requested changes becomes too large. These small requested changes—"death by a thousand cuts,"—was a form of execution used in China beginning roughly AD 900.

International projects usually have long value chains and multiple organizations, as we have seen. With this in mind, consider the next scenario. A value chain organization in Panama that is doing medical record transcription finds that the software provided by the medical records service provider in Chile will not accept magnetic resonance imaging (MRI) information. The lead project manager is informed by the service provider that a change in the contract is required, and the lead project management firm located in Brazil must provide the missing functionality. Is this a change? The answer takes a bit of description.

The contract between the two key organizations, the service providers in Chile and Brazil, and the U.K. customer has a clause that says "Provide all software to enable the transcription of all medical information formats utilized in the U.K." Clearly, MRI information is required. When the service provider in Chile wrote the contract for the value chain organization in Panama, it did not copy the specifications from the U.K. customer but rather summarized them into a few pages that said simply "Transcribe all information provided." What is needed could be a change, if implied in the contract with the value chain organization was the idea that the organization in Chile would furnish any software not readily available in the Panamanian market. Our point is that knowing if a change has occurred is complicated by the fact that scope documents are not perfect and also by the fact that the documents provided down the value chain may not be 100% of the primary level scope. Our advice on this point is to flow down, or pass along, the same scope descriptions to the value chain. Doing this will make the wording consistent and eliminate some of the difficulties.

Staying with this example, assume that the Chilean service provider passed on exactly the same scope documents to the Panamanian value chain organization. Is the Brazilian organization then in the position of furnishing all software to all of the value chain for the Chilean service provider? A quick visit to the scope split document reveals that this level of detail is not addressed. If each service provider is responsible for its own scope of work, and the work is clearly transcription, then who owns the change? This is one reason we recommended the buildup of a CPE WBS. It will not catch everything, but likely many similar issues were discussed, and that provides guidance about intent. In any case, would this be a customer change? What if the MRI files were from first-generation machines that stopped being used 10 years ago and none of the software on the market used for transcription will work? Imagine what it would cost the Brazilian service provider to write the necessary software.

The lead project manager must sort out such issues. In Spanish, the word *enredo* roughly means "entangled," "complicated," like being snarled in a fisherman's net. This term helps to describe just how difficult it can even to determine if a change has occurred. Our advice is for the project managers, and particularly the lead project manager, to do the necessary research of the written scope and the discussions that occurred at the planning sessions. Then he or she should pass this information along up the value chain with the request. Demonstrate respect for the organization and project manager who is your customer, and be equitable.

The lead project manager must establish a change control process for the CPE and communicate it to the participants very early in the initiating and planning process. The change control system needs to consider the conundrum we just described and be built with problems like this in mind. The change control process begins with identification and continues with analysis of the change. Analysis includes determining if the change impacts the project management plan, how, to what extent, and what can be done, if anything, to mitigate the effects. The most important aspect of this process is consistency. Have a checklist of items to review for each change, and follow it. We have used the next checklist successfully.

- *Change description*. What is the change, and what conditions of the contract scope are altered? Answering this simple question will help mightily in explaining why the lead project manager believes it was not part of the original scope. Defining the change will also flush out any misunderstandings on what was requested.

- *Requested by*. Who is requesting the change specifically? It could be a partner or a value chain organization. It could also be the customer's technical department or the sponsor.
- *Impact on time*. Put an activity into the project schedule, link the logic, and see what happens. Then see if crashing or fast tracking can mitigate it, and, if so, whether there is an associated acceleration cost to stay on schedule.
- *Impact on cost*. In addition to a possible time impact, there could be added material or service costs. See if there is adequate contingency to cover the costs, or if the project budget at completion will have to be changed.
- *Impact on quality*. Will the additional work result in the dilution of resources, or overstressing the resources? Mistakes are more likely when longer hours are worked. Potential technical impacts also must be considered with each change. Assess them, and provide options to mitigate.
- *Impact on risk*. Some changes can change the risk profile of a project. On one project in Miami, the contractor bid the project to complete the foundation work before the wet season. A piece of foundation work was added, but the contractor did not consider that it would push the remaining work into the wet season. It did, and the one week of added work resulted in two months of delay due to flooded foundations. A risk assessment must be performed for any added scope.
- *Written approval*. Many, if not all, international contracts contain a clause that forbids the service provider(s) from making changes to a project unless they are approved in advance by the customer. Such clauses are there for a reason, and we recommend that they be followed. This is yet another reason to have a change control process that includes an approval process. Such a process names the people who are authorized to execute changes and their limits of change authority. For example, the lead project manager may have an authority level of ¥100,000 on a ¥100 billion project, so a change request of ¥200,000 would have to be approved by the sponsor. In a consortium like Figure 6.2, service provider B would also have to

sign off, and the authority level of the project manager would again be an issue. The same is the case for the customer.

- *Disputes*. There will be occasions when the customer and the service provider(s) cannot agree that a change has occurred or on one or more of its impacts. On a project in Turkey, the customer convinced the service provider to go ahead and do extra work to "keep the project moving" while the change was discussed. The service provider did as requested, the Turkish government changed, and the verbal agreement to negotiate the change evaporated. Then the government reverted to the no-change-without-written-authorization clause and denied the extra work, which by then had been completed. Having a process in place to deal with disputes is essential to keep a project moving forward while the disputes are being negotiated. In Chapter 5, we touched on this issue at the project level. Here the need for clarity in the process needs to be at a basic and easy-to-follow level.

We suggest that organizations prepare a format of their choosing to capture the checklist items and track the changes to the CPE WBS. We recommend each change be added to the CPE WBS. In some cases, the change may need to be a separate work package. In some cases, it can be added to an existing work package. The important thing is to make sure that the changes are recorded in a disciplined and consistent way. We have seen hundreds of projects start with good intentions and aspirations but fail due to lack of a change control system. Too often projects try to reconstruct the changes that have occurred once the budget is in jeopardy. This is simply too late, and there is no reason for it. Even if a customer is difficult on the issue of changes, it still deserves to have a professional project manager provide the facts about changes as the project progresses.

To return to our sailing metaphor, each component of project management requires establishing planning a course, sailing, checking periodically to determine the actual location, make course corrections, and replanning the future course. The *PMBOK* calls this plan-do-check-act, and it was based on the work of Shewart and Demming (*PMBOK* 2009). The "act" part consists of forecasting, adjustment, and replanning. The process applies here to scope. The scope baseline is predicated on the scope understanding at the time it was created. As the project progresses, the CPE learns more about the implied needs of the customer, and changes occur. Each progress update provides knowledge that adds accuracy to the plan. Our advice is not to believe that once the baseline plan is complete, planning is complete. Completion of the scope baseline represents the first course, not the final one. Planning begins day 1 and continues throughout the project.

In this section, we have covered the WBS. Estimated costs were shown in the examples for illustrative purposes. As we said in the introduction, the WBS should precede the estimating in terms of sequence. We do not mean that a planning session should be convened to do WBS and another to do estimating. Rather they both should be done in the same meeting, but focus on the work packages first, then on the estimating. The reason is to save time by keeping focused. Likewise, when in the CPE WBS joint planning session, focus on the work packages first, then the estimating, then the risk. This sequence does not need to be adhered to dogmatically; use it as a guideline to keep the group focused on one thing at a time.

Chapter 7

Cost and Progress Management

> Honesty is the best policy—when there is money in it.
>
> —*Mark Twain*

> Whenever there is profit to be made then think of honesty.
>
> —*Chinese proverb*

In this chapter, we discuss how to manage cost across organizational boundaries. As with scope and time, cost is one leg of the so-called triple constraint, and it is easy to utilize. As we point out, however, the metric "within budget" is not as easy and obvious as one may think. We think of funding as fuel; it enables the lead project manager to mobilize resources to complete the work. It is a limited resource, and therefore precious, and must be utilized effectively and efficiently. How to accomplish these things is the subject of this chapter.

7.1 COST AND CURRENCY STANDARDS

On international projects, working with different currencies is fundamental. The lead project manager can help the collaborative project enterprise (CPE) avoid difficulties by setting forth basic guidelines during the joint project planning meetings:

- Overall project reporting will be done in a single designated currency.
- Project documents must show a currency symbol or description.
- Project documents must clearly indicate the exchange rate utilized.
- Project documents will use a single convention for displaying amount—for example, €1,000.00 versus €1.000,00.

On large international projects, the risk for currency devaluations relative to another currency must be managed. As we discuss in Chapter 8, these risks need to be identified and assigned to the party that has the best ability to manage them. In this way, the risks may be mitigated by putting into place the currency hedges necessary to manage the risk at the CPE level. If these risks are left on a to-each-his-own basis, some firms will suffer because they are not sophisticated enough to manage the risks, and some firms will build an arbitrary contingency into their price to compensate—a contingency that is likely to be either too large or too small. Neither of these methods is efficient and, in a lengthy value chain, either can add significant costs to a project needlessly.

In speaking of buffers in the critical chain method, Goldratt (1997) argued that value chains magnify the effect of addressing risk by increasing the time necessary to complete a project. The same is the case with estimates for managing risks. As we discussed in Chapter 6 and will see in Chapter 9, the challenge on international projects is creating a CPE culture of trust so that the participating organizations are comfortable sharing information. Goldratt proposed the critical chain as a method of pooling risk, and that is exactly the approach that will yield the greatest utility on international projects.

The challenge for the lead project manager is to build a culture of trust in the CPE quickly. The remainder of this chapter is premised on the assumption that this can be done. To start, the lead project manager must lead the way by divulging. The trick is to divulge enough without providing access to proprietary or competitive information. That is an art form, and relies on the lead project manager's experience, intuition, and cross-cultural leadership intelligence (XLQ). It begins with one of the most difficult issues, contingency.

7.2 PROJECT CONTINGENCY

Figure 6.5 provided a view of a CPE work breakdown structure (WBS), and we discussed the importance of candid and transparent discussions regarding contingency, and the reason for which contingency was created. Figure 6.11 showed a CPE contingency fund, and we discussed how it was compiled. The critical issue for creating a CPE contingency fund is open and honest communications and transparency.

Imagine that a customer, the World Health Organization (WHO), has requested quotations for a new project with a goal of reducing the

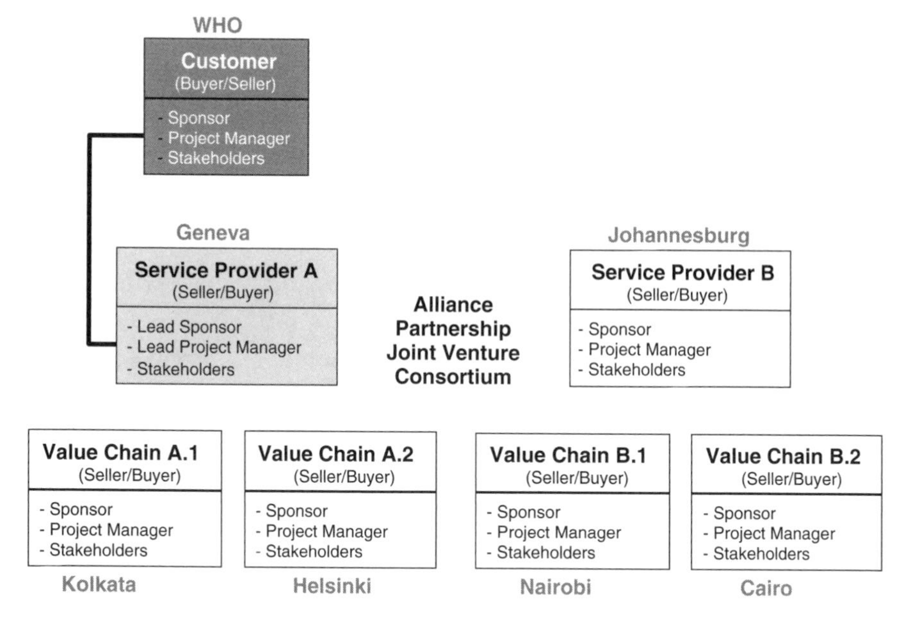

Figure 7.1 Tendering

incidence of malaria in Africa by 50% in four years. Service provider A in Figure 7.1 intends to submit a bid to the WHO. Imagine you are the lead project manager for service provider A and know the work well, but you also know that you will need to engage a partner who can provide the netting and medications that will be required. The bids are due in four weeks, and that is not adequate time to negotiate with a potential partner. All that can be done is to engage in some general discussions about rough order of magnitude estimates and scope splits.

You do have a tentative bid from one of your value chain, predicated on some assumed clarifications on scope from the customer. As service provider A, you decide to submit a proposal, with the intention of filling in the necessary partner and value chain once you are successful.

If service provider A is successful, now it must go to the marketplace and solicit formal bids. First question is, does service provider A divulge the contingency that was built into the price to cover the lack of precision? For example, service provider A estimated that the cost for the work required of service provider B is €120,000,000, but that included a contingency of 50% since there was a great deal of uncertainty over the current market conditions. Your contract value is €250,000,000, and service provider A included a 30% contingency in its portion of the work since some of the countries involved are not politically stable. Therefore, the breakdown looks like this as service provider A begins the search for a partner:

- Total contract value—€250,000,000
- Service provider A's share of the total contract value—€130,000,000
- Service provider A's contingency—€39,000,000
- Service provider B's share of the total contract value estimated by service provider A—€120,000,000
- Service provider B's contingency estimated by service provider A—€60,000,000

Should service provider A divulge any or all of these numbers during the negotiations with service provider B? Consider the extremes.

Service provider A divulges the total contract value of €250,000,000, and the estimates for service provider B's work. Since you are going to be partners, you decide to build a foundation of transparency. One possible disadvantage is that you provide your partner with information that it can use to its advantage in your initial negotiations and later if problems arise. Thus, you could undermine your organization's profits on the project. One possible advantage is that you demonstrate your earnest desire to be open and transparent in your dealings with your partner. At the other extreme, you decide to divulge nothing and plan to explain yourself if the partner learns the numbers later. The advantage is that you do not initially provide a possible advantage to your partner. The disadvantage is that is shows your desire to withhold information.

The real question is if transparency and the trust it engenders (not guarantees) is worth the potential loss of profits to your organization. Based on our experience, trust is worth many times over the potential added profits, even on projects where the partnership is a transactional one, not a relationship. The reputation of an organization does make a difference on international projects. We have seen price increases over time when firms choose to maximize short-term profits at the cost of transparency.

How the lead project manager deals with this issue of contingency and transparency is critical to the success or failure of a project. To keep a project under budget, it is essential that all of the project managers share information candidly. The lead project manager cannot predict an estimate at completion (EAC) within a reasonable range of accuracy

(+/-10%) if the project managers of the partner and value chain do not disclose their contingencies. In addition, the lead project manager should not expect that others would do what he or she is unwilling to do.

Once the lead project manager divulges, it is almost certain that the customer will learn the numbers. Therefore, better to divulge to the customer first and gain trust. We can hear the complaints already! It is like playing poker where the other organizations can see your hand but you cannot see theirs, you say. We agree. If it is games you desire, do not divulge. In a nuclear arms race, who would be willing to divulge their strengths first? If no one is confident enough to do so, the race continues, and failure can be catastrophic. The same holds true on international projects. We do not make this recommendation lightly, and we make it in all seriousness. Experience shows that leadership is a key ingredient in the success of international projects, and trust is the hub of leadership. It takes a lot of courage to do this and may result in short-term losses. We recommend as much transparency as is possible. It is not a guarantee that others will follow, and it may not have the desired effect on accuracy. However, it will change the dynamics on the project, and it will have a positive effect on the reputation of the organization. In our experience, it also has a large positive impact on customers, even if they choose not to be transparent.

Start with a fair cost estimate for the entire project, including the customer's scope, the partner's scope, and the value chain scope. When building the CPE EAC, challenge each participant to defend his or her estimate and his or her contention that there is no contingency built in—you will likely hear this more than once. It is a lot of work to build an estimate, but it provides the lead project manager with a baseline for asking tough questions in a polite and thoughtful way. It also demonstrates the commitment and concern the lead project manager has for each participant organization. Besides, it is the job of the lead project manager to understand the overall project, not just his or her organization's portion. Imagine the empathy you demonstrate if you catch a mistake on the part of a value chain participant and mention it.

There is one last practical aspect of this process. Imagine that in Figure 7.1 there are three added levels of supply chain organizations below value chain A.1. If each level of value chain adds 10% to the lower level price to account for uncertainty, that adds 30% to the CPE budget at completion (BAC) for the value chain work. What if there are 10 levels of value chain? If the CPE is a public-private partnerships

(PPP), the value for money metrics can be skewed as a result. Since most such agreements are negotiated, it only makes sense to remove as much padding as possible. The other alternative is to competitively bid the value chain work, and assume that the lowest bid is both adequate for the work and has been stripped of contingency since it is low. We prefer not to speculate but rather to engage and ask. This issue of trust and transparency underlies all of international project management, and certainly the individual cost estimates.

One last item before we discuss the EAC. We prefer to use a consistent application of contingency across all WBS work packages. That means if our control limits are +/-10%, we apply that to all work package estimates. The reason is statistically the probability of any one being wrong is equal. Like the Program Evaluation and Review Technique (PERT), it is an attempt to manage the numbers and remove as much bias as possible. There may be occasions where added or reduced contingency may be appropriate, as when a fixed price is received for equipment supply, but those should be handled on an exception basis. Remember, in our terms, contingency is the imperfection of estimates.

7.3 ACCURACY OF EAC

International projects are tendered in many different ways and can range from competitively bid to PPP. It is important for the lead project manager to understand the maturity of the estimates for each participant organization, and a logical starting place is the terminology. Terms vary widely across industries and countries, so it is vital that the lead project manager set the standard. As we have indicated, knowing where each organization is in the process is important for the lead project manager, and making sure everyone speaks the same estimating language is vital. Here are some suggested definitions to keep the language clear.

- *Rough order of magnitude.* The participant organization does not have the details of the scope, has not had enough time to prepare a detailed WBS, does not have the expertise and will need to find a partner, and so on. For example, this is an estimate prepared in a hurry for a tender. Accuracy +/-50%.
- *Preliminary estimate.* The participant organization needs a detailed scope explanation and clarification, more time to prepare a detailed

WBS, but has acquired the expertise necessary to do the work. Accuracy +/-40%.

- *Detailed estimate.* The participant organization has completed its detailed WBS, has had its scope questions answered, and has thoroughly reviewed its estimate. The participant organization can define its scope boundaries. Accuracy +/-25%.
- *Confirmed estimate.* The participant organization has met with the CPE and the lead project manager, reviewed their estimate in detail, and has established an agreed-on contingency reserve for its work. Accuracy +/-20%. This is assuming the accuracy of the organizations estimate is +/-20%. In practice, it could be more or less, and the 20% is only illustrative. Also, the concept here is that if the participant organization is uncertain about scope and has added, say, 50% contingency for an item, the discussion may enable that to be reduced back to the organization's normal accuracy range.

  Note: Risk should *not* be included in this contingency fund. We will discuss more on risk in Chapter 8. The risk estimate for the project needs to be included in the baseline estimate but kept separate from the contingency for estimating errors. This estimate should be at the work package level and able to be summarized easily. More on this in the next section.
- *Baseline estimate.* The participant organization has completely reviewed the CPE plan in comparison to its confirmed estimate and it comfortable with its scope and that of the other participant organizations that may have an effect on its work. Accuracy +/-10%. This is also called a budget at completion.

We also recommend discussing the basis of the estimates provided. Were they analogous from other projects, from a corporate database, or from corporate heuristics (€/kilowatt for a power project), from a published database (e.g., R.S. Means), from a brainstorming session, a combination, or PERT. These questions should not take the form of an inquisition but rather an attempt to understand the approach taken and for the lead project manager to feel comfortable with the estimate's level of accuracy. There is nothing wrong with probing your partners and with your partners probing you. It is healthy and natural, and can expose problems. When the inquiries are greeted with anger or indigence, it may be time to consider working on the relationship before pursuing financial details.

The steps or sequence outlined in Chapter 5 should be followed. A durable project EAC is simply a summation of dependable WBS EAC's from the participant organizations. It is a lot of work but creates a solid foundation.

7.4 CPE WORK PACKAGES

According to the *Project Management Body of Knowledge* (*PMBOK* 2004), a work package is "a deliverable or project work component at the lowest level of each branch of the work breakdown structure." Another way to think about it is as the lowest level needed by the project manager of the participant organization to manage work on the project. In Figure 7.1, it is the level of detail required for value chain A.1 to manage its work. Obviously, service provider A would not need this level of detail to manage the overall project, but that does not mean it should not be requested, and provided. Service provider A must understand as much of the details of a project as possible, and then use her or his judgment to determine which detail is needed to manage the project—just as you would do for your own organization. It does not mean that the detail will be used to control or status the project.

By requesting the work package detail, the lead project manager is showing interest in the work of the other organization and in helping it to be successful on the project. Once an understanding of the detail is achieved, the lead project manager can summarize the work packages to a level appropriate for control. After the baseline estimate is completed and progress updates begin, the CPE work package provides an extremely powerful tool to assess why an activity estimate has a variance. A variance is a deviation outside of established control limits, not a lack of precision. It may be difficult to pry the information from participant organizations before the project begins, but it is virtually impossible to do so after execution begins.

The common complaints about this approach are that it takes too much time, that is not the business of the lead project manager, or that the estimates are not to that level of detail. All this should leave the lead project manager with a sense of impending disaster. It is most certainly the job of the lead project manager to have the project virtually assembled in detail before execution. If the estimate is a preliminary one, it should not be used to plan how the project will be executed. If required to be used because of the structure of the agreement or the circumstances, it must be made clear to the CPE that it is highly inaccurate. If this estimate is given to another value chain organization to use for its work, it would be wrong not to say that it is speculative and inaccurate.

From a practical viewpoint, assume that the project manager for each participant organization spends 10% of his or her time on progress reporting. Then imagine that each project manager now has to keep three WBS estimates: one for the value chain, one for the internal sponsor, and one for the buyer. Now the time required becomes much more than 10%, and it places the organization's project manager at risk of it being discovered that there are actually three, or more, estimates for the same project. Having multiple estimates really undermines credibility and destroys any trust that has been created.

As described in Chapter 5, the process of developing a WBS is time consuming and needs the detailed input of each participant organization. The lead project manager shoulders the added task of assembling all individual WBS's into CPE work packages. The sequences described in Chapter 5 for scope and in Chapter 8 for time are the same as for the creation of the CPE work packages: Each participant organization develops its work packages, then meets with value chain participants and aggregates the work packages. The lead project manager then meets with the partner(s) and again aggregates the work packages, and finally the lead project managers and partner(s) meet with the customer and user to aggregate their portions of the project into CPE work packages.

The aggregation of the work packages will lead to a CPE WBS at the work package level that can be summarized to provide the level of detail needed/desired by each participant organization. The CPE WBS with work package level detail, and the established contingency fund, provides the lead project manager with the financial tools to create a baseline EAC once the schedule has been completed—more on that in Chapter 8. During the course of the project, the lead project manager must report on the status of the costs, to which we turn next.

7.5 COST UPDATING AND CHANGE CONTROL

At this point, the CPE WBS has been completed, along with the contingency funds, and the metrics for progress have been established and agreed on relating to the EAC baseline. Remember, the BAC is the budget for the project; the EAC is the estimate of the total project costs by the lead project manager at each progress update. One way of measuring

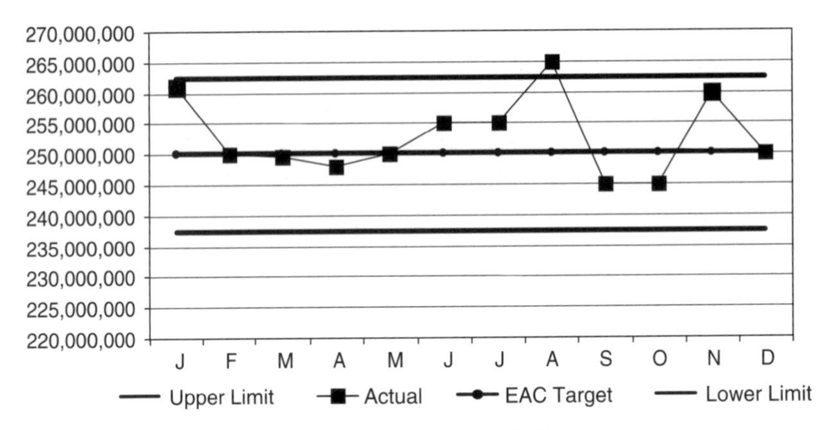

Figure 7.2 EAC Control Chart

success on projects, and one we suggest to our readers for their consideration, is to use the size of the contingency fund as the measure of success. This provides a single measure for the customer, and variances outside a preestablished range require explanation, lessons learned, and mitigation measures.

For example, assume that the range of accuracy on the estimates and contingency is established at +/-10% of the EAC. Then the control limits for variance might be set at +/-5% or tighter if potential cost overruns are a critical component to the customer. Figure 7.2 provides an example of the approach. In the first month, the figure shows that the EAC has increased from the baseline of €250,000,000 (using the same example as in Section 6.2) to just above €260,000,000. The deviation is within the +/-5% variance limits so would not need to be explained, and no mitigation plan would be required.

In the month of August, where the EAC has exceeded the control limits, an explanation of the reason and a mitigation plan would be required. It is also possible to attach bonuses and penalties to variances, to encourage the CPE WBS engagement we discussed earlier. These incentives can then flow down to the participant organization that caused or contributed to the creation of the variance. This will serve to encourage honesty and candor in the creation of the WBS, work packages, estimates, and contingency for each participant organization. The deviations in the other months, not perfect but within control limits, are acceptable accuracy. The customer will be most interested in knowing if there is a single firm causing the deviations, or if the deviations are random. The customer will of course want an assessment of all variances as to the cause, the effect on the BAC, and the mitigation measures to be undertaken.

Any changes that have occurred in the project can manifest themselves as variances if the baseline has not been changed. The change control process, described in Chapter 5, must include the manner in which changes are to be reflected in the EAC and the BAC. Staying with our previous example, if the BAC is €250,000,000 and a change request for €2,500,000 is implemented, there may not be a need to change the baseline BAC; a change of €25,000,000 might well necessitate a new BAC. In Figure 7.2, the variance in August could well be due to a change request of €2,500,000, in which case no mitigation plan would be required. The other issue to be addressed in the change control plan is when the baseline should change. If a €2,500,000 change is made, the new contract value will be higher by that amount, but this alone may or may not necessitate a change in the baseline. If the character of the project is changed, the baseline needs to be changed.

The analogy we suggest is placing apples in a basket. If the base contract requires 100 apples to be put into the basket in 100 minutes at a cost of £0.10 per apple, adding 10 apples could likely be accomplished in the same time, at perhaps a small cost increase. The EAC would change, the BAC would change, but it would not be necessary to change the baseline. If the change was instead 50 apples, the character of the project would have changed, and a new baseline would likely be warranted. The lead project manager needs to make this decision in consultation with the CPE participants, especially the partner and customer, after assessing the situation. Therefore, we suggest that the EAC and BAC can vary, and can vary independently, and that changes in these do not alone require a change in the project baseline. That is a judgment call that must be made by the lead project manager.

The key in control is to make all of the participant organizations in the CPE aware of what will be measured, BAC and EAC, what each means, and what the control limits on variances are to be. This information can be used to lead in to discussions on the WBS exercises and the contingencies.

Once the CPE WBS has been established, the next task is to assess risk. Remember our closing for Chapter 5, where we mentioned that the sequence recommended in this book is designed to save time, and the risk plan should be built in the same planning sessions as the scope and estimate, just after them.

7.6 PHYSICAL PROGRESS

One of the most challenging aspects of international project management is determining progress achieved, or your actual position in our sailing metaphor. We are sad to say that in our experience, most projects suffer from a lack of reasonably accurate physical progress estimates.

Physical progress is a measure of the physical work completed, and the time and resources used to achieve it. Some now use earned value (EV), and we discuss that shortly. Physical progress begins with the design of the WBS and the level of detail of the work package. We recommended 1% of the project duration as minimum work package duration and one reporting period as the maximum. The key is to have actual information that is equal in detail to the plan, and vice versa. The contingency is for inaccuracy in estimates, not for inaccuracy in cost account practices. It can be, but that requires a far greater level of financial sophistication.

If an organization requires people to account for their time hourly every day, the granularity of the actual costs will support a work package duration of one hour. Many organizations only require a monthly or yearly accounting of time. If the actual cost information is monthly, and the work package duration is five days, the project manager will be forced to estimate the amount of time a person spent out of the month on the project.

In exploring the cause of a cost deviation on an internal organizational work package, the project manager could attribute it to any or all of these issues:

- Misestimating the actual time the person worked on the project because of the difference between the level of detail of the work package and the level of detail of the accounting system.
- Lower productivity of the people who worked on the task. Productivity is output divided by input, so if the work package granularity is exactly the same as the cost accounting system, the question is how efficient were the people working on the work package. They could be unfamiliar with the work, inadequately trained for the work, ill, and so on.
- Misestimating the productivity rate used for establishing the baseline for this work package. If there is not a direct connection between the work package and the accounting system, you can see how complex estimating will be.
- Change in local priorities. International projects utilize people from numerous country profit centers. Each profit center has priorities,

and they shift. The project manager should build a baseline plan and get buy-in from the functional managers at each profit center. Equally important is to establish an early-warning system with the functional managers so that changes in local priorities are known as soon as possible.

- Changes in scope. Evaluations for scope changes must consider the impact, if any, on the work packages. Therefore, if the work package required three people for three days for eight hours, and the change adds one hour but extends the work into the subsequent week, where a priority change is to occur, the impact may be much more than one hour.

The goal is to reduce or eliminate as many of these causes of variation as possible internally. Then there is the problem of determining physical progress. Imagine laying a new gas pipeline across Turkey. There are 1,000 kilometers, to be completed in 1,000 days, and a cost of TRL 100 billion. The project manager wants to know the actual physical progress at day 100, to compare it to the baseline plan of, let us say, 100 kilometers. If the pipe is complete, tested, and backfilled, clearly the work is 100% complete, so the progress from a linear measure is 10%. Assume the actual cost to do the work is TRL 10 billion, and the estimated cost is the same, so from a cost perspective 10% of the work has been completed. However, imagine that the pipe is not tested, backfilled only 20 kilometers, welded for 70 kilometers, lying on the ground for 30 kilometers, all at a cost of TRL 12 billion. What is the physical progress?

If the work package was "Furnish and Install gas pipe—marker 0.00 to marker 100.00" with a duration of 100 days, how would you tell the physical progress if the customer wants reports every 30 days? The answer is guess, most likely, based on best judgment. Our suggestion is to make activities that are no longer than the reporting period first. Then break the activities down into components that are somewhat more measurable; in other words, use judgment at a lower level of detail. For our example, the idea is to break down the installation of the gas pipe into these parts:

- *Fabricate and deliver gas pipe—marker 0.00 to 100.00.* Likely a value chain organization or two will do this work. Assume the cost is TRL 5 billion, and the duration is 100 days. Here the project manager will need to know the physical progress from the value chain and the way it is determined. Also, remember our point earlier of not becoming a slave to detail and being artful in knowing where to stop with the work package. This activity could be split into three tasks to meet the heuristic of limiting the duration of a work package to a reporting period. That would increase the accuracy of the physical progress estimate but would require people to account for their time in more detail. We would make it more detailed, but to keep the example simple, assume it was not. In such case, the project manager could check to see how many kilometers of pipe had been delivered.

Even if the analysis is broken down into smaller lengths, there is still the question of physical progress resources. The pipe may have been ex works and payment for materials net 30 days after it was on the loading dock at the fabricator. It has not been delivered, by a different value chain organization, and the transportation costs have not been paid. Should you go to a level of "Fabricate gas pipe—marker 0.00 to 30.00," or stay at a higher level of detail? If this were a critical activity, we would recommend more detail. The point is that there is a balance among the level of detail of the work package, the level of accuracy of the physical progress estimate, and the level of accuracy of the EAC.

- *Unload and string gas pipe—marker 0.00 to 100.00.* Assume that part of the work here is that of the value chain. Assume the cost is TRL 1 billion, and the duration is 20 days. The same considerations apply as described earlier. We would break the work packages down further. This breakdown also improves the accuracy of the lessons learned in the kaizen process of continual improvement of the estimating accuracy. At a lower level, the physical progress estimate is more accurate and therefore will be of more use when projecting productivity and the EAC.
- *Align and weld gas pipe—marker 0.00 to 100.00.* Same as above.
- *Test gas pipe—marker 0.00 to 100.00.* Same as above.
- *Backfill gas pipe—marker 0.00 to 100.00.* Same as above.

Project managers who do not break down the "Furnish and Install gas pipe—marker 0.00 to marker 100.00" will still have to consider each of the detailed steps noted and then make a judgment call as to how much weighting to assign to each. It is a less expensive approach and requires less effort, but it is less accurate. Imagine the activity has float of 500 days. It might be more prudent to break down for management of the value chain into the major components so that the procurement documents can adequately describe the needs and can be tracked back to the project baseline. There are a number of very good reasons for doing this for partners and the value chain:

- The same internal considerations listed: misestimating, lower productivity, misreporting of the productivity rate, change in priorities, and changes in scope. Here the dilemma is the more general the schedule the more difficult to anticipate problems, and the more difficult to recover if anticipated progress is not achieved.
- The interfaces between different organizations are often delicately balanced. Imagine value chain A.1 is a pipe supplier, value chain A.2 is a pipe fabricator, value chain A.3 is a pipe shipper, value chain A.4 is a pipe handler, and value chain A.5 is a pipe welding organization. If the level of detail is "Furnish and install gas pipe—marker 0.00 to marker 100.00," all four of their scopes reside in a single activity with no way to anticipate the results of an internal change in one organization on another.
- If the productivity of one organization, say value chain A.4, in stringing out the pipe falls beyond, value chain A.5 may suffer productivity losses. If those continue, value chain A.5 will protest the added costs, and a dispute will arise where there is no detailed baseline to rely on.
- Changes in scope can be convoluted. If the scope split document requires participation of five organizations and a change affects one of those organizations, how will the impact on the others be evaluated? In practice, it is a judgment call, but that is a fairly subjective way to manage a complex project. More detail would help to accelerate the process significantly.
- Transparency and XLQ. The lead project manager can increase both by moving farther down in the detail. Examining the detail provides multiple opportunities to demonstrate XLQ in all of its attributes.

Bewildered yet? Now think about an information technology project where such things as "Write code" or "Test code" are work packages. How could you break these down further? As we have said, it takes some experience to estimate physical progress and to build a baseline

CPE WBS. You need a sense of the cost/benefit ratio for detail to know if it should be developed at all, to understand the accuracy it will provide, to appreciate the criticality of the work package, to anticipate problems, to forecast the future, to perform the check and act portion of plan-do-check-act, and to lead the CPE.

There are a number of ways to determine physical progress, some more precise, some less. Here is a brief review of a few.

- *Our approach.* Make work package durations no longer than a reporting period and detailed enough to lead the work. On many projects, this will result in work packages that are a week long at a minimum and four weeks long at a maximum. At worst, this method provides a 66% probability of being correct even if the project manager only guesses. If the project manager has thought through the details as suggested, accuracy increases significantly. If, as is often the case, the work packages are one week in duration, it is fairly easy to see if the work was actually completed. In general, we prefer a bit more detail and generally craft a CPE WBS that has activities measurable on a weekly basis.
- *The 50-50 rule.* The idea is to assign 50% complete at the start of a work package and 50% complete at the end. This is a probabilistic approach that will be more accurate the shorter the duration and the more work packages in the project.
- *Earned value.* Earned value is the budgeted cost of the work performed, or physical progress. It considers the cost of achieving progress. To assess earned value, it is necessary to know actual costs, budgeted cost of the work planned (baseline), and the budgeted cost of the work scheduled. It is also necessary to know planned and actual in some measurable way, such as lineal meters, kilograms, pages, lines of code, number of patients treated, number of judges trained, and so on. Think back to the example for the gas line in Turkey, and knowing how many meters were installed, how long it took, and how much it cost. Using this technique requires knowing actual cost contemporaneously and knowing the quantity of physical work planned and performed.
- It is not especially accurate at the beginning of a project; studies have found that the cumulative value of the cost performance index (CPI) stabilizes by the time a project is about 20% complete. After that, it does not vary by more than 0.10 from the value obtained at

the 20% point, and the accuracy tends to worsen from that point of stability until project completion. A number of books that detail the process for those of you interested in learning more (Bower 2007; Fleming and Koppelman 2000; Kerzner 2006; Turner 1999).

- *Divination, just guess.* We have seen too much of this method. Due diligence is required to manage the details at a reasonable professional level. Judgment will still be required, but guessing at the CPE level is not the way to do it.

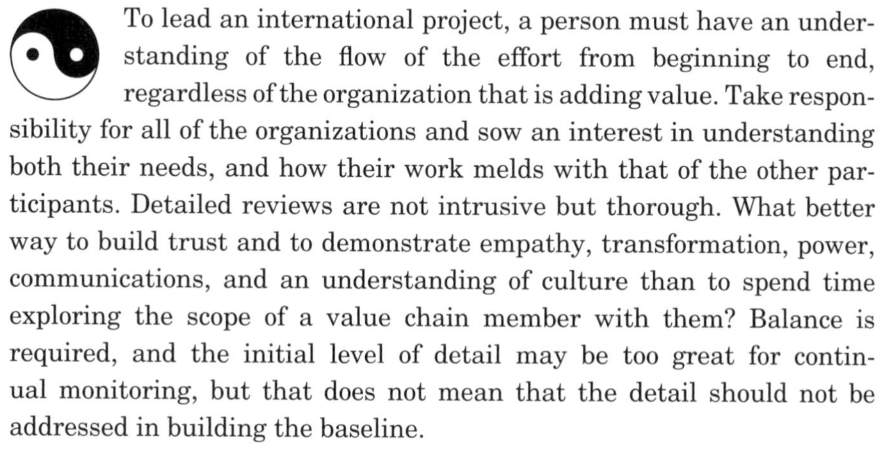

To lead an international project, a person must have an understanding of the flow of the effort from beginning to end, regardless of the organization that is adding value. Take responsibility for all of the organizations and sow an interest in understanding both their needs, and how their work melds with that of the other participants. Detailed reviews are not intrusive but thorough. What better way to build trust and to demonstrate empathy, transformation, power, communications, and an understanding of culture than to spend time exploring the scope of a value chain member with them? Balance is required, and the initial level of detail may be too great for continual monitoring, but that does not mean that the detail should not be addressed in building the baseline.

For physical progress, we strongly recommend that the lead project manager push the work packages level down as far as necessary to measure progress adequately. As we have indicated, it is critically important to know the physical progress when forecasting the way ahead. The forecast may prove that a revised baseline plan needs to be enacted and may require some organizations to mitigate the failings of others. This is what teams do. It also provides yet another opportunity for the lead project manager to demonstrate her or his XLQ.

Chapter 8

Risk Management

> The policy of being too cautious is the greatest risk of all.
>
> —*Jawaharlal Nehru*

> God sells knowledge for labor, honor for risk.
>
> —*Dutch and Saudi Arabian proverb*

In this chapter, we discuss risk management from the viewpoints of the participant organizations and of the collaborative project enterprise (CPE). From experience, the key to effective risk management is to actively manage a limited number (we suggest no more than six) of risks thoroughly. The risk management attitude we suggest is the same that major international cities have for catastrophic occurrences such as earthquakes, cyclones, and floods. There should be a risk plan written down, complete with a budget, schedule, process, and preapproved response. Think of a risk plan as a mini project plan, and you have the idea. If the risk occurs, all that remains to be done is to implement the plan. This means the risks need to be integrated into the schedule, budget, and scope of the project—transparently and purposefully.

8.1 ORGANIZATIONAL PROJECT RISK PLAN

Planning for risks begins at the corporate level of each participant organization. The first question is: What is the risk attitude of the organization: risk seeker, risk neutral, or risk averse? A risk seeker would be willing to gamble a possible loss ¥10,000,000 of gain or profit of ¥1,000,000—or a profit/loss ratio of 0.10. At the other extreme, a risk-averse organization would be interested in projects with a profit/loss ratio of perhaps 0.90. There are innumerable risk attitudes between

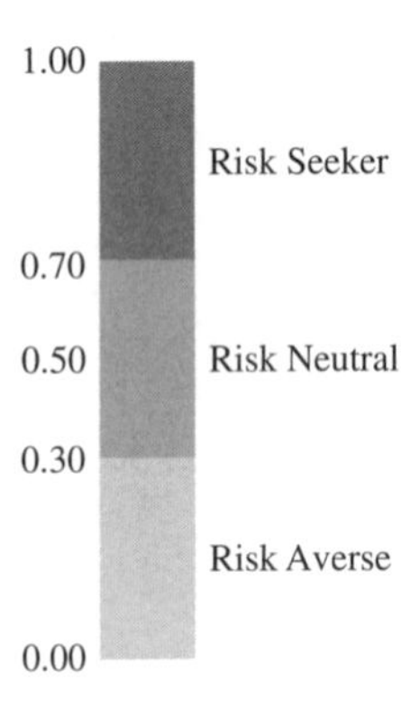

Figure 8.1 Corporate Risk Attitude

these extremes. The project manager of each organization in the CPE must know, guess, or divine the corporate risk attitude first. Figure 8.1 provides a graphical view of a corporate risk attitude scale.

The risk plan, for each organization, must also address how the project risk profile will be accumulated and assessed. We recommend a numerical system that utilizes probability times impact (P × I) with both dimensions ranked from 0.00 to 1.00. A probability of 0.0 would mean no chance of occurrence; an impact 0.0 would mean no impact. If a project has a risk ranking that is assessed at 0.75 and the corporate risk attitude is 0.45, the risks would be above the corporate tolerance level and would need to be actively managed. Conversely, a project risk ranking for the same organization that had an evaluated score of 0.30 would need to be passively managed. More on the details and active and passive risks later. For now, the main idea is that the project manager knows the risk attitude of his or her firm.

The next issue that must be addressed in preparing a risk management plan is a method of filtering the risks. For a participant organization on a project, the risks need to be divided into business risks and project risks. Examples of business risk could be shortage of staff, shortage of funding, political risks, market risks, and currency risks. Examples of project risks could include changes in technology of the product, weather, delivery delays, and government regulation changes that impact the project product. In Figure 8.2, the light gray items are project risks and the dark gray items are business risks, some of which need to be actively managed. In the center of the chart are black boxes. These are risks that

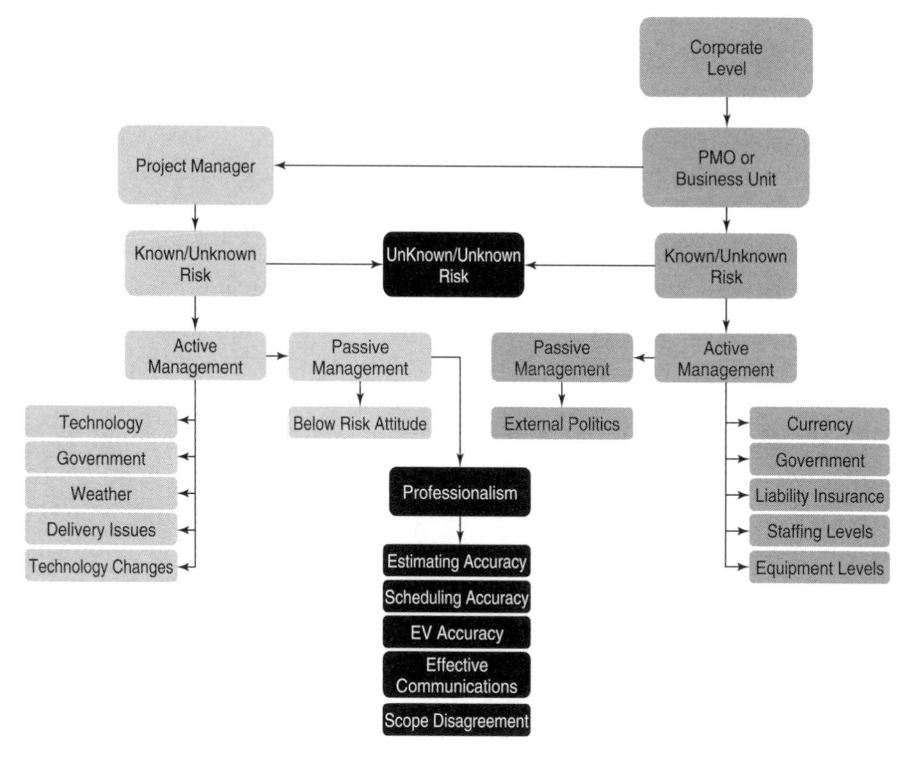

Figure 8.2 Risk Filtering Part A

we recommend are passively managed by the project manager in the case of project risks, and by the business manager in the case of business risks.

An actively managed risk must come from the filtering process, must have a P × I greater than the corporate risk attitude, must have a mini project plan constructed for it, and must be preapproved by the key stakeholders. Then it is managed the same way a work breakdown structure (WBS) activity would be managed. Passively managed risks are identified during the risk filtering process and have a P × I less than the corporate risk attitude. For these risks, a mini project plan does not need to be created; nor is key stakeholder approval needed.

Actively managed risks are those that require identification, probability assessment, impact assessment, qualitative and quantitative analysis, estimating, scheduling, ownership assessment, trigger event description, response strategies, and reporting. In other words, they are the risks that require a significant investment in time, cost, and planning. The risk of a project manager being less than perfect in estimates, activity durations, cost accounting, and understanding of the scope are *not* the type of risks that should be actively managed. They should be addressed by contingency. Think of it this way: If you would not be embarrassed to present a plan that shows how you are going to manage the risk that your communication plan is fraught with potential miscommunications, you should consider managing such risks actively—especially if it is presented to your customer. If not, train project managers to reduce the error rate, build in contingencies to cover the balance, and manage project management risks by having people who are trained and do their due diligence. That is what a customer expects. This is not the view, by the way, taken in the *Project Management Body of Knowledge* (*PMBOK* 2004).

In our experience, one of the reasons that risk management has not gotten much traction in the project environment is that there are simply too many risks to be managed. So few are managed because there is not adequate time to do so. Confronted with such a dilemma, project managers will develop an ad hoc process and try to pin down the risks with the highest visibility; in doing so, they may get lucky, or they may not. If you are serious about managing risk, you must winnow down the list and focus the process on the risks that require active management and those that exceed the organization's rick attitude.

Another major issue with risk is ownership. We contend that the organization most capable of managing a risk should own the risk. Imagine you are the project manager for a water project in Peru. You are planning to subcontract the excavation work to a local Peruvian organization. The customer, a financial firm from Sao Paulo, and the user, a Peruvian governmental agency, have written a contract that places the risk of archeological discoveries in the right-of-way on the service provider, your organization. Your organization has never done a project in Peru, and your value chain subcontractor is from Lima and does not know the local conditions where the waterway is to be built. Clearly, the customer and user are at least concerned about the risk, or they would not have written it into the contract. Who would be in the best position to manage the risk?

While you are pondering that question, let us consider another difficult issue with risk management planning, and that is transparency. On international projects, the process for risk is sometime like the game of musical chairs, where 12 children walk around 11 chairs while music is playing. When the music stops, whoever is left standing is out of the game. To extend the metaphor to risk, the person left standing gets the risk. If a customer has provided the right-of-way for our Peruvian project, and has reason to believe that there could be a burial ground on or near the property, it can choose to divulge that information, withhold the information, or write a contract clause to transfer it to the service provider. It is hoped that the sales teams for the lead organization will know that such a risk exists, regardless of the customer's action, and addresses the risk in its tender. The lead organization also has a choice to make with the value chain: Divulge the information, withhold the information, or write a contract clause to transfer it down the value chain. It is tempting, from the customer's viewpoint, to say nothing about the possibility of burial grounds, particularly if it is not certain that they exist. The concern is that the professional organizations that will be tendering will include any evaluated costs for this perceived risk in their pricing.

We suggest that the organization in the best position to manage this risk is actually the customer. The reason is that if a burial ground is discovered, other governmental agencies will need to be involved, and it is likely that other rights-of-way will need to be acquired and a redesign of the pipeline undertaken. In taking this approach, the organizations submitting the tender know the risk exists, know that the customer will take the risk, and can pass it down the value chain. If the customer chooses to withhold the information or to assign it to the tendering organization, the lead organization will have to undertake the cost of design changes, governmental agency coordination, property acquisition, and a significant amount of time. In the best case (for the project), the tendering organizations will build the cost of the risk into their price; in the worst case they will not and the legal system probably will need to be engaged. Once the legal system is engaged, costs could far exceed the cost to remedy the problem.

Let us return once more to the issue of filtering. We cannot overemphasize the importance of filtering out the risk to arrive at the six risks that exceed the organization's risk attitude. Figure 8.3 provides a graphical view of some typical categories of risk that an individual organization would face.

Starting at the center is the CPE; on the right is the individual organization's business, and on the left is the project. We recommend that the business internal and business external risks be passively managed by the organization and not be addressed as part of the risk management plan. Only the CPE external risks should be actively managed. The figure provides a standard list of risks that can be used as a template for brainstorming.

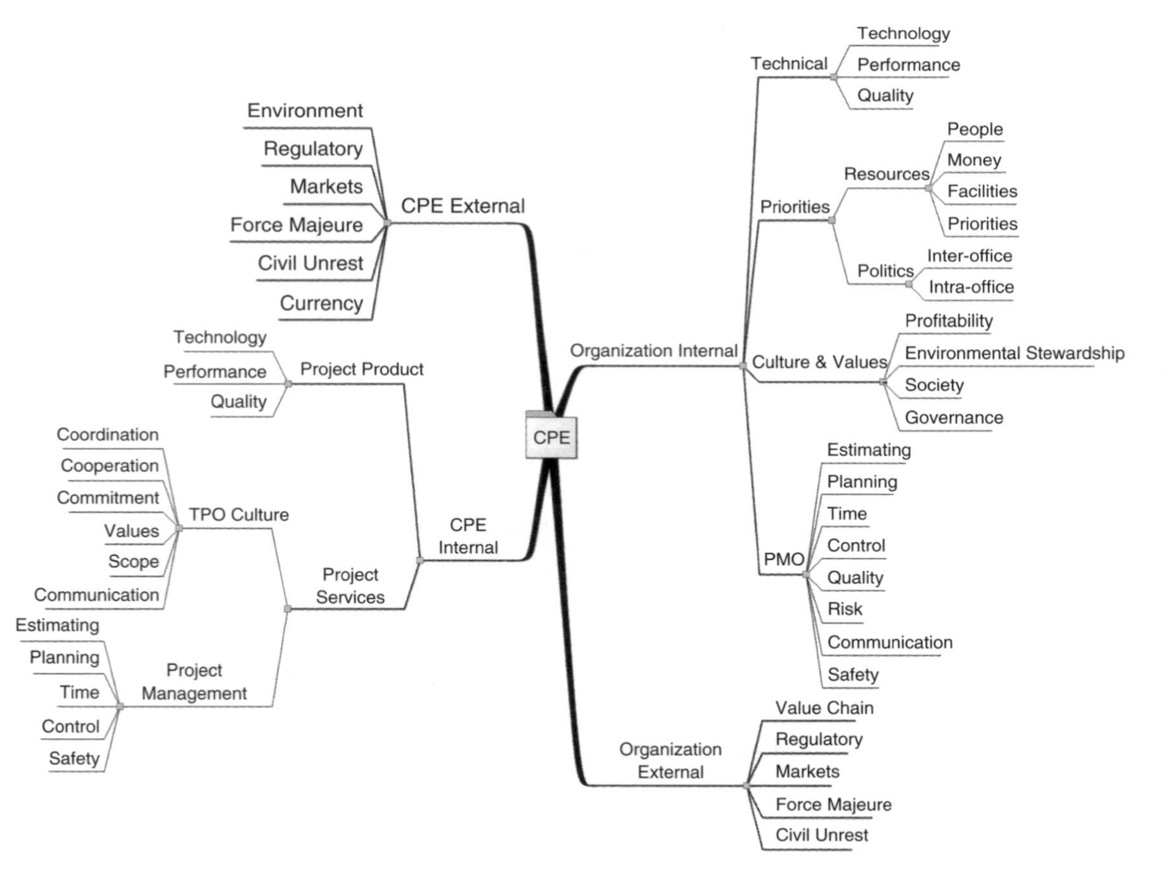

Figure 8.3 Risk Filtering Part B

Therefore, in Figure 8.2, the risks to be actively managed are those in the CPE external category of Figure 8.3. In other words, these risks must be evaluated as to whether they are better managed as part of the project or part of the organization. For example, staying with the Peruvian project, if there is a risk that the equipment to be imported for the project is lost at sea, should the project manage the risk? If the business has blanket marine coverage for operations, it may prove more efficient to have the project added to the business insurance coverage. If there is no business coverage, then insurance should be acquired for the project. The same would be the case for currency hedges. If the risk is to be managed by the organization, there will be a cost incurred to extend coverage to the specific project, and that would need to be priced and paid under the project but actively managed under the organization.

There is more, however. Most marine policies have different coverage premiums for cargo shipped on deck and in the hold. Here coordination must occur between the CPE and the lead project organization. Shipping costs will increase by shipping in the hold, and each partner and value chain covered by the policy must adhere to the insured option. If the customer has an umbrella policy for the project, however, the same flow-down of requirements would need to occur.

Under business internal, risks such as inadequate organization resources, transparency issues, poor communication, or changes in internal business priorities are not risks to be actively managed by the CPE. Likewise, CPE internal risks of making a nonperfect schedule, having poor communication, not understanding the scope, and failure to inspect the product for quality go to the professionalism of the project manager and should not be actively managed. These things need to be managed by the project manager but not managed as active risks. As we suggested earlier, one easy way to separate active from passive risks is to determine if you would want to put them into your schedule and report on them each period to the customer and the CPE participants.

This filtering of risks is the easy part. Each organization in the CPE has to determine the risks associated with its work and check to determine if the risks are above the organization's risk attitude. Then the lead project manager needs to confirm that the risks are truly CPE external risks and determine who would be in the best position to manage them. Remember, this is the perspective of a collaborative environment where the lead project manager is directly involved from the beginning of the initiating and planning phase. Consistency by the lead project manager is essential, for different organizations have different risk attitudes, and the attitude of each needs to be respected.

The ownership of risks flows directly from the discipline just described, assuming that there is open and transparent knowledge sharing. If the Peruvian project was tendered, and the non-Peruvian firm won the tender without a local partner, the process can still be followed, but it may not be possible to give the risk to the organization in the best position to manage it (i.e., the Peruvian government). One potential impediment to assigning risk to the firm most capable of managing it is the type of contract environment used. If the organization that won the tender won via a competitive public bid, it is unlikely that transparent discussions on risk and risk ownership occurred between the customer and the service provider. It is possible that this agreement will restrict the potential for transparent communications with the value chain and any potential partners. We discuss the reasons in more detail in Chapter 10.

The organizational project risk plan must include a process for risk management that addresses risk identification, analysis, response, and ownership. The process should be prepared at the corporate level by the organization's risk management department, its project management office, or an executive-level manager with access to the corporate strategy discussions and the project manager. We have seen it done in an ad hoc manner by each project manager, but the results, as you can imagine, are inconsistent and potentially dangerous to the organization and to the project. For the CPE risk plan, we recommend that the lead project manager develop a risk attitude for the CPE, outline the processes described earlier, and then seek concurrence from the key stakeholders. This plan then provides a framework into which will fit the individual organizational risk plans.

The CPE risk plan is the high-level plan for the entire project. Once the key stakeholders accept the plan, the detailed risk planning may begin. As with all other aspects of international project management discussed in this book, planning for risks begins with the individual organization and is accumulated and codified by the lead project manager.

8.2 PROJECT RISK IDENTIFICATION AND ANALYSIS

On international projects, the identification of risks can become a lifelong career. Too often in business, the sheer number of potential risks can overwhelm a lead project manager and cause the process to be abandoned to a more expeditious approach—subjectivity. As we have indicated, the goal is to get to the six most significant risks by using filtering techniques. Those techniques begin from the point where the risks have been identified. Now we need to talk about the identification process itself.

Although each organization and each project within the organization will be different, generally there are three general approaches to the identification of risks: expert opinion, brainstorming, and an approved checklist, such as that shown in Figure 8.3. Bias exists in all techniques, and the project manager must exercise a healthy dose of skepticism.

Expert opinion can come from a variety of sources, internal and external, and may be expensive if you are forced to rely on consultants. The advantage of internal experts is that they have direct personal knowledge of how the organization conducts a particular type of project. The disadvantage is that they may not know anything about the customer, partner, contract terms, or geographical location(s) involved. We believe that internal experts are an excellent source for international project managers, if the experts have in fact managed international projects. If not, be more cautious and scrutinize their input.

Expert opinion can also be gained through external experts. The advantage of external experts is that they likely know something about the customer, partner, contract terms, or geographical location(s). The disadvantage is that they may not know anything about how the organization conducts a particular type of project. There is also the Delphi technique for soliciting expert opinion. This technique reduces bias through anonymity. For example, if you have selected three experts, you ask them all to provide their opinions without seeing or knowing who the other experts are. The opinions are gathered, combined, and then shared with each expert. Still not knowing who the other experts are, each considers the summary of opinion and can change or confirm his or her initial opinion. This process then continues until a consensus is reached.

In using expert opinion on international projects, we strongly recommend that project managers understand the knowledge context of the experts by asking them about their experience:

- Size, complexity, and duration of previous projects
- Customer knowledge
- Partner knowledge
- Value chain knowledge
- Geographical knowledge

- Cultural knowledge
- Success of projects
- Lessons learned on risk

Expert opinion also comes from the organizations participating in the project. As with the WBS and scope split effort described in Chapter 5, identifying risks requires the accumulation of efforts of each individual organization.

Brainstorming is a process where a project manager assembles a team to create a list of potential risks for the project. All ideas are simply recorded, and no judgment is made on the validity or importance of the ideas. All ideas are equal. Once a list is created, it must then be filtered and prioritized.

We explore the process of risk identification in the next section, so we need an example project to make the points more clear. Imagine we have a project in South Africa to construct a new toll road, and the structure of the contract environment is a public-private partnership (PPP) as shown in Figure 8.4. The customer is a financial consortium led by an investment bank in London and a sovereign fund in Dubai, and guaranteed by the government of South Africa. The user will be a toll road operations firm based in Spain. You are the lead project manager with an international engineering and construction firm from Brazil, with a local construction joint venture (JV) partner that will do the property acquisition and some of the construction, and of course a global value chain.

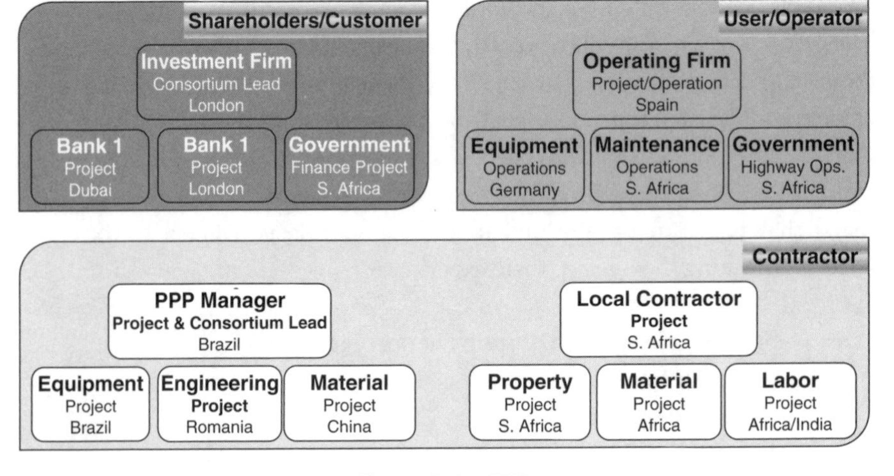

Figure 8.4 PPP

The toll road is not popular with everyone in South Africa, but it is desperately needed. The project has a budget of ZAR 3 billion and duration of 36 months. There is a penalty of ZAR 300,000 per day if the project is not ready for operation within 36 months. Now let us build a risk plan.

8.2.1 Organizational Risk Register

The starting point is for each organization to construct its own risk register, as was done with the WBS. The lead project manager can brainstorm the potential risks for all of the organizations, talk to some experts, and utilize a checklist of risks for similar projects as a starting point. In this way, she or he can create a richer understanding of the risks that confront the project and thus anticipate the work of the other organizations. By so doing, the lead project manager can demonstrate concern for the project and for the work of the other organizations. Such empathy and consideration, servant leadership, begins to build a foundation of trust—more on that in Chapter 11. Figure 8.5 shows an example of a risk register. A risk register is a listing of risk including the probability and impact, and P × I. For active risks it also includes the trigger event, the response strategy, the time horizon, the costs, and assumptions.

The first column is an arbitrary sequential numbering system to organize the risk, followed by a description of the risk itself. As with the WBS, precise descriptions greatly aid in the communications process. In the figure, public outcry is not very illuminating for it could represent a small group that is vocal or a large one. Some frequency, such as daily protests or protests that reach the government level, would be more helpful in thinking about how to manage this risk.

The "Probability" and "Impact" columns are subjectively assessed on a scale of 0.00 (low) to 1.0 (high). The probability that Risk #1 will occur is relatively low (0.40), but the impact on the project if it does occur could be relatively high (0.60). The product of these two assessments is shown in the "P × I" column as being 0.24. The "Trigger" column provides a judgment of the occurrence that will trigger, or begin, the agreed-on risk response, or one of the modules of the risk response.

The "Response" column describes the actions to be taken to minimize the probability and/or impact of the risk on the project. In the figure, Risk #1 shows two steps to be taken. The first is to engage the community

						Avoid Miktigate Transfer Accept		
Risk #	**Risk Description**	**Probability**	**Impact**	**P x I**	**Trigger**	**Response**	**Owner**	**Assumptions**
1	Public outcry about allignment	0.40	0.60	0.24	Public protests	Multiple	Multiple	Despite efforts project will be 3 months late
2	Property rights unclear	0.90	0.70	0.63	Public records search provides no ownership information	1. Accept. Re-allign roadway. 2. Accept. Utilize eminent domain	South African contractor	1. Assume that allignment can be altered cost to redesign 30,000. 2. Assume cost to litigate 20,000
3	Global financial crisis	0.90	1.00	0.90	Stock markets tumble	Mitigate	London financial group	Sell bonds in Rands, and put currency hedge in place. Cost is cost of hedge

Figure 8.5 Risk Register

early in the project in order to demonstrate concern and to elicit reaction. On a number of our previous projects, this approach helped to bring the community's objections to the project early on and enabled community members to select a group of representatives to interact with the lead project manager or the CPE spokesperson. In this way, the overall impact of an unpopular project can be diminished. On our real project, the public condemnation of the project continued despite all of our efforts, but the project continued to completion. In the example of our South African roadway project, the hiring of a local public relations (PR) firm and a campaign to engage the public would be one possible way to mitigate the risk. The outline of the program and the estimated costs are shown in the "Assumptions" column, which captures the overview plan of the risk.

Now let us look a more detailed view of Risk #1, provided in Figure 8.6. The risk number is followed by a detailed description of the risk, and a recap of the probability, impact, and the probability times the impact, and the trigger event. Following this, the strategy and tactics are described in detail, and the assumption that despite the efforts o mitigate, there will be a residual risk.

The initial plan is to hire a PR firm, and undertake monthly meetings. The plan also includes setting up a board, which will require five preliminary meetings. The figure presents an example of how to allocate these risks and the financial exposures for each participant organization. The lead project manager wants to recommend that the risk be given to the organization that most capable of managing it. In our example, the overall risk needs to be coordinated by the lead project manager, but the costs are more properly divided among the participant organizations. First, the South African firm should handle the hiring of the PR firm, as it has intimate knowledge of the market, and any disputes will be directly linked to its acquisition activities.

The creation of the board will require the participation of all participants, including the U.K. financial firm and the South African government. The South African firm would take the costs for the meeting venue, and the other costs are estimates of what each other organization would incur to attend the meetings. Similarly, the monthly meetings will require the participation of all organizations. The more interesting issue is the penalty. When done properly, penalties, often called liquidated damages (LDs), are included in contracts to pass along all damages that will be incurred if the project is not completed on time. Penalties can include such things as LDs, loss of revenues, loss of business, and forgone

Description	Probability	Impact	P x I	Trigger	
Risk #1					
Public outcry about highway allignment. Public outcry stops or significantly delay the acquisition process, which in turn delays the construction and the completion.	0.40	0.60	0.24	Public protests	
Strategy					
Risk is possible due to the property right issues that exist in South Africa. Strategy is to be proactive and build property right sensitivity into the acquisition process, and to engage the public support from the start of the project.					
Tactics					
1. Public support should begin with assistance from the JV partnership and the government to form a board of governance that consists of elected public representatives and members of the TPO.					
2. Monthly meetings to review the status of the acquisitions, the plans to ensure equity and fairness, and the status of the design and construction. % represents the responsibility to participate					
Assumptions					
Assumption that despite all efforts the project completion date will be delayed by three months or 90 days.					
	Ownership				
	Brazilian Firm	**South African Firm**	**UK Financial Customer**	**South African Government**	**Total**
Strategy Costs					
1. Hire PR firm		ZAR 700,000			ZAR 700,000
2. Board creation and meetings (5 each)	ZAR 100,000	ZAR 200,000	ZAR 200,000	ZAR 50,000	ZAR 550,000
3. Monthly meetings (36 each)	ZAR 360,000	ZAR 720,000	ZAR 120,000	ZAR 360,000	ZAR 1,560,000
Residual Costs					
1. Penalty (ZAR 300,000 per day for 90 days)			ZAR 27,000,000		ZAR 27,000,000
Totals	ZAR 460,000	ZAR 1,620,000	ZAR 27,320,000	ZAR 410,000	ZAR 29,810,000

Figure 8.6 Risk Register Detail

opportunity costs. The losses could be "hard," or receipts for added expenditures, or "soft," such as loss of business. In most legal systems, such costs that are "estimated fairly" in advance, such as LDs, are recoverable as long as both parties have agreed to them in advance by signing the agreement. Let us assume that the penalty in our case was actually included in the value for money (VFM) that the investment organizations used to acquire funding for the project. Thus, the bondholders purchased the bonds with an AA risk rating, in part due to this government guarantee. As an aside, on the issue of VFM, a good deal of the research into PPPs, such as our project, has found that the majority of the VFM is in the transfer of risk. Most of these risks come from the operations of the facility once completed.

Let us assume that you as the Brazilian lead have completed your due diligence scheduling effort and find that, using the Program Evaluation and Review Technique, you will be 90 days late with the project. While it is possible you could do better, it is also possible you could do worse. As shown in Figure 8.6, the exposure is ZAR 27 million, so who should take this risk? Our suggestion is that the organization in the best position to manage this risk should accept it; in this case it is the government of South Africa, as it has domain over property rights. From experience, this is the last organization that will take it.

What if this risk register is being developed after the signing of the contract, rather than before? In Chapter 10, we discuss some contract strategies to help ameliorate this issue, but for now assume it is after the contract and that the Brazilian/South African JV did not include it in the pricing for the project. Also, assume that the South African government wants nothing to do with either the PR problems or the cost risk, and the U.K. organization has no intention of modifying its VFM assessment. We recommend that the JV actively manage this risk to the best of its abilities and report on it at every progress meeting. The lead organization must attend to the risk in any case, but the ownership in this case should specifically reside with the South African JV partner.

As you can see with this one example on this one project, risk management takes time, and it can be quite complex. That is why we recommend filtering the risk down to a manageable number, no more than six, and actively managing them. The risk of being an imperfect project manager should be managed passively, with training, support, and strong lessons learned processes.

If each organization prepares its own risk register by thinking openly about risks:

- It greatly facilitates an atmosphere of trust among the members of the CPE. The team sees that transparency and honesty about difficult issues is the norm, not something to be feared. Thus, small problems are far easier to deal with. It shows leadership on the part of the lead project manager. One of the aspects of cross-cultural leadership we discuss in Chapter 11 is fearlessness.
- It greatly reduces the negative impact of risks on the project. While in our example the South African government may not be willing to take the risks, it may be more than willing to work in the background to mitigate. It is in its best interest politically.
- It helps build cohesiveness in the CPE. It begins the process of imbuing a culture of cooperation.
- It exposes risk as a task to be managed rather than as an emotional quagmire to be ignored.

Now let us look at the risk register for the JV partnership.

8.2.2 Partner Risk Register

Before we discuss how the JV partnership might deal with this risk, there is a question of timing. On some international projects, as we have mentioned earlier, the JV lead organization may have had a long-term relationship with the U.K. firm that designed the VFM for the project. In fact, it is quite possible that the Brazilian firm participated in the initial planning and marketing of the project for the South African government and the investors. Let us assume this is the case and that the Brazilian firm is now searching for a partner. What we mean by this is that the contract was already signed among the U.K., Brazilian, and South African government organizations. Therefore, they would have already created a risk register for the project. We will assume that it was done at the level of detail suggested and that it is transparently shared.

The Brazilian organization would then include the risk register in the negotiations with the South African organizations bidding on the work. This is one of the many tests of leadership for the lead project manager and her or his organization. If this information was disclosed, we would expect the bids of the South African firms to reflect the penalty in their bid. They of course may think it is not a problem because they have political connections, or they may think that they can push it off on the government, or they may think that the penalty is going to be larger or smaller. What if the value of the contract work for this firm is ZAR 270 million, with an anticipated profit of ZAR 15 million? If the risk occurs, this organization would bear a disproportionate financial burden. It is hoped that it would increase its price to offset this potential loss, which would mean that the Brazilian firm would bear the cost of the risk. What if another South African bidder decides to take the risk and does not include it in its price, and does not include the cost to mitigate—a savings of ZAR 1.62 million? Unfortunately, sometimes this is the math.

In our experience, transparency is the best approach, and if a proposition appears to be too good to be true, it usually is. This sequencing problem is difficult to deal with because of the variety of ways that projects are developed in practice and the amount of time it takes to complete the initiation and planning. In Chapter 10, we set forth a process that suggests some options on how to deal with this issue. For now, suffice to say that the Brazilian organization should make the risk register part of the negotiations with the South African bidders and should demand that they prepare a risk register of their own before a price is submitted.

Once both partners have compiled a joint risk register, then the discussions regarding who should own and manage the risk can be productive. In some cases, such as the example of the risk of public outcry, the risk should be shared among all the organizations involved in the project. The risk register and risk register details shown in Figure 8.5 and Figure 8.6 provide a format for recording these decisions. As an aside, look back at Figure 6.8 for a moment. It is critically important to keep the risk estimates separate from the contingency. The risk estimates may need contingency added to them to account for uncertainty in the estimates, but do *not* confuse or intermingle risk estimates with the contingency fund for the project.

The reason for not intermingling them is as follows. The contingency fund is a hedge against a project manager or lead project manager being imperfect. It should be established, as we indicated earlier, in a consistent way, and is intended to cover the fact that we are all imperfect in our estimates. That is quite different from the cost for a risk that is actually a work package. If risks are left in the same pile as contingency, they will be treated as such. The cost of dealing with risks is separate, and must be managed like any other work package activity on the project. Think of it this way: Contingency is one bank account, work package costs are another, and risk costs are another. If they are intermingled, you will lose control.

The steps are the same as with the WBS and budget at completion (BAC) with one important exception. In the case of risks, the JV partners need to have built a preliminary risk register for the project and have assigned ownership between them and the customer and user. The reason is that some of the value chain may be completely unaware of political risks. In our example, to do this, each JV partner needs the schedule, WBS, and risk register that has been created by the JV. It unlikely that the Romanian organization would know about politics in South Africa at the level required to imagine risks. The lead project manager must educate the value chain to some extent about the vision of the entire project. This is true in all aspects of leading the CPE, but especially here with risks, for they are more nuanced.

8.2.3 Value Chain Risk Register

Each JV partner must begin with his or her own internal make-buy decisions: which of the WBS work packages should be procured, how much time and cost is estimated for each, and what risks should be passed on to the value chain. One of the great benefits to having done the JV risk register is that most of the difficult decisions in procurement have already been made. Procurement under uncertainty is a complex and difficult undertaking, and the process we are discussing with help to remove much of that uncertainty.

In practice, many of the make-buy decisions are made during the WBS, BAC, and JV risk register process. Now each organization can go to the market for pricing with a very clear idea of the scope, timing, cost, and risk associated with each work package. This process yields more precise and lower pricing in general and enables quick decisions to be made on pricing appropriateness, because there is a fair cost estimate for each work package. It enables transparency in the value chain transactions, and that builds trust.

Having been on the seller's side many times, we can tell you that trust often leads to lower pricing. Jeff Immelt, the chief executive officer of General Electric, says that "green is green," meaning to be a green is good business and will make the organization money. The same concept applies here: Transparency is good business and will lead to increased profits.

That is not to say that it will always lead to success on a specific project, and certainly it will not be easy, especially when it is not reciprocal. However, in our experience, transparency pays off long term.

On large, complex international projects, it will not be possible for the JV partners to pass along 100% of most risk costs, simply because of the project scale. That means that, during negotiations, a decision must be taken with the value chain organizations on how much risk they can tolerate for the price. For example, consider the penalty for being late of ZAR 300,000 per day. For a value chain subcontractor that will do property appraisal work valued at ZAR 1 million, a delay of four days will exceed the entire value of the contract. Imagine a simple issue like a property owner showing up with a gun and prohibiting the appraiser to enter the property for one week until an officer can accompany him or her. In such seemingly intractable situations, we recommend returning to the basic question: Who is in the best position to manage the risk? If the answer is the customer, but the customer refuses to accept the risk, the highest level of political power should take the risk—in this case, the JV partner. However, the risk trigger recognition clearly should be the responsibility of the value chain organization. Failure of the value chain organization to provide immediate notification should carry a penalty that is proportionate to the value of the contract—this is another case of applying the XLQ descriptors of equity and fairness.

As you have seen so far, this level of effort takes time and resources. That is one reason why we strongly recommend that international project managers must filter the risks into active and passive categories and deal only with the active ones. Some project managers may feel uncomfortable being so transparent, and we certainly can empathize, but the effort pays big dividends on the leadership scale. Fearlessness, transparency, truth, equity and fairness, open communications, being vulnerable (you practice transparency but your value chain colleague does not), and empathizing *demonstrate* leadership—more about that in Chapter 11.

As all project managers know, this process is iterative. If you are transparent, you may learn about other risks that were not on the JV risk register but should be. Once the JV partners have a fully populated their internal risk registers, including that of the value chain, then the customer and the user need to be consulted again.

8.2.4 CPE Risk Register

On most projects, the customer and the user develop a project and then invite organizations to provide technical and price proposals, unless the PPP approach is utilized. Therefore, there is obviously some level of interaction on scope, cost, and time at the beginning of the process. If the customer and the user practice transparency, the JV partnership, in our example, has better information and can reduce the costs of tendering and the time required. In this section, we are discussing the detailed level of discussions.

Discussing risk with customers and users is not a task for the timid. When we were a customer and a user, we did not want to hear about risks that we should have considered; particularly if we had overlooked them and they would cost us more money and/or time. Painful experience has taught that this questioning is actually a blessing in disguise, for it is better to manage risks rather than have risks manage you. In our experience, a well-considered risk register with details and active risk management for a well selected and limited number of risks will help convince a customer to listen. What helped to convince us was to make each active risk a work package, and put the packages into the schedule. In most legal systems, the so-called means and methods—meaning how the work is conducted—is the sole domain of the service provider. If the JV partners believe a risk is important enough to manage actively, as a work package, the customer does not have the legal right to insist it be removed. Informed customers know this; if they are uninformed, it is your job as the lead project manager to educate them. Part of the job of a lead project manager could well be to protect customers from themselves.

Returning to our example case, imagine that the property appraiser value chain organization has informed you that in one particular township, there has been violence over property rights, and some appraisers have been injured in attempting to do their job. Here is a risk that transcends the basic contractual agreements and is one that the South African government must help mitigate. If it does not, the project will most certainly come to a stop, and people may be injured. It is pointless to ignore such a risk, and the lead project manager must do her or his diligence in building the strongest, most compelling case to persuade the customer to proactively assist in mitigating the risk as part of the risk response plan. The JV and its value chain cannot correct a political issue that has existed for decades. If the customer chooses to ignore the risk, we recommend that the lead project manager firmly, but respectfully, insist that the risk remain a work package and that it be managed accordingly.

The assumption for our example is that the ownership for the risk can be established in negotiations before the agreement is signed, using a collaborative structure. If the agreement has already been signed, the CPE risk register still needs to be produced; however it will require a stronger lead project manager. If the financials and VFM have been preestablished, then the transparency issue is more of a challenge. In our example, the JV partners would have a price for the project established and, it is hoped, would have built in a risk fund for the property issue. If they have, divulging it to a customer may seem nutty, but if it is not divulged it will not be managed—at least not in the manner we have suggested above. Think of it this way: If the certainty of violence is 100%, the choices would be to:

A. Divulge it and have the risk fund exposed.
B. Divulge it and conceal the risk fund.
C. Ignore it until the problem manifests itself and then claim ignorance.
D. Ignore it and hope it does not happen.
E. Ignore it and try to convince the customer that it is their problem and that it was not known during the bidding if it happens.

Options B, C, and E are at best disingenuous and are possibly an ethics violation. Option D is not a wise option for your organization, for you are eliminating any possibility of mitigating the risk. Usually—not always—such risks are covered by generic legal clauses in the general conditions of the contract that place any unidentified risks on the JV. An argument can be made that the risk of violence was not known by the JV partnership at the time of bidding and that the customer and user should have known it, and they should have divulged it. This would follow option A, and we prefer this approach. It will not eliminate the problem, but it will expose it for discussion early on when there is still time to act. In the words of Sun Tzu: "He who excels at resolving difficulties does so before they arise, and he who excels in conquering his enemies triumphs before threats materialize."

Our best advice is to identify transparently risks on a project, determine those to be managed actively, and get them into the plan as a work package.

8.3 PROJECT RISK FUND AND MONITORING

The monitoring is the same for risk as for the other aspects of project management. You do you plan, your check to see where you are, and you adjust your course—as in sailing. Risk, if managed actively as work packages, should be reported on at each progress meeting as part of the agenda. We prefer to have a single member of the lead project manager's team responsible for monitoring a specific risk, evaluating it each reporting period, and providing a status report on it. This could be a value chain project manager or a staff member for one of the JV partners. The report would reassess each of the items in the risk register that placed the risk on the active list (e.g., probability, impact, trigger, response). Some risk characteristics will cause the risk to become passive; others will cause them to become active. For this reason, ongoing diligence and attention are required. In the risk filtering, we separated project risks into active and passive categories, and recommend managing the active ones as work packages. It is entirely possible that risks placed in the passive category could become active, and vice versa, so make sure to keep the initial listing. It is also possible that unanticipated risks arise that require evaluation. Again, as with all aspects of project management, this should be a plan-do-check-act iterative periodic review.

In our example, let us assume the risk has been negotiated in the agreement, and the ownership and mitigation plans are as shown in Figure 8.6. The owner of the risk would review the risk and the mitigation measures undertaken, and report on the evaluation of their effectiveness. It may be necessary to increase the public meetings, replace the board, or reduce the meetings, depending on community response. The estimated costs should be revisited as with any other work package, and the progress must be evaluated. It may be that the risks in one community have been mitigated, but another risk has arisen in an area not previously considered a risk. In short, the risk needs the same due diligence and attention as any other aspect of project management.

One metric than can be used to assess the success of the risk management effort is the use of the risk fund. Measure the risk fund at the beginning and end of the project, and see what the difference was. If the accuracy control limits for the project are set at +/-10%, did the risk fund vary outside these limits? If so, a graduated scale can be established to provide benchmarks. These then provide a numeric way to capture lessons learned, along with some specific bullets to describe the risk and the methods of mitigating it. For the CPE, the knowledge gained from the lessons learned process should be shared with all. The lessons learned can help reduce costs on future projects for each organization.

Chapter 9

Time Management

Better to sell time than to buy it.

—*Finnish Proverb*

He that looks not before finds himself behind.

—*Maltese Proverb*

In this chapter, we define some of the primary issues surrounding time. Time is one of the most critical components of a project, for it determines cost and it establishes the sequence of events for a project. This sequence of events enables the lead project manager to anticipate and to lead the team around potential problems.

9.1 SOME ESSENTIALS

In the project management profession, we hear people use the terms "schedule" and "plan" and "scheduler" and "planner" interchangeably. In our view, a scheduler is a person who understands the software and the basic rules of logic. A planner is a person who understands processes and how the processes must be linked to complete a project. A planner might tell a scheduler that the information technology (IT) department must finish writing the code before the code can be tested. A scheduler would then translate the planners' statement into an activity, "start code," and insert a finish-to-start restraint with activity, "test code." Planning requires far more experience and knowledge than does scheduling. A project plan includes a time management plan that we all call a schedule, but a schedule is *not* a project plan.

Now we need to define some technical terms, beginning with critical path. The critical path is the longest sequence of events from start to finish of a project. It also is the shortest time required to complete the project. Some people also define the critical path as the path of zero float. That may be true, but it can also mislead for there is also a near-critical path on each project. We have asked hundreds of project management professionals around the globe how accurate their estimates of cost and time really are. The consensus forms between 70% and 90%, so let us use 80% accuracy to demonstrate the point. Imagine you have a project with a critical path that is 500 workdays long. That means that the project could prove to be 600 days long (+20%) or 400 days long (–20%). In this case, an activity with a float of +100 days (600-day duration) could actually be critical, but it is not yet known to be. When we talk of critical path, we mean this bandwidth of accuracy rather than precision.

Float can be negative as well as positive. When a project critical path is longer than the time permitted for performance, the float will become negative. In the case just given, imagine that the critical path logic developed by the team comes to 700 days; the project would have a –100-day critical path, provided the end date was fixed in the software. Negative float is a useful tool in analyzing schedule networks, particularly when it is necessary to evaluate the impacts of make-or-buy options, overtime, extended workweeks, accuracy range in the durations themselves, or of course logic.

As a quick refresher, Figure 9.1 shows the fundamental logical relationships, the concept of float, and the concept of elapsed time. Other books (Kerzner 2006; Lester 2007; Turner 1999) address scheduling in great detail, and we encourage those who are not experienced in scheduling to turn to one of these works. You must understand planning to lead a project, and a knowledge of scheduling will enhance your abilities.

From experience, we have a few scheduling heuristics that might prove useful. To save time and improve quality, standardized processes are a great help, and the heuristics are offered in that vein. These are not hard-and-fast rules, and come from lessons learned over many projects. The first three are repeats from the work breakdown structure (WBS):

- Shortest duration—Work packages should not have durations that are shorter than 1% of the duration of the project. Normally a week for projects over a year in duration.
- Longest duration—Work packages should not have a duration longer than the reporting period.

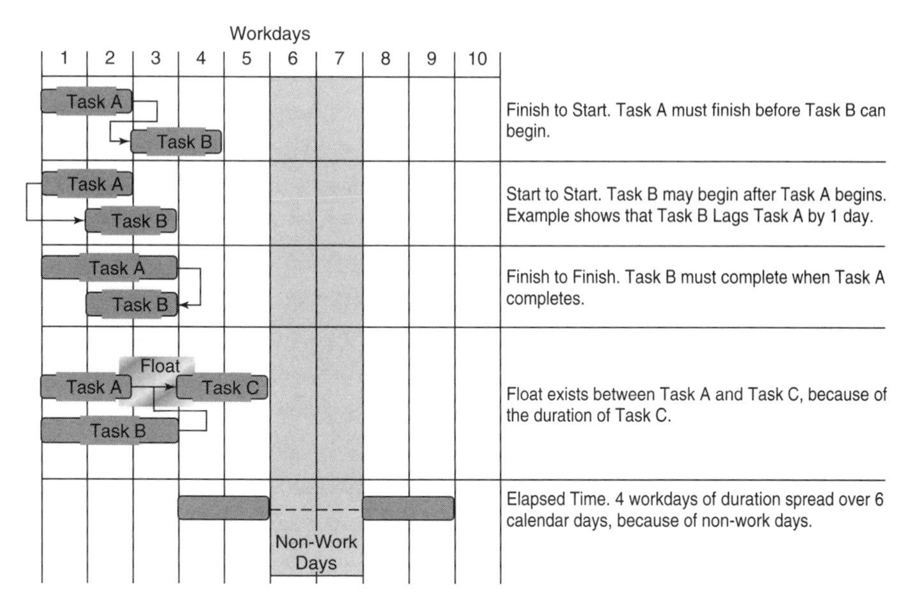

Figure 9.1 Schedule Terminology

- One work package = One chart of account. A chart of account is, you will recall, one component account within a corporate accounting system such as SAP.
- One work package = One schedule activity. A work package is, you will recall, the lowest level of detail in the work breakdown structure. The *Project Management Body of Knowledge* (*PMBOK* 2004) suggests that it is appropriate to break down a work package into multiple schedule activities. This may be required on occasion, but it is not recommended. The reason is that work packages have resources assigned to them, and if they are broken down further, a project manager loses the ability to manage resources. If more detail is required, then the team has not yet found the greatest level of detail required, or work package.
- Use finish-to-start relationships when possible. These relationships are easier to analyze when the changes occur, and they will. However, the overriding test must be that the logic mirrors reality and reflects the way you actually intend to prosecute the work.
- Avoid using leads and lags. Leads and lags make schedule analysis arduous. Unless it is essential to include them, we strongly recommend that the logic be developed to avoid their use.

Now let us shift the focus to time zones and non–work days. International projects must utilize multiple time zones and calendars, and a lead project manager must pay particular attention to these issues. Cross-cultural leadership includes the ability to empathize and to communicate effectively. What better way to demonstrate empathy than to show knowledge of and concern for local holidays and work hours? These also have a direct impact on synchronous communication patterns.

One of the first issues that all international projects face is the difference in time zones. Fortunately, numerous software tools available today address this issue; one of many is www.timeanddate.com/worldclock. We recommend that a standardized format be placed on a project server and included in the welcome package given to each person joining the project team. An example that we have used is shown in Figure 9.2.

The time map graphically shows when people are asleep, waking, or at work. It is easy to develop and must be refreshed when daylight savings time changes. The white shows when people are working. We show this period as a 10-hour workday, but that will vary depending on the location. The black is when people are sleeping, and the gray/white transition is when people are waking or slowing down from their day. The lead project manager, in consultation with the team, selects "prime time" (best time)

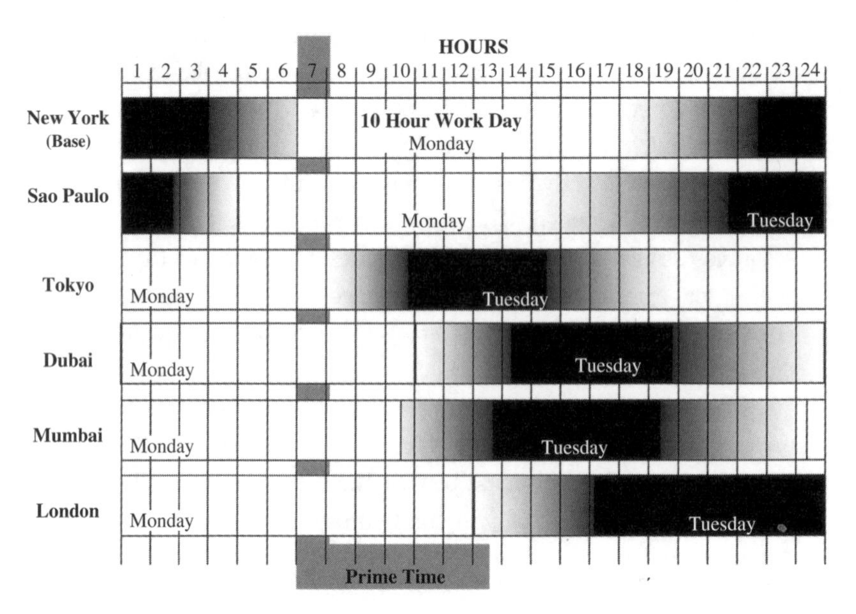

Figure 9.2 Time Map

for people to communicate synchronously. "Synchronous" means communicating with others at the same moment, although in different time zones. The use of such things as telephones, meeting software, Skype, chat rooms, and text messaging would be examples of the technology. "Asynchronous" means communicating with others at different moments using such things as discussion boards and e-mail. Figure 9.2 was constructed using June 2, 2008, as an example, when daylight savings time was in effect in the United States and the United Kingdom.

Imagine for a moment that a work package has a duration of five days, and the accuracy, using Program Evaluation and Review Technique (PERT), is +/-20%. In other words, actual duration could be six days or four days. If the work was to be completed in Tokyo on June 2, 2008, if it was completed at 2300 hours New York time, was it on schedule? Yes, but at the upper control limit since it was Tuesday in Tokyo. We point this out to demonstrate the endless complexity possible and to show that a lead project manager must avoid getting to this level of detail, unless of course you have an outage or an international IT "go live."

Figure 9.2 is provided to help demonstrate empathy and consideration for the members of the collaborative project enterprise (CPE). When we were working in Asia, 10- to 12- hour workdays was not uncommon. The team in the United States would want a conference call to discuss the project beginning at 2300 our time so that they could have coffee and get some e-mail out of the way. It was a convenient time for the U.S. team but incredibly inconvenient for us, and this situation persisted for months. We recommend scheduling synchronous meetings around prime time. If that is not possible, rotate the meeting times to be considerate of others.

Another aspect to consider on international projects are non-working days, like weekends, shown in Figure 9.3 that being regular nonworking days. In this example, we compare the nonworking weekend days of Saudi Arabia to those of the western world. Neglecting holidays, the normal interaction time of an international project with teams in these two locations is reduced to just 3.5 days each week. If there is a seven-hour time difference from New York to Jeddah, that leaves about three hours each day, as shown in Figure 9.2, for synchronous communications.

Thursday is lost because the team in Jeddah has already started the weekend when the New York team comes to work. Let us do the math:

- Assume that there are 2,600 work hours available every year: 52 weeks × 50 hours per week.

Figure 9.3 Weekends

- Assume that there are 260 work hours available every year for synchronous communications, taking into account the differences in time zones: 52 weeks × 5 hours per week.
- Assume that for the team in Saudi Arabia, there are 156 hours available every year for synchronous communications when taking into account the difference in nonworking days: 52 weeks × 3 hours per week.
- Assume now that local holidays are considered. Holidays vary by country and by practice. A few Indian national colleagues and I attempted to make a list of the holidays in India: official and unofficial holidays, and a truly remarkable number of religious holidays. In fact, we discovered that almost every day there is a holiday somewhere in India.
- The lesson learned is that the listing needs to be specific to the location of the team within the country; do not generalize, especially in diverse cultures. While business hours are less variable, they can require special attention in Spain and Mexico, for example, because of extended breaks for the midday meal, or in Saudi Arabia,

where prayer times must be respected. Numerous sites on the Internet (see, e.g., www.wholeworldcalendar.com)that list holidays in different countries, but we recommend checking with the team in country.

- If there are 20 nonconcurrent holidays per year for the CPE, the number of prime-time synchronous hours available drops by another 20 hours (20 days × 1 hour), to 136 hours per year. Said another way, about 2.5 hours per week is available for synchronous communication in the CPE.

Our point is that the time to talk to the CPE team is extremely limited on international projects; therefore, the effective use of nonsynchronous techniques is critical. Synchronous meetings time should be reserved for progress meetings and crisis situations. Accomplishing this is a challenge but also a great learning tool for it teaches people how to do a better job of communicating. More on that in Chapter 11. For now, let us turn to another leadership issue, float.

9.2 FLOAT, ETHICS, AND LEADERSHIP

Our research shows that effective cross-cultural leadership requires trust. Trust is "the willingness of a party to be vulnerable to the actions of another party based on the expectation that the other will perform a particular action important to the trustor, irrespective of the ability to monitor that other party" (Mayer, Davis et al. 1995, p. 712). Trust is a fragile commodity; it takes time to build, but it can be lost in an instant. Trust is a precious for commodity project managers, and particularly for lead project managers. Let us illustrate by considering a common occurrence on most projects.

Imagine you are the lead project manager in a partnership to build a new airport in Oviedo, Spain. It is a 30-month contract, with liquidated damages of $300,000 per day. You have completed the planning using a consistent PERT approach and find that you will need no more than 28 months to complete the work. Remember the PERT approach for time is the best-case duration, plus the most likely case times 4, plus the worst case, all divided by 6. The formula, again, is:

$$\frac{\text{Best Case} + (4 \times \text{Most Likely Case}) + \text{Worst Case}}{6}$$

Now the decision needs to be made on how to display, or not, these two months of float. Consider these questions:

- Do you show it to the customer, your partner, and the value chain?
- Do you show it to your internal sponsor, resource managers, and team members?
- Do you spread it back into the activities so that it is not visible?
- Do you create two schedules, one for distribution and one for the lead project manager?
- Would your approach change if you knew that in most legal jurisdictions float belongs to the project?
- What if you make your schedule and your partner does not because it did not know about the float available?

This dilemma probably has confronted every project manager since the pyramids, and there are no pat answers in part because it is an ethical issue. Is it appropriate to mislead others to protect yourself and your organization? In 2007, the Project Management Institute issued a revised code of ethics and professional conduct with an answer. Under "Honesty, Mandatory Standards," Section 5.2.1, the code states: "We do not engage in or condone behavior that is designed to deceive others, including but not limited to making misleading or false statements, stating half-truths, providing information out of context or withholding information that, if known, would render our statements as misleading or incomplete" (p. 5).

We have posed this sort of scenario to international project managers in training sessions. Easily 95% of the time these managers want to bury the float in the activities for the customer so the critical path is 30 months long and then keep other versions of the schedule for different purposes. Then the students suggest a 28-month schedule for the project manager, a 26-month schedule for the project manager's team, and a 24-month schedule for the value chain. In our experience, if a project manager can get a durable resource-loaded baseline schedule and keep it updated, it is amazing; juggling two or more is not possible, and it ignores the ethical issue.

The students were not unethical; rather they have learned that float can encourage predatory practices by others. Customers use up float for changes without paying for extended overhead costs, value chain organizations use up float to rearrange factory floor time for their own benefit, matrix employees use up float to work on other projects.... The students have learned to shield themselves from the risk of being transparent, not because they do not want to tell the truth, but because they fear the consequences of doing so. The question is a simple one: Is it ethical? According to the *PMBOK,* it is not. We think lying represents the seeds of long-term catastrophe and that it cannot coexist with leadership. As was said by the Persian poet Sa'dī, "A falsehood is like the cut of a saber; for though the wound may heal, the scar of it will remain."

If you do not disclose float, you must amortize it over all of the activities or lengthen some select activities to use the students' approach, described above by keeping multiple schedules, for the customer. It will be discovered eventually, and the message to the CPE will be that it is acceptable to hide information. On large complex projects, it is easy to forget where the float was hidden. When it comes time to evaluate progress and do lessons learned, it can actually harm your organization and the CPE. How? Your lessons learned might be that it only requires five months to do the code for a project, when you had estimated six. The lessons learned may not capture the fact that there was one month of hidden float in the work package.

From our experience in executing, we recommend that you identify float as float so that you can lead the CPE. It is always in the best interest of an organization to manage its resources and risks as effectively as possible, and having a single plan that is transparent facilitates this. We have seen many project managers attempt to cloak float and to maintain multiple schedules, but seldom have we seen either approach work. Neither strategy has a high probability of success. To add one more layer of complexity, imagine you have decided to divulge the float and to be transparent with the planning. You go to the joint planning session with the customer and find that not only does the customer have a hidden agenda, but also so does your joint venture partner. This is the test of leadership, and everyone will be watching.

From our experience in forensic analysis and dispute resolution, the best course is also transparency. As a professional, do your diligence; disclose information that is needed by others to manage their work. The return on investment for taking this approach is trust. The CPE will know that there are no hidden agendas with the lead project manager and that she or he is fearless even when it would be easier to prevaricate. The CPE cultural norms and values set by such actions are that truth, honesty, integrity, dependability, and character are the goals. We have argued (Grisham 2007) that people copy the behavior of a leader, both knowingly and physiologically. Many empirical psychological and sociological tests have demonstrated that people will copy the behavior of those they admire and those to whom they give power. If so, then what behavior would you like to see manifested in the lead project manager on a complex international project: honesty or deceit?

Jack Welsh, the former chairman of General Electric, described the following benchmark test: Do not write or say a thing unless you would like to see it on the front page of the *Wall Street Journal*. We have a less sophisticated suggestion. You come home from a hard day as the lead project manager, and your six-year-old asks you what you did that day. Would you be comfortable telling the child that you misled your team and that it is okay to do so? We have heard, and earlier in our career constructed, many reasons not to disclose float. Over the long term, we have learned the hard way that this is not a sustainable approach. The lead project manager will demonstrate the ethical standards for the CPE by how candid she or he is with the team. This issue of candor lies at the root of many project management dilemmas. Consider a piece of proprietary information that belongs to your organization. Certainly you should not divulge it to others for it is one of the competitive knowledge advantages for the organization. However, you can divulge that it is proprietary information and will not be disclosed.

Imagine that the lead project manager knows the customer is demanding, very knowledgeable about the product of the project, and has a reputation for being nontransparent. The lead organization could withhold this information from the value chain to reduce the costs of bids. Is it ethical to withhold this information, if revealing it would increase the cost to the organization, and it is not possible to pass these costs along to the customer?

A major dilemma on international projects is that organizations and people have different values and norms. Figure 9.4 provides a way to think about how organizations prepare their estimates and schedules and how they communicate them. A firm that seeks risk (because it can manage risk) and has a strong corporate ethic requiring transparency in all dealings will use a PERT approach and divulge the float to the CPE. At the other extreme, a firm would tend to withhold or hide the float. On international projects, the issue of transparency and risk vary widely across organizational and social cultures. On a single project, it is almost certain that a lead project manager will find organizational cultures in all of the four boxes shown; the dilemma is that you may not know which organization is where during the initiating and planning phase. Worse

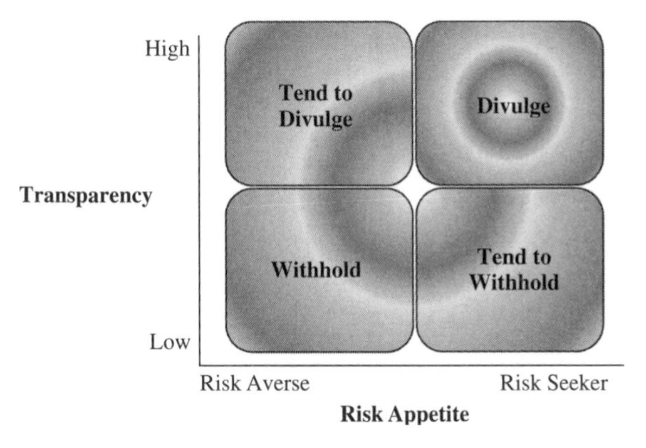

Figure 9.4 Sharing Information

yet, within an organization, the project manager may have a different set of values from the organization he or she represents.

There are no easy answers to the dilemma. The lead project manager has a range of choices on offer bounded by the two extremes. The first is a so-called race to the bottom: set the transparency standards so low that everyone can meet them. The second extreme is to set the transparency standards high by divulging and then by encouraging others to follow this lead. We prefer the latter approach, as we are convinced that it will lead to more frequent project success. However, there are dangers in divulging and times that it is not appropriate. Our suggestions in the collaborative environment are:

- *PERT.* The lead project manager requires a PERT assessment of each activity from each organization, not because it will guarantee that a project is completed on time but because it will give everyone an indication of how the organization prepared the schedule. It will also give an indication of where the organization may be in Figure 9.4.
- *Due diligence.* The project manager and the lead project manager prepare their own estimates for the work of their organization. For example, the contract says the work must be completed in 12 months. The due diligence schedule effort is the unbiased view from a professional's eyes of how long it should take. Then the two are compared.
- *Positive float.* If the comparison results in positive float, recognize it in the schedule as such. Then make it clear that the estimate

at completion (EAC) for the project either includes or excludes the requirement for the organization to expend funds, specifically overhead, to be on the project for this additional time. The concept is relatively simple: If there are no cost risks, why not divulge the float? Be warned, however, that, in general, if the ownership for float is not defined in the contract, it is the property of the project, not the organization. Our recommendation for the project manager in this situation is to offer two options. First, the shorter critical path and an understanding that extended condition costs will accompany float days used by any other organization in the CPE. Second, the longer critical path where float can be used by anyone, but the extended condition costs will be added into the EAC; this option provides transparency, and it mitigates risk. The lead project manager can optimize project resources with the knowledge of float.

Imagine that value chain A.1, a software provider, has 30 days of positive float on its critical path and that value chain B.1, a hardware provider, has 30 days of negative float on its critical path. If the sequence is software ready, load onto hardware, and test, then from the CPE perspective, if value chain A.1 hides the float and value chain B.1 hides the problem, the CPE schedule actually is wrong by 60 days. The 30 days of positive float are not really needed and if used actually will exacerbate the 30 days of negative float.

- *Negative float.* If the comparison results in negative float—not enough time to do the work—recognize that fact. In this case, the organization's critical path is longer than that required. Here the project manager for the organization should provide a critical path that meets the requirements with a list of steps, and costs, that were necessary to bring the work within the project needs. For example, if it was found that 3,000 person-hours of overtime will be required on the last three work packages, share that information with the CPE; this is transparency with mitigated risk.

Our recommendation is to be transparent and truthful. Share your professionalism, and inform the other organizations of your plans to deal with the positive and negative findings of your planning. We are not recommending giving away organizational time or resources without equitable compensation. No one is perfect, and that is why float and contingency are commodities that need to be recognized, estimated, and managed. Having confronted the issues of float, we are now ready to develop a project schedule.

9.3 DEVELOPING A CPE SCHEDULE

We recommend that the WBS, cost estimating, and risk register be developed before the schedule. Creating a schedule requires a work package level understanding of the scope, duration, resources, and assumptions of the CPE WBS. The scheduling process then can focus on establishing the logic of the project. One useful tool in developing schedule logic is Post-it notes.

On complex international projects, teams are always diverse in terms of culture, technical expertise, political acumen, position, and scheduling knowledge. For example, on a large power project, we had a room full of craftspeople (ironworkers, pipe fitters, electricians, superintendents, etc.) who had no idea of what a predecessor or successor was, or the difference between start-to-start or finish-to-finish relationships. However, they knew exactly what work needed to be done, and in what sequence. By putting the work package descriptions on Post-it notes and a time scale on a white board, they could move the work packages around until they had the sequence they knew they needed. The planner can lead such a discussion by posing questions and facilitating creativity. Then the planner inserts the logic, confirming the relationships with questions such as: Does this activity have to be completed before this one can begin? The craftspeople provide the sequencing, the project manager provides the planning, and the scheduler inputs the information into the scheduling software.

The sequence of events in building a CPE schedule is the same as it was for the WBS exercise, four steps that begin with the internal organizational schedule.

9.3.1 Step 1: Organizational Schedule

The first step builds on the WBS that has already been created. We recommend that the project manager strive to do the WBS first and not concurrently stray into risk, quality, cost, and schedule. It saves time and keeps the processes on track and the team focused on one thing at a time. If the strategy is to complete the project plan during one joint planning session, then we recommend finishing WBS, the estimating effort, and the risk register first, and then taking on the scheduling. The reason is that many of the questions needed to complete the schedule flow from these other processes. The schedule template can be built from the WBS dictionary with the only input remaining being the logic. Assuming that the goal of the joint planning session is to complete the entire project plan, the sequence of efforts that we recommended in Chapter 4 bears repeating:

- Scope management plan
- Cost management plan
- Risk management plan
- Schedule management plan
- Quality management plan
- Procurement management plan
- Diversity plan (human relations and communications in the *PMBOK*)

We find it useful to print out the WBS work packages on file label stickers, as shown in Figure 9.5 and then stick them onto Post-it notes. The WBS#, duration, risk #, work package description, total cost, contingency, and resource names come straight off the WBS. Before you start, create vertical timelines to demark weeks, months, or quarters on a white board so that people can see where they are in time as they think about the logic. Many technical professionals have a sense of how long things take; this format helps them move from the heuristic level to the work package level.

Then using the Post-it notes, arrange the activities in logical order by time. As this is being done, a scheduler should be keeping track of the

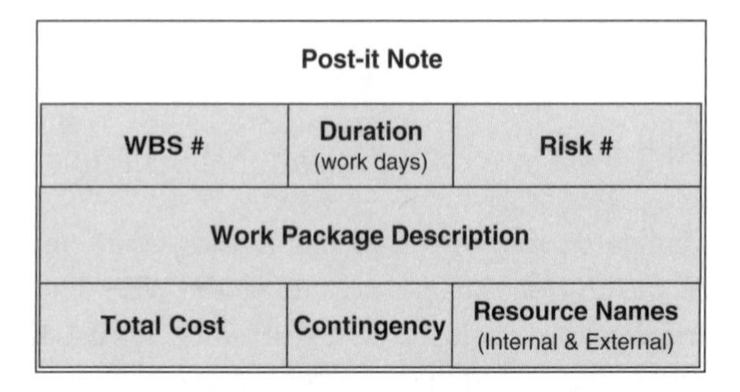

Figure 9.5 Schedule Tool

comments, assumptions, and changes that will surface. Once the Post-it notes are arranged in time, it is easy to see where potential resource constraints might exist and if resource constraints need to be added. In our experience, most project managers schedule based on limited, or constrained, resources. We prefer unconstrained resources, because it does not force logic into the network that may be artificial. Permit the software to help you identify periods of overallocated resources. Then it is relatively easy to consider the changes that must be made on an exception basis to remedy the problem.

When all of the work packages and any added ones have been arranged in this manner, it is time to add the scheduling logic (finish-to-start, start-to-start, finish-to-finish). Here a scheduler or someone who knows scheduling will need to assist the team in drawing the sequence correctly. In defining the logic, more discussion will ensue, and likely the need to shift the work packages around again will arise.

Once the diagram is completed, it will look like a precedence diagram (PDM). Now the information must be loaded into the schedule template that already contains the WBS work package activity information. Then the schedule can be completed overnight and ready for the next morning meeting, where the duration of the project developed is tested against the contract duration. This is what we called due diligence earlier. Then the issue of float must be discussed, and the disposition determined. If, for example, there were positive 20 days of float on the critical path to the required end date, one option would be to add a work package called "Float Reserve" with duration of 20 days as the last activity in the network. At the other extreme, if the template shows a negative float or shows resources overutilized, the team can methodically undertake an analysis to bring both back to the required completion date, being careful to record the steps taken to do so, such as adding overtime.

The last step is to decide on the summarization levels that are necessary for the key stakeholders and the value chain. As with the WBS, the issue of transparency must be addressed. It is also important to provide only the level of detail needed for each organization to perform its work effectively. Too much detail will only overburden the other members of the CPE. Our best advice is to provide what you would want if you were in the other person's position. Be respectful of others' needs and of their time. This balance may be achieved by establishing milestones to highlight the interface points and then being open to provide added details if requested.

Once again, we want to emphasize the importance of keeping just *one* schedule. As with the WBS effort, the next step is the value chain schedule.

9.3.2 Step 2: Value Chain Schedule

The preparation of the value chain schedules requires experience and judgment, and we suggest using the same approach as for the organizational schedule. At the beginning of the joint planning sessions, the buyer of the service and the seller of the service should have already have traded schedules and become familiar with them. In these meetings, the key for the buyer of the services will be to get sufficient information to enable it to manage the project without being buried in detail. The key for the seller is to be able to provide the level of detail required by the buyer without having to keep a separate schedule. Both schedules need to be integrated so that the seller schedule rolls directly into the buyer's schedule.

When the seller and buyer have work that is intertwined, we recommend using the Post-it note approach. When the seller is simply providing a commodity that is a one-time delivery, clearly this is not required. Imagine a distributed control system vendor as the seller for a new pharmaceutical organization buyer. The buyer would likely want more that a "system ready" milestone from the seller, such as a system witness test, training, and the like. The buyer may want to require "hold" points, such as the system witness test, where the seller cannot continue with the work until these points have been observed and approved by the buyer.

The project manager for the buyer must take this opportunity to check the approach taken by the seller in establishing their durations. Did it use best case, worst case, likely case, PERT, or did it rely on previous experience? The question is whether float exists in its network, and is it based on normal hours and days? We recommend asking about shop loads for manufactured products so that the buyer can get a sense of the factory's capacity and what other projects are scheduled to be ongoing concurrently. Likewise, for services, ask how deep the seller's organization is and how much other work is in its pipeline. Knowing this, the buyer's project manager can select some events that occur before the required milestone or hold point to provide early warning of potential delays.

It is important for project managers for the service provider and the value chain organizations to carefully consider the need to automatically collapse and extend the level of detail in the schedule. There will be detail in the value chain schedule that the service provider simply does not need, and there are activities in the service provider schedule that are not needed by the value chain schedule. The key is to identify those areas and create milestones and activity logic that enables both organizations to update their schedules and reintegrate them effortlessly. On some projects, it is easier to keep all of the activities in every schedule; normally this occurs when the activities are interlinked to the point that separating them requires more effort than the benefit derived.

The project manager for the service provider has the additional obligation to share with the value chain schedule information from other value chain organizations and other project participants that impact on a particular value chain organization's ability to perform its work.

9.3.3 Step 3: Partner Schedule

The considerations and tools for building a partner schedule are the same as for building a value chain schedule. The key issue here is the level of detail that needs to be included. Service provider B, the partner shown in Figure 6.2 will want, and perhaps need, more information about the work of service provider A and its value chain. The intimacy of the relationship between the service providers will play a part in determining the level of transparency. In an alliance, the partners will likely share almost everything, including in some cases margins and overhead amounts. In a consortium, there would likely not be the same level of information sharing. In a perfect world, both partners would openly share information, but in the real world, this is often not the case. In Chapter 5, we discussed the levels of detail in the WBS and of the work packages. The guiding principle again is that the lead project manager, working for service provider A, needs enough detail to manage the project but not so much as to become a slave to the detail.

On almost all projects, the value chain for service provider A will need information from the value chain for service provider B, and vice versa. For example, imagine a project to digitize an oncology center where service provider A is responsible for the physical structure and service provider B is responsible for the IT systems. The value chain for service provider A would need to know the physical characteristics and wiring requirements for the intelligent equipment, and the value chain for the intelligent equipment would need to know mounting requirements and physical equipment characteristics, such as room temperatures.

Is it better to have the value chain people talk directly, or to have all communications run up-across-down-up-across-down?

Depending on the intimacy of the partnership between the service providers, we recommend that value chain organizations be included in the partner joint planning sessions when they play a critical role and their technical needs are better communicated directly. If this is the case, we prefer to have the service providers meetings first so that each partner can acquire the necessary knowledge to place their work in the proper context.

In our experience, the true characteristics of a partnership become evident during the partner joint planning session. There is a natural caution and a tendency to withhold information in all relationships, particularly those that involve risk. There is a saying in Alaska: "I don't need to outrun the bear, just you." If a partnership has this as a foundation, competition will rule the day. Information is advantage, and divulging it requires trust. This hesitancy to share knowledge is on display at most partner joint planning sessions we have attended. The individuals are getting to know one another and are discovering whom they can trust and whom they cannot. The scheduling is important, but not as important as the sociological and psychological dance that unfolds. It is here that the project manager and the lead project manager earn their salaries.

In an article on trust, Mayer, Davis et al. (1995) define trust as being "the willingness of a party to be vulnerable to the actions of another party based on the expectation that the other will perform a particular action important to the trustor, irrespective of the ability to monitor that other party" (p. 712). Building and nurturing trust is job one for an international project manager, and the joint planning sessions are where the seeds are planted. It begins by being willing to divulge information yourself and by fulfilling your commitments.

The problem with trust is that it takes a long time to mature but can be lost in an instant—it is a fragile commodity. Imagine you are the lead project manager for service provider A, and your value chain has told you that delivery of the custom cabinetry for the reception area and servers will likely, 80% probability, be late in arriving. The value chain has presented an option to improve the delivery by doubling the price, but it will reduce the probability only to 40%, so there is a residual risk. It is not certain that the delivery will be late, so what does the lead project manager share with the partner? Does the lead project manager tell the value chain organization not to divulge the potential problem? What if that the lead project manager decides to inform service provider B that the delivery date is not yet known? As we recommended, it takes courage to divulge, but honesty will serve you well.

Now imagine that service provider B has a similar problem with some hardware but has decided to pay for an accelerated delivery date. Over coffee, employees of the value chain organizations for both service providers are talking about sailing. A quick bond forms and they start talking about their respective value chains. Global value chain delays force both of them to work longer hours, and this cuts into their time for sailing. Then they discuss the project issues, and the facts escape. If they do, the lead project manager may lose all trust created, and perhaps for the remainder of the project. Excuses can be made, patches placed on relationships, promises to do better in the future, but the default will be mistrust on every issue that arises.

Current research on brain chemistry shows that people physiologically copy the actions of others (Grisham 2007), that our brains actually wire themselves to mimic the behavior of others. In apes, it even extends to copying what is imagined to be associated with the action—not just seeing another eat a banana, but the taste itself. While the research is quite young, it points to the importance actions. If the lead project manager withholds information, everyone else will find this action acceptable. By withholding, the lead project manager would be creating a culture of secrecy.

We have emphasized the issue of trust here, because it manifests itself most clearly during the early parts of a project in the initiation and planning phase. Now it is time to include the customer and user.

9.3.4 Step 4: CPE Schedule

In general, customers are purchasing services externally because they do not have the technical expertise, resource capacity, or attitude toward risk. Although some international customers do form alliances to perform multiple projects, this is the exception rather than the rule. Most often a customer is attempting to fulfill its needs as quickly as possible for the least cost and risk. If the customer desires a transactional relationship that is in an adversarial environment, this combination tends to suppress the open sharing of information in part because it could expose ambiguities in the scope or in the customer's ability to support the project.

Imagine that the government of the Democratic Republic of Congo wants to build a road from Kinshasa to Kindu. It has managed to obtain a loan from the World Bank for the physical installation, but funding for the capacity building of the highway ministry is still being processed. The road is desperately needed, so the government decides to proceed with the hope that the highway personnel will be available when needed. Therefore, the government writes a contract that requires the service provider joint venture to train the Congolese highway personnel by assigning them as staff to the service providers. The service provider partnership will reduce costs accordingly and will be responsible for the actions, and inactions, of the Congolese highway personnel.

As the lead project manager for the joint venture, your ability to be successful will be tied directly to the government's ability to provide enough qualified people when they are needed. There are three philosophic approaches this dilemma:

1. *Ostrich approach.* Ignore it, not your problem; deal with it if it happens. You might get lucky!
2. *Legal approach.* Get legal involved in the joint planning sessions. You are 80% certain that the government will not be able to support the efforts. It has a staff of 6 in Kinshasa, but you estimate it will need 60 spread over the entire route, and the project begins in a month. Plan for failure and claims.
3. *Leadership approach.* Build a plan that helps the government, and you, achieve success. It is best to deal with the problem at the joint planning session rather than later during execution. Likely the local value chain can provide intelligence on the current capabilities of the government and of the market to provide staff, and the training and housing required. The EAC can be adjusted with a plan to phase in government employees and train them in a rational way.

Now imagine that your organization has already signed the contract for the project, and you are the lead project manager in the joint planning session with the customer. Assume that the customer is wildly optimistic about its ability to get its people but that it really does not want to discuss any timetables, and certainly does not want to make any hard-and-fast commitments. The customer will probably not be willing to display its deficiencies and may in fact want to follow the ostrich or the legal approach. A Congolese proverb says, "Little by little grow the bananas." Thus, the customer may prefer taking a slow approach and a wait-and-see attitude. However, international projects are like sailing, and a small change in bearing at the start of the journey can cause a huge error in the final destination. International projects are also like supertankers; because of the political momentum, it takes time to correct the course once a deviation is recognized.

How you begin a project is critically important; therefore, we prefer the leadership approach. The lead project manager can build trust and begin to grow the bananas by leading this effort. It is difficult, requires emotional and financial sacrifice, and will probably not bear fruit immediately, but it is the course with the highest probability of success in the end. In the joint planning session with the customer, it is important to divulge as much of the thinking, planning, and expertise of the CPE as possible. We recommend taking critical value chain organizations that will play a pivotal role in the project to the meeting so that the customer can feel comfortable with the depth of expertise and planning that has transpired. It is wise to take time to walk the customer through the process we have described so that it understands the effort and attention devoted to planning the project. It helps build trust, demonstrates empathy, shows respect and concern, and displays the professionalism of the CPE.

The scheduling process itself is the same as for the partner schedule. The trick is to make the customer feel comfortable enough to engage in the process, even if it is not yet willing to divulge information. A potential initial dilemma for the lead project manager, if the project was tendered, is that the sales team oversold the project during the negotiations. By "oversold" we mean chose to overlook issues, such as existing staffing levels in the Congo example. Once a contract is executed, it is a promise to perform. The lead project manager must do his or her due diligence first but then must fulfill the promised obligations. In this context, by "due diligence" we mean review, before the customer meeting, the contract for responsibilities, assess the current levels of customer capability, assess the customer's plan for adding resources, review based on the budget at completion (contract amount) and the estimated time for completion.

Then sit back with a cup of tea and consider the probability of the customer being ready when you need resources. If your judgment tells you that the customer will likely fail, build a mitigation plan with time and cost estimates for presentation to the customer, user, and key stakeholders. The alternative mitigation plan may be rejected, and the burden of the costs borne by your organization, but that may happen anyway. It is better to be a proactive leader, and build confidence and trust in you, by broaching the subject straight away. This is one way to snatch victory from the jaws of defeat.

Do your diligence. It is your job to plan the project using your expertise and experience without being biased by a contract. You must then reconcile the contract with your due diligence. We strongly recommend that a project manager plan the contract without being dogmatically restrained by the contract conditions. Project managers and lead project managers are paid for their expertise, so use it. Be creative, proactive, and transparent.

9.3.5 CPE Example

Look back at Figure 8.4 and recall the design of the public-private partnership. To illustrate the steps just discussed, we have constructed a partial schedule for early portions of the project. Figure 9.6 shows a few of the early activities that would be required related to the joint planning meeting. We have shown some durations that are 40 days, beyond our recommendations for reporting periods, but only for the reason of brevity. Activity numbers are in the far left column.

Activity #10, "C. 1 Advertise for Operator," would clearly need to have more detail in order for it to be managed properly. The items in bold are called hammocks. Hammocks represent an earliest start, latest finish for those activities that are indented below them. In this way, the value chain details can be included in the plan but the details suppressed for summarization. For example, activity #9 is a hammock and has a duration of 120 days, and that duration includes all of the subactivities #10 through #13.

Likewise at the next level down, activities #27 through #29 are for a supplier to a value chain organization. In this case, value chain organization A.1 has a value chain supplier A.1.1. Devising a basic numbering system will greatly facilitate knowing the organizations that are participating in the project and through which parent organization they report. So, a second-tier value chain organization for service provider A would be A.1.1. If this second-tier organization had two value chain organizations as the third tier, they might be A.1.1.1, and A.1.1.2. Having a consistent numbering system helps when reading a schedule. You can of course look at resource views, but having a mnemonic in the activity description saves time.

In Figure 9.6, activity #3, "C.1 Design Initial Financial Plan," is one of the customer's activities. Remember that the financing for this project was from a consortium of investors, so they need to be numbered in a similar

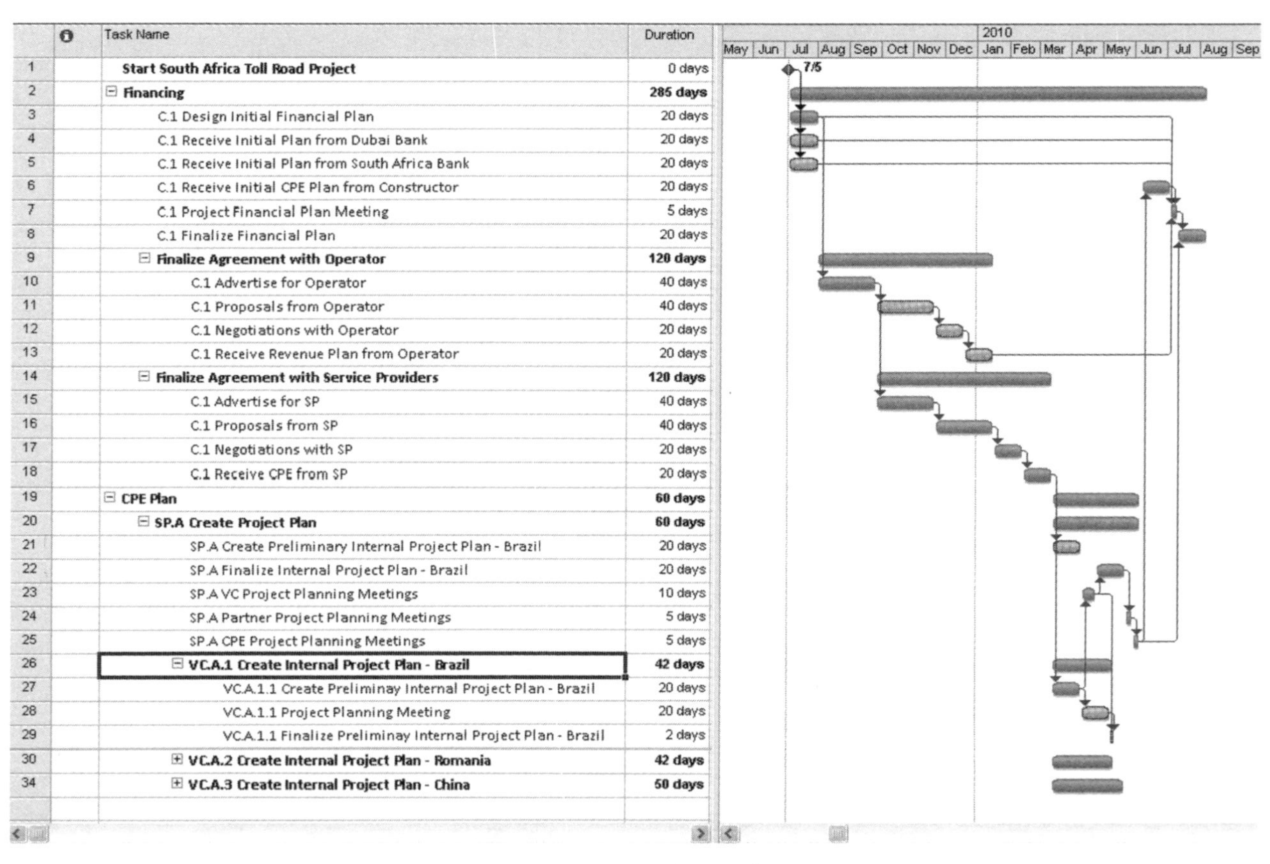

Figure 9.6 Partial CPE Plan

manner. The advantage of having a numbering system permits anonymity for some of the investors without losing track of their contributions and obligations. Notice that there is a constraint between activity #25 and activity #6. Financing cannot be completed until the financial consortium understands the cash flow requirements during execution and the revenue stream predictions during operations. Once built into the financial plan, that information may—likely will—necessitate a change in the CPE plan.

Notice that activities #11 and #15 are running concurrently. That will shorten the time required but is far less than optional, for the operator will have very specific ideas of the facilities required, their durability, and the costs of maintenance. If done sequentially, it would add 120 days to the project, and that may put the London Interbank Offered Rate in a range that is unacceptable for the project. The CPE plan provides a way to evaluate decisions under uncertainty and to optimize time and cost.

At the end of this process, the lead project manager will have completed a CPE baseline schedule that has been assembled with the input of all organizations and approved by the key stakeholders. Building the project plan is the easy part. What follows is the true challenge for international project managers: updating the schedule and change control.

9.4 SCHEDULE UPDATES AND CHANGE CONTROL

Schedule updates are performed most often at predetermined intervals, dictated by the customer. On very large projects, updates may occur quarterly; on large projects, monthly; and on small ones, weekly. For our purposes here, assume we are describing monthly reports. The meetings may be virtual or face to face, and may require the participation of value chain organizations or external organizations (e.g., community groups) if appropriate.

A few basic points are in order before we begin. First, we recommend that the lead project manager never bring up bad news or problems for the first time at such meetings. Those should be discussed between the key stakeholders long before the report is issues and before the meeting occurs. Never put key stakeholders in the position of being surprised or of possibly losing face. We are not saying, or suggesting, that a project manager or lead project manager should be timid or avoid difficult issues. What we are saying is that she or he should extend the courtesy to the key stakeholders to discuss it with them first off line.

Second, make certain that the information needs of the key stakeholders are perfectly understood. What format do they want, what level of detail, how far in advance of the meeting do they need the information, what other organizations might they share it with? If the information is to be provided to an organization that is providing financial support, it may need a different format, for example. More on this in Chapter 11.

We address four different but related issues in this section: getting started, changes in logic, changes in performance calculation, and scope changes.

9.4.1 Getting the Project Started

From the viewpoint of the customer, user, or partner, unannounced changes in logic or duration can be a source of considerable concern and consternation. Approval of the project plan and baseline is a time-consuming affair, and each organization invests significant time and expense fitting efforts into the CPE plan. As we have discussed, the implementation and planning period is critical, for it is the time that the lead project manager is building the team culture and trust in her- or himself. Consider the impact on the lead project manager's credibility and trust if the very first schedule update comes back unrecognizable. The reward for such an occurrence is trepidation and a lack of trust all round. Many, if not all, disputes begin or are fueled by this single issue.

Another problem we see in disputes on adversarial contracts is that the lead project manager takes a month to sort out the CPE schedule and get it to the customer. By this time, the first monthly progress update is due, and the customer gets two schedules with different activities, different durations on the activities, different logic connecting the activities, and different floats. One is the original baseline schedule from the contract, and the second the revised baseline schedule, the schedule created after discussions with the CPE. It is tempting to supply schedules that have the same completion dates, so often the revised baseline schedule must be arbitrarily adjusted to fit within the contract boundaries. In our experience with disputes, lead project managers on such adversarial contracts often find themselves in the position of planning the project, building the CPE culture, and seeking approval on the CPE baseline schedule concurrent with submitting the first progress update.

Imagine you are the customer and your team of professionals, the low bidders, provides you with a plan that is clearly fraught with problems after you have waited patiently for a month to know where the project is headed. What we see is serious concern and demands for immediate action to create a believable plan. Often payment is connected to the submission of the CPE plan; in such circumstances, customers are inclined to hold the payment hostage. More pressure on the lead project manager results, from above and below. At this point, the lead project manager is sailing without a course, does not know where he or she is, and must start working on the next progress report while trying to construct a baseline plan. We have seen numerous projects where the customer, even at the end of a project, never accepts the baseline plan. This is the case almost always on adversarial projects, not on collaborative ones.

This is one reason we strongly recommend that organizations have lead project managers involved in the tendering process and that they attempt to negotiate adequate time to complete the CPE baseline schedule before the first update is submitted. This will allow the lead project manager to assemble the schedules from all of the team, gain approval from the customer, and then provide copies of the CPE baseline schedule to the entire team. Then the team provides its progress measured against this schedule. Customers want a CPE baseline schedule early in the project, and rightly so. Yet forcing the schedule to be done too quickly is entirely counterproductive. Here are some suggestions for customers and lead project managers on international projects:

- *Risk.* Recognize that a risk exists if detailed actions are taken on plans that have not yet been finalized. For example, if a development bank is funding a project, it will want to know the cash needs for each month of the project. The bank will then set its currency hedges in place based on this cash flow. Hedges are not inexpensive on large international projects, and they cannot be revised without added costs. If the hedge is based on a preliminary milestone schedule, it probably will need to be changed once the detailed baseline CPE plan approved. Decide who will take this risk.
- *Adversarial environment.* Require the CPE baseline schedule to be submitted, and approved, within 45 days of formal contract award. Require the first progress update 60 days after formal contract award, based on the approved CPE baseline schedule only. While 45 days may be too tight on very complex projects with large dispersed value chains, this timing can be adjusted as necessary. If it is adjusted, the first progress report should be adjusted as well.
- *Collaborative environment.* Require the submission and approval of an approved CPE baseline schedule prior to formal contract award. Depending on the gestation period of the selection phase, this schedule may need to be a high-level milestone schedule so

that the necessary financial instruments can be put into place. If so, the timing will need to be adjusted accordingly. Likewise with a fast-track project, where the implementation and planning phases overlap with execution. Otherwise, in no event should execution of the project begin before a detailed CPE baseline plan is approved.

It is critical to separate the approval of the CPE baseline schedule from the first progress update. They should be two individual and distinct activities, for there will almost certainly be changes once the schedule is updated. The next section addresses this issue.

9.4.2 Changes in Logic

As all planners know, building the logic for a schedule is an art form. The closer the logic and durations match reality, the more stable and durable the schedule will be. As it is not possible to model a project perfectly, changes in the logic and duration are inevitable. There are numerous reasons that logic and duration are changed in practice. In this section we are focusing on non–scope-related changes. From our experience, here are some changes that frequently occur:

- *Incorrect durations.* The estimate of the CPE work package duration is not based on historical practice, has been manipulated for transparency reasons, or is simply incorrect. These changes are not related to occurrences like scope changes or force majeure.
- *Incorrect sequencing.* The logic in the CPE baseline does not reflect the way the work is actually being conducted.
- *Incorrect resources.* The estimate of the resources is not based on historical practice, has been manipulated for transparency reasons, or is simply incorrect. The change could be due to a lack of resources, improperly trained resources, overcommitted resources during a period, or waiting on a predecessor activity that has not yet completed.

Many customers include language in the contract documents requiring that an explanation of any logic changes be provided with each the schedule update. Consider these examples:

- Activity 1000—Evaluate Environmental Concerns. Changed duration from 20 days to 30 days. Reason: The baseline duration estimate was based on higher productivity.

- Activity 1000—Evaluate Environmental Concerns. Changed logic from a finish-to-start with Activity 1010—Prepare Environmental Impact Report, to a start to start. Reason; The baseline sequence was not the most effective way to proceed.

In our experience from the customer side, we have found that despite such contract requirements, the lead project manager often either refuses to report them or constructs an argument that blames the customer or a third party for the changes without analysis. Sometimes it is difficult for people to report their errors, and not all organizations will support a project manager who willingly divulges the reason for the change is an error in duration, sequencing, or resources. It takes courage to admit your mistakes but doing so demonstrates truth and fearlessness and is the foundation of trust.

We have often heard from project managers: "It is not the business of the customer or our partner, and certainly not of our suppliers to know why the schedule was changed—that is our business." We think not, unless the organization can produce the entire project without external help. If an organization needs others, it is the business of all the participants. Each participant relies on the others to perform in a timely manner and to know if errors will affect its work. As we indicated earlier, the lead project manager should require a PERT assessment of each work package activity from each participant. Then when updating for progress, the lead project manager should require an explanation of changes in duration, logic, or resources with each schedule update.

As an aside, software packages such as Primavera have been developed to identify all activities where changes have occurred since the last update, as a backup plan for lead project managers. Next, let us look at the issue of changes in performance of an activity.

9.4.3 Changes in Performance Calculation

Numerous things can impact the performance of an activity. In this section, we want to focus on network calculation techniques. The examples come from the forensic side of international projects.

A Brazilian organization was the service provider on a large project and found itself in a dispute with the customer for a variety of reasons. The Brazilian organization were expert in the use of Primavera, which is popular scheduling software on large projects, and knew that it could change the calculated completion date of the project by leaving the logic and durations intact and by using the "retained logic" or "progress override" software functions instead. Our observation was that huge amounts of effort went into avoiding changing the logic or durations, although the project was being executed in a completely different manner. The lead project manager actually told us that his company only was using the published schedule to satisfy the customer's reporting requirements and that it was building the project from its own internal schedule: one schedule for the people executing the work, and one for the customer. People will revert to protectionism in adversarial environments, as we have said.

As with the changes in logic, the CPE baseline schedule updates should be performed on a consistent basis so that the baseline measurements are comparable. We are strongly in favor of staying with the logic on progress calculations. If the logic does not represent the way the lead project manager intends to execute the work, it needs to be corrected. How else would the other organizations in the CPE integrate their work? One schedule is difficult and expensive to maintain properly; more than one leads to confusion, extra cost, and delays. This is easy to say, but think back to our discussion on float. It takes courage to tell the truth.

On another project that had become mired in disputes, the lead project manager changed progress percentages on subsequent updates without explanation. For example, progress during June was recorded at 60%, and then changed to 40% in July, 70% in August, and 50% in September. What we discovered in analyzing the data was that the lead project manager changed the baseline each month, which obviously altered the progress. Because the baseline increased and then decreased, no one, including the lead project manager, knew where the project was. Using our sailing metaphor, it was like having four different marine charts, with no course on any of them.

Adjustments to progress sometimes are required to correct mistakes, but when they are required, an explanation is mandatory and is an essential part of lessons learned. It is also essential in building trust. Customers will not like the news, we know, but they will respect the candor and honesty. The participants in a project have enough changes to cope with without having the baseline changed or the calculation method manipulated. CPEs look to the lead project manager to provide consistency they can trust.

Before completing this section, let us take a look at scope changes and how they are addressed in the schedule updates.

9.4.4 Scope Changes

As we discussed in Chapter 5, when scope changes are recognized, they must be analyzed to determine if they have any impact on the project critical path. Change procedures normally require written authorization from the customer before the lead project manager undertakes the change, and before such authorization is provided the change control procedure normally will require a documented analysis. Often changes must be made quickly to keep the project work moving forward, and documented analysis under such conditions is essential.

We recommend that potential change activities be placed in the CPE current baseline schedule and linked with appropriate logic. After calculating the network, it can be determined if the activity causes a change in the critical path. If it does, then the lead project manager can assess options for crashing (adding resources, hours, or days), fast tracking (making activities parallel or concurrent), or extension of the end date to determine the cost to overcome the delay caused by the change. If the change does not affect the critical path, then the question of dilution of resources and quality should be addressed as well. In either event, we recommend leaving the change in the CPE baseline schedule, for it is part of the project scope.

If the scope control procedure is followed and the changes are incorporated into the network, a complete history of the project showing potential cumulative impacts of changes on the project as a whole is available. The so-called death by a thousand cuts or cumulative effect of multiple small changes may not impact the critical path itself, but the overall affect usually will become visible if all changes are put into the schedule. On many projects we have seen, the forensic determination of what caused a project to fail emerges quickly when the changes are combined into the CPE baseline schedule. Oddly, in many of these cases, that was exactly what was not done during the life of the project. We recommend that *all* changes be inserted into the CPE baseline schedule.

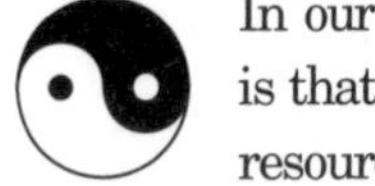

In our experience, one of the primary reasons that project fail is that the CPE baseline schedule is not prepared openly, is not resource loaded, and the logic is not kept up to date. It is like sailing without a sextant and charts. Building a CPE baseline schedule is not quick or easy. It requires hard decisions to be taken and compromises to be made among the value chain, partners, customer, and user. It may point out serious gaps in a firm's tender, gaps that may be embarrassing if exposed to others. Our best advice is to disclose whenever possible, for problems will become known anyway. Transparency in all facets of project management is our recommendation.

Another facet of transparency is evident in Chapter 10, which focuses on quality and customer satisfaction.

Chapter 10

Quality Management—Customer Satisfaction

It is the quality of our work which will please God and not the quantity.

—*Mahatma Ghandi*

By the work, one knows the workmen.

—*Jean de la Fontaine*

10.1 WHAT IS QUALITY?

In this chapter, we discuss the elusive commodity called quality or, in our terms, customer satisfaction. To begin, look back at Figure 2.8, which shows customer satisfaction as the central goal for international project managers and for lead international project managers. With this in mind, we provide a few definitions of quality:

- The International Organization for Standardization (ISO) ISO9000 defines quality as "the totality of feature and characteristics of a product or service that bears on its ability to satisfy stated or implied needs" (IPMA 2006).
- The *Project Management Body of Knowledge (PMBOK* 2004) defines quality as "the degree to which a set of inherent characteristics fulfills requirements" (p. 371).
- The International Project Management Association (IPMA) defines quality in two ways: "Project quality is defined as fulfilling the requirements agreed for the project. Project management quality is defined as fulfilling the requirements agreed for the management of the project" (Section 1.2, p. 2).

Quality is subjective. It is not necessarily perfection, and its aspects should be defined, or agreed to, by the customer. On many projects, the reason a service provider is hired is that the customer either does not have the necessary expertise to plan and execute the project or does not want to take the risk. In the case of expertise, the customer may not know the quality standards for the industry, but it definitely has a idea of what it thinks quality means. How would a customer define the requirements, and how would it imply its needs? What a lead project manager needs are strong communication skills (see Chapter 11), patience and persistence, and a continual dialogue with the customer on what she or he thinks quality is. To know the implied needs, the lead project manager must develop a relationship with the customer.

As we have said throughout this book, scope specifications are never perfect. Thus, the customer's explicit needs cannot be known from reading the scope documents alone. The lead project manager will develop an understanding of the ambiguities in the documents through the processes we have discussed in the last eight chapters. The implied needs come from a relationship with the customer and user. The relationship includes attitudes, the way they process information, their political position, their candor, their vision, their aptitude, their willingness to take risks, their personality, their culture, and more. Some customers are driven to meet the project deadline, some the budget constraints, some the quality of the product, some the quality of the service, some the political visibility of the project, and some all of these things together in different proportions. The lead project manager must listen actively and frequently to the voice of the customers and understand their words from their perspective; in other words, the lead project manager must empathize. We therefore this definition for customer satisfaction:

> Customer satisfaction is achieved by fulfilling the implied and stated needs of the customer in both the services and products provided. It is the aggregate of all the touch points that customers have with the collaborative project enterprise (CPE).

Consider an information technology (IT) project where the government of Indonesia wants to integrate its SAP system with a new tourism platform. The new platform will enable tourists from around the world to select a language, seek information, purchase crafts, make reservations, and download documents and photos. The government has IT professionals on staff that were trained in Indonesia and have only a domestic view of government IT platforms. The potential service providers are a consortium, brought together only for this project, of organizations from Finland, India, and the United States. The organizations have expertise in each of the areas needed for the complete platform. There is no international standard that would clearly define a universally accepted definition of customer satisfaction for such an endeavor, so the lead project manager would be forced to construct one.

Let us assume the government of Indonesia recognizes this as an issue and wants to make its best efforts to describe what it needs. Accordingly, it hires a specialty IT consultant who has managed a similar project for the government of Sri Lanka. The consultant subject matter expert (SME), who was trained in Dubai, knows precisely what the Sri Lankan government came to accept as quality for the past project, and has drafted a specification from those lessons learned for the Indonesian government. The consortium, with clarification questions asked and answered, then bids on these specifications in an adversarial environment. You can well imagine that the picture of quality is likely quite different in the minds of each participant. If ISO definition of quality is used, the consortium knows it does not know what is implied when it bids—unless of course negotiations preceded the bid.

Even if the stated needs are explicit and the SME does a good job of describing them, the needs still can be misunderstood. Assume the specifications require 99% reliability defined as 99% of the year the system will be accessible by users, travelers, and service providers. A service provider can read this as 99% of a year, excluding maintenance. A customer can read this a 99% of a year, including maintenance. Both parties understand that quality is not perfection, but the understanding of the metric placed on availability varies. Both are quality, just different levels. To achieve 99% of every minute may indeed require a backup system. If the Indonesian power grid goes down, is that time deducted from the calculation, or is the service provider to provide a backup generation capability? If the customers' IT staff makes a mistake and the system goes down, does that mean the seller delivered a nonquality product, or one that should have detected input errors?

Imagine that the lead organization is based in India. Assume that the project manager for the U.S. firm has an arrogant and demanding approach with the customer, bordering on disrespectful. But the project manager is excellent at what he does and is held in the highest esteem by his organization. Now imagine being in the joint planning meeting with the customer to discuss scope, and it is discovered that a portion of the scope availability is not mutually understood. As the discussion progress, it becomes clear that the customer must have 99% of a year including maintenance and that this is in the scope of the U.S. organization according to the scope split document. An argument breaks out where the U.S. project manager becomes angry and storms out of the room. Disregard the issue of who is right or wrong on the issue; the point is that the customer likely will not be satisfied with this encounter and will not be looking forward to the remainder of the project. The implied need is respect and personal consideration for the customer, and it will not be written into the scope document. As the example illustrates, the perception of quality is woven into the fabric of a project and touches all aspects.

Our experience is that no matter how complete a scope specification is, there will always be a question about what exactly constitutes customer satisfaction and quality and the big missing component of implied needs. Figure 2.8 shows that there are two components to customer satisfaction: product and scope. The product of the project, in the example a new tourism IT platform, will have measureable requirements described in the scope specifications: throughput, availability, ease of use, error free, compatible, and so forth. In our example, the scope specification would describe practices that are generally accepted by people in the Indonesian IT industry and from the experience and expertise of the SME. The consortium will clearly see the scope documents from its context of experience and expertise. It should be no surprise that the views of the consortium, the SME, the customer, and ultimately the user will be different.

Far more difficult than the product, though, is the problem of describing the stated and implied needs for the service component of the project. By "service," we mean the value added by the lead project manager, such as planning, communications, integration, scheduling, cost, risk, leadership, quality assurance and control, procurement, and more. In the past, we have seen most customers describe the symptoms of professional project management in the general conditions, not the scope specifications. For example, it is not uncommon to see a five- to six-page description of the scheduling requirement for a project. However, that description is usually a list of to-do items, not a description of a quality plan or schedule. More recently, customers have taken to including the requirement that the project manager or lead project manager be a project management professional (PMP) or be IPMA Level B certified. These professional standards are then relied on as the minimum aspects of quality service. Remember, however, that these are standards of care, not definitions of quality. An international lead project manager with 20 years of experience will produce a project plan of different accuracy and durability than will a first-time project manager who has just passed the PMP.

To repeat, the description of quality for the product and the service can only be imperfect and cannot be assumed to be bounded only by what is written. In addition to the implied and written needs, there is also the consideration of cost. Mercedes-Benz makes a range of automobiles that most people would consider quality products. A 500 series has more features and creature comforts than a 200 series, but both are quality products. The 500 series, however, costs much more, so it is easy for customers to confuse, purposely or not, added features with quality.

Even in six sigma organizations, with 1 defect every 3.4 million times a project is performed, there are control limits. If you are a customer in Bogota and you discover under the hood that your new Mercedes-Benz was manufactured in Brazil rather than Germany, would you still see it as a quality car? What if the "made-in" sticker said Germany, but most parts were manufactured in Brazil, shipped to Germany, value added, and then shipped back to Bogota? Do you feel better about your purchase? Would you be willing to pay the added cost to get the made-in Germany sticker? Global sourcing requires the adept management of complex value chains to keep cost low and quality to the same standards. In this case, it is the Mercedes-Benz logo on the hood; that is the definition of quality, for if it fails, it is fixed regardless of where it was assembled.

Global sourcing is a difficult but not impossible challenge. In our experience, the customer must confirm what quality is through serious, detailed, patient, and persistent questioning and input from the lead project manager, and doing this takes time. The customer may be satisfied with a product or service that the service provider thinks unacceptable, or vice versa. The judge and jury need to be the customer, because it is she or he who will decide if the project is accepted or not. We do not mean to imply that a customer should get free services, free product features, or a level of quality not available in the industry. It is the job of the lead project manager to educate the customer and engage in a dialogue to make certain there is a common understanding of the meaning or quality for the project and of the metrics to be used to verify that it has been provided.

Each discipline will have its own standards for international projects, with some being promulgated by the United States, some by the European Union, some by Australia, and some ad hoc. Product quality standards depend on the industry; they vary for healthcare, energy, and IT, for example. For project management services themselves, a few international benchmarks follow.

10.2 PROJECT MANAGEMENT QUALITY STANDARDS

The ISO Web page (www.iso.com) provides a listing of the various ISO standards. In addition to ISO, there are also the American National Standards Institute (ANSI) standards for project management.

- ISO 9000: 2005. The ISO Web page provides an abstract:

  It describes fundamentals of quality management systems, which form the subject of the ISO 9000 family, and defines related terms. It is applicable to the following:

  a) Organizations seeking advantage through the implementation of a quality management system;
  b) Organizations seeking confidence from their suppliers that their product requirements will be satisfied;
  c) Users of the products;
  d) Those concerned with a mutual understanding of the terminology used in quality management (e.g. suppliers, customers, regulators);
  e) Those internal or external to the organization who assess the quality management system or audit it for conformity with the requirements of ISO 9001 (e.g. auditors, regulators, certification/registration bodies);
  f) Those internal or external to the organization who gives advice or training on the quality management system appropriate to that organization;
  g) Developers of related standards.

- ISO 9004: 2000 Quality Management Systems—Guidelines for Performance Improvements. The ISO Web page provides an abstract:

  This International Standard provides guidelines beyond the requirements given in ISO 9001 in order to consider both the effectiveness and efficiency of a quality management system, and consequently the potential for improvement of the performance of an organization. When compared to ISO 9001, the objectives of customer satisfaction and product quality are extended to include the satisfaction of interested parties and the performance of the organization. This International Standard is applicable to the processes of the organization and consequently the quality management principles on which it is based can be deployed throughout the organization. The focus of this International Standard is the achievement of ongoing improvement, measured through the satisfaction of customers and other interested parties. This International Standard consists of guidance and recommendations and is neither intended for certification, regulatory or contractual use, nor as a guide to the implementation of ISO 9001.

- ISO15188: 2001 Project Management Guidelines for Terminology Standardization
- ISO 10006: 2003 Quality Management Systems—Guidelines for Quality Management in Projects. This document supplements ISO9004. From the ISO Web page, the abstract says:

  It gives guidance on the application of quality management in projects. It is applicable to projects of varying complexity, small or large, of short or long duration, in different environments, and irrespective of the kind of product or process involved. It is not a guide to "project management" itself. Guidance on quality in project management processes is discussed in this International Standard. Guidance on quality in a project's product-related processes, and on the "process approach" is covered in ISO 9004.

- Section 5.2.2 of ISO 10006:2003 states:

  [O]rganizations depend on their customers and therefore should understand current and future customer needs, should meet customer requirements and strive to exceed customer expectations.... Satisfaction of the customers' and other interested parties' requirements is necessary for the success of the project. These requirements should be clearly understood to ensure that all processes focus on, and are capable of, meeting them.

- ANSI. The Project Management Institute (PMI) is a member and standards developer under ANSI. ANSI is a founding member of ISO.
- Project Management Institute Project Management Professional. According to the PMI Web page (www.pmi.org), "the PMP credential

recognizes demonstrated knowledge and skill in leading and directing project teams and in delivering project results within the constraints of schedule, budget and resources."

- IPMA certification is available in four levels, according to the IPMA Web page (www.ipma.org).
  Level A: Certified Projects Director able to manage complex project portfolios and program.
  Level B: Certified Senior Project Manager able to manage complex projects. Minimum five years of experience.
  Level C: Certified Project Manager able to manage projects with limited complexity. Minimum three years of experience.
  Level D: Certified Project Management Associate able to apply project management knowledge when working in a project.

As you can see, there are a number of standards for project managers, and each emphasizes different aspects of the profession. IPMA and ISO focus on leadership and personal skills where PMI and PRINCE2 focus on process. In addition, the boundaries of IPMA, PMI, and ISO do not cover many international structures for CPEs. ISO does in a generic way but not in enough detail to call it a specification. Furthermore, the certification processes that PMI and IPMA have should increase the probability that a project manager has the requisite skills and experience to understand what quality project management means in general. Then there are the customers and users who certainly know what they think quality project management means. Many customers have strong business backgrounds and their own set of expectations of what is quality management; they may know nothing about project management but may have expansive expectations. Our advice on considering what represents quality project management services can be summarized in these points:

- Have a proven project management process. ISO requires that an organization must have a process and, in order to keep its certification, must, through metrics and testing, confirm that it is following that process. Unfortunately, the quality is in the process, not in confirming that the process was followed. The metrics provide the consistency that will enable a kaizen process to make the work more efficient over time.

  A side note is in order here. Recently the term "agile project management" has been used in more project management publications.

Our definition of this term is project management performed under general guidelines rather than strict processes. The idea is to enable project managers to exercise their expertise and skills in a guided but flexible environment. Internationally this means having processes that can be utilized anywhere the organization works but that are flexible enough to allow for the necessary adjustments. We fully agree, but think that *all* international project management must be, and is, agile.

- Have project managers who are certified by PMI and/or IPMA. This will help provide consistency.
- Have a selection process that matches the experience, skill level, and cross-cultural leadership intelligence of the project manager to the complexity and difficulty of the project.
- Ensure partners and value chain organizations have project management processes that are at least as stringent as yours. If not, recommend the use of your standards for the project if you are the lead project manager, and garner acceptance from the other organizations in the CPE. Quality is not free, on an absolute basis, but rather free if compared to the cost of poor quality.
- Educate the CPE on the project management processes you will use for the project. You must respectfully explain the dimensions of quality for your version of project management, and seek others' concurrence. This negotiation process is essential.
- Most important, share all of this with the customer and user. Then, using agile project management, be prepared to adjust it all.

Now let us have a look at some ideas about how to implement these ideas for the CPE.

10.3 PROJECT MANAGEMENT QUALITY PLAN

The starting point of the project management quality plan is a quality plan that addresses the product of the project, and the project management services for the project. The plan must begin with the quality standards for the product and the services, metrics, and dashboards as we have discussed. Each of these needs to be designed for the project after discussions with the key stakeholders and especially with the customer and user. For this section, let us explore the extreme of having as a starting point a performance specification for the IT project in Indonesia and that the contract will have a negotiated fixed price structure. First, the easier task: the product.

10.3.1 Product Quality Aspects

As any good trial attorney will tell you, you never ask a question you do not know the answer to. Arbitrators occasionally must deal with a case that involves one pro se party (someone not represented by an attorney). In such cases, the arbitrator must educate, not advocate, by informing the pro se party of the potential pitfalls of self-representation and of how the process of arbitration works. The lessons learned from the legal side of projects are: anticipate, prepare thoroughly, and educate the customer on the options and the benefits and disadvantages of each. Doing this requires a lot of effort, but it is better to spend the time before you sign the contract rather than after. Again, we prefer collaborative environments, during which this discussion can take place and be priced.

For IT products, we know of few recognized international standards. The same applies in other industries. Some industries, such as power vessels, do have an international acceptance, but many others are country standards that some adopt and others do not, such as U.S. codes or European codes. If there are standards that are somewhat applicable, they can be used as the starting point for discussions. For our example, let us assume that there are no standards. As we said earlier, the lead project manager must understand the stated and implied needs of the customer *and* the user. One way to do this is by performing a limited needs analysis of the business.

To understand the conversation with the customer and user, there must be a common context. The lead project manager can provide that context for the IT portion, but the business and political atmosphere needs to come from the local partner and the business perspectives from the customer and user. This exercise will expose a number of scope and time issues from the implied category and will demonstrate empathy. We are not suggesting a full-blown technical assessment but rather a summary-level investigation.

In our example, we recommend that the lead project manager assemble a checklist of quality dimensions for discussions with the customer and user in close consultation with the organizations in the consortium. For the Indonesia project, the checklist might look something like this:

- CPE work breakdown structure (WBS), budget at completion (BAC), risk plan, and schedule
- Existing hardware and software efficacy
- New hardware and software grade
- Existing user competence level
- Employee competence level
- Integration of software platforms
- Speed of system
- Availability of system
- Ease of use
- Upgrades
- Security
- Others

The last item is for the customer and user to fill in any other items that they believe to be aspects of a quality product. When reviewing the checklist, the goal must be to establish metrics that are measurable and agreed on. Doing this will limit the subjectivity and potential disputes over intention. For example, new hardware can be of different grades, faster throughput, better warranty, and better support, and of course will likely cost more if it is. The metric here is tangible, money, and can be established reliably only if the customer and user know what they are getting for their money.

Ease of use is at the other end of the scale. Here we recommend the use of a dashboard. Engage the customer and user in a discussion about what they believe the aspects of ease of use are. They may say that the system should be usable by Indonesians who have a *Sekolah Menengah Pertama* education (equivalent to a middle school education in the West, or about nine years of education). To provide some reaction time, this aspect can be tested on different modules of the new platform as they become available. The lead project manager could suggest that a testing program be established in the Jakarta school system for students with an academic standing of average or better. The system would be considered excellent if 90% of the students found it "easy to use." At less than 90%, there could be either a financial penalty to be paid, or a redesign of the software. These shortfalls could be offset by the potential advertising gain of having future users testing the software. In Indonesia, as in most places around the globe, children are the advocates of technology with their parents and in their social groups.

As you can see, the structure of the contract, the costs, the time, the risks, and the quality are all intertwined. This is one of the reasons why we recommended starting with the WBS, for it established a foundation for structured discussions. In practice, the use of the nominal group technique works well in facilitating an expeditious completion of the quality plan. The nominal group technique is a process to prioritize brainstorming sessions.

The idea is to do a brainstorming session and create a listing of ideas. Then each participant is given, say, 30 adhesive dots that they can invest in any and all ideas, based on how they see their importance. Quickly a few ideas will emerge as clear winners. Then repeat the brainstorming process with the prioritized items, and go again. This system can also be used with an Ishikawa diagram.

Above all, make sure you establish metrics for each aspect of quality. The devilishly difficult ones deserve your special attention. Table 10.1 provides a template for recording the dimensions of customer satisfaction. The columns capture the importance to the customer and which of the CPEs are responsible to participate in the process. The descriptions in each column should be adjusted to fit the structure of the CPE, so "respond" would be service provider A if the contract agreement requires single point. Exploring this one issue to this level of detail will pay large dividends when outsourcing the code to a value chain organization. No haziness about ease of use means you can write and manage a contract with far less ambiguity, clearer goals, more appropriate pricing, better timing, and less risk. It also means that you have laid the groundwork for better communications among the stakeholders. The time spent will pay huge dividends.

The foundation for the quality plan for the product then is the aggregation of these aspects and metrics of quality identified by the lead project manager and confirmed by the customer and user. We will discuss more on the plan soon, but first, let us look at the more difficult part of a quality plan: services.

Table 10.1 Quality Metrics

Item	Customer Quality Descriptions	Customer Rating (0–10)	Service Provider A	Value Chain A.1	Value Chain A.1	Service Provider B	Value Chain B.1	Value Chain B.1
1	Prompt & complete response to inquiries	9	Respond	Input	N/A	Consult	Input	N/A

10.3.2 Service Quality Aspects

Service quality aspects are standards for project management quality, but they are only generic. The final measure of quality is that of the customer and user, not that of the PMI, IPMA, or ISO. There is a tendency for organizations that manage numerous projects to insist that their internal processes provide, and define, quality of service. This may be true to the extent that it is measurably consistent against a goal or aspect that has been established by the organization, but that is as far as it goes. For example, a standard goal for a project management communication process might be the timely and satisfactory response to customer inquiries. The organization may believe that "as soon as possible" is adequate, or at the other extreme, a metric that requires response within 48 hours. The customer and user may say they want "prompt and complete" responses. Will the existing process fulfill the stated and implied needs of the customer and user? If not, is the organization willing to change the project management processes to meet the stated and implied needs? If the adjustments are minor, probably; if they are major, expensive changes, probably not. If the project manager agrees to make the changes, how will he or she know that they worked?

As another example, think about the scheduling aspect of project management. Let us assume that for our IT project, the lead project manager has done six similar projects in the past for different governments in Malaysia, China, India, and Germany. In addition, let us assume that they were of similar financial size, similar frame, with the same organization, using the same generic processes. Each project finished within 20% of the original BAC, within 20% of the estimated completion date, and experienced scope changes of 10% of the BAC. Each reporting period, the deviation from the plan was within the control limits, and occasionally there was a variance in scope, time, and cost. Think about the last time you knew this about previous projects, especially IT projects, and the last time you told a customer and user you were not perfect. When we were the customer, we would ask the service provider project manager if being inaccurate by 20% was standard in the industry.

All of the leading writers on business management and most organizations readily accept the concept that lessons learned are essential in reducing cost, improving quality, and increasing revenues. In our experience, this is truly a global principle. The trouble is, few actually do it, and fewer still do it well. Most project managers attempt to avoid the issue of past performance, because the knowledge is not harvested and because there are no international metrics for benchmarking quality; there are at best just guidelines. What do you say if the customer and user broach this issue with you as the lead project manager?

It is not a perfect world, and project managers are certainly not perfect. It is important for customers to understand this; that is the responsibility of the lead project manager. The trouble is suggesting that one is imperfect can alarm customers and users, and some partners, for that matter. Without tangible lessons learned evidence, the lead project manager must get political support and set forth control limits for the project even if they are arbitrary. Having a process and following it does not necessarily mean you get a quality result; it means you get consistency within a range. Look at Figures 10.1 and 10.2 and notice the difference between the two figures. In Figure 10.2, while there are no variances, the standard deviation is large and clearly there is a lot of deviation in amounts. Now look at Figure 10.2 which has two periods with variance, but the remaining periods show little deviation from the expected. Now imagine you are the customer: Which is the quality effort, and in which do you have more confidence? Although we like Figure 10.2, the customer may not, even after we explain why.

Control charts are one way to present project information that has its roots in statistical analysis. Variances are excursions of the data

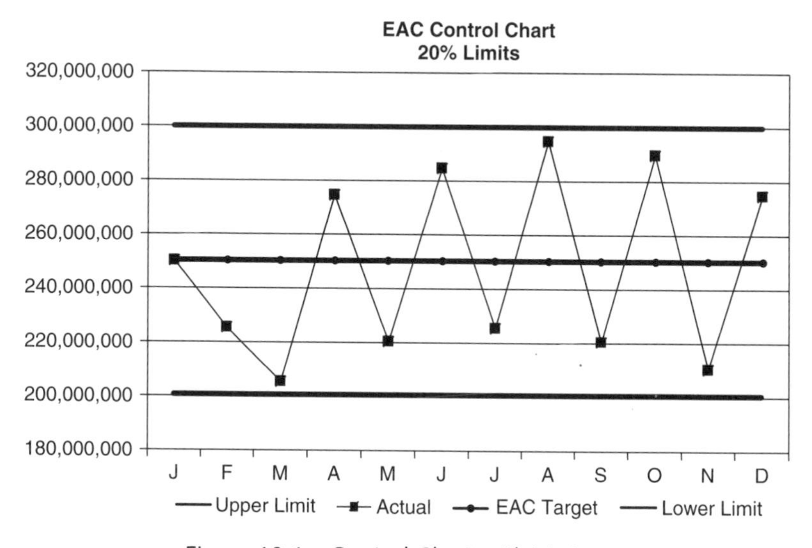

Figure 10.1 Control Chart with Variance

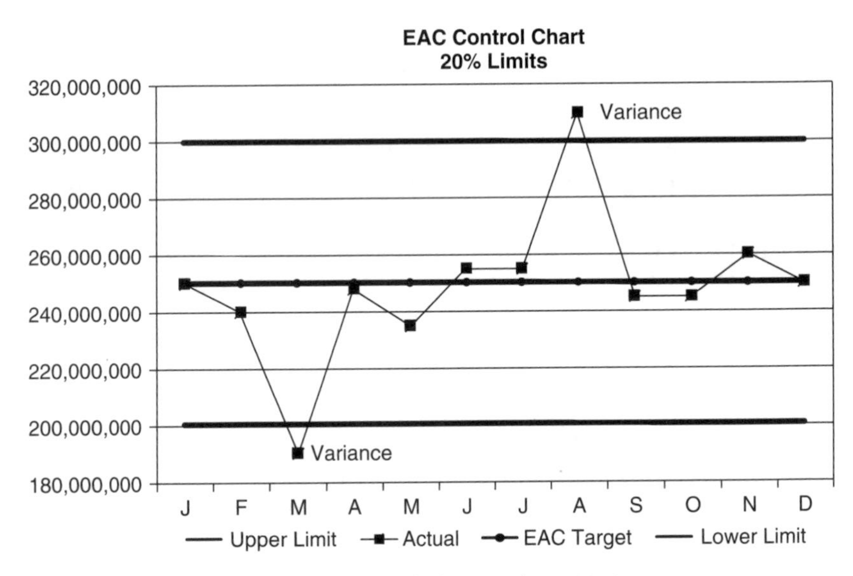

Figure 10.2 Control Chart without Variance

outside the control limits. Professional international project managers must assess the reasons for variances, determine how to fix their underlying causes, and then report to key stakeholders. Control charts provide a way to establish the range of quality that is acceptable; in the examples shown, that is +/-20%. Imagine that you suggest +/-20% and the customer insists on +/-10% for the range. In Figure 10.1 there are variances to explain almost every month, but in Figure 10.2 there are only two. If the customer had established +/-10% accuracy as the metric for quality in estimating/control, Figure 10.1 shows a near complete lack of quality, and Figure 10.2 shows that the quality metric was met for the majority of the project.

In our research, some organizations have developed evaluation checklists for project management quality, but the questions are subjective evaluations rather than metrics, and they tend to be generic, such as: "Rank the quality of the project management services from 1 to 5." Furthermore, our research shows that many do the evaluation at the end of the project rather than at each reporting period. We think both of these methods need improvement for international projects. Metrics help to focus the CPE on what quality is and help to provide tangible values for lessons learned. Lessons learned—some call it knowledge management—are useful only if they are easily harvested and easily accessible. Harvesting a qualitative assessment of 3 on a scale of 1 to 5 for estimating does little to help an organization learn. Incorporating the data from a control chart into a corporate database provides useful information that is easily harvested and accessed, especially if it is in a consistent format. This is one place where a PMO can add significant value.

In six sigma organizations, the goal is 1 defect in every 3.4 million opportunities, so the control chart limits are effectively perfection, with no deviation. This level of detail requires a huge amount of investment in training, planning, measuring, and control. Although many firms find it to be outside their reach and interest, six sigma offers some good ideas:

- Establish processes for the services and products you provide and metrics that define what quality means.
- Establish cost-effective systems to capture the numbers that make up the metrics.
- Establish a process to relate the stated and implied needs of the customer to the organization's established internal processes. In addition, what is the organization's policy when these two are in conflict?
- Establish a communication system that provides feedback to the customer on each progress report or quality metric, and lessons learned for the organization.
- Establish processes for evaluation of variance, such as Ishikawa diagrams, nominal group technique, statistical analysis, and benchmarking to the lessons learned database.

Another graphical technique that customers tend to like is the use of dashboards to display quality information. Returning to our example IT project, imagine the customer tells the lead project manager that quality communications means "prompt and complete" responses to e-mail questions. Each adjective needs a metric to define what the customer's stated and implied needs are. Imagine that the lead project manager determines that what the customer needs is a response within four hours and that an acceptable response is not "we received your e-mail and will get back to you with an answer." Through further questioning, the lead project manager finds that a response within eight hours may be acceptable sometimes but longer than that is unacceptable. First a system must be established to measure the response time on each e-mail. For those that do not provide an answer, the time of response is established when the response is actually given.

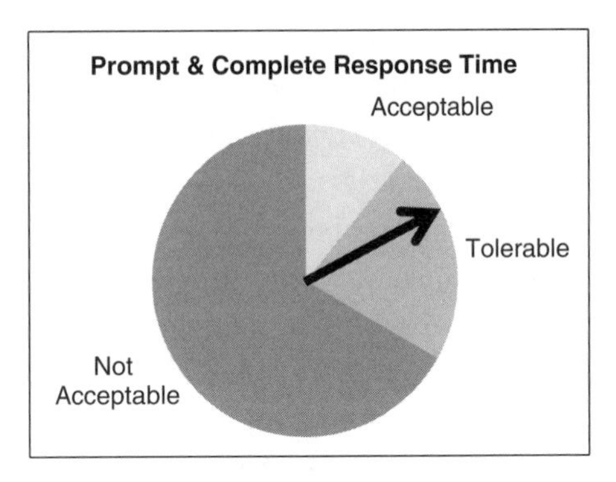

Figure 10.3 Response Dashboard

If the lead project manager then measured the performance for the first reporting period and found it to have a mean of 4.7 hours, the dashboard would look like Figure 10.3. The pointer represents the actual measurement mean for the reporting period and shows that more work is required to bring the customer satisfaction metric into the acceptable zone. The other benefit of having a quantitative measure such as this is the ability to harvest it for lessons learned. The dialogue with the customer can be "We have an existing process for prompt and complete responses to inquiries that we have been monitoring for the past five years. The process produces a mean response time of 4.7 hours. Would that be acceptable as a measure of quality for you?"

The key is to explore the aspects of quality with the customer and then get metrics on each one. Without metrics there will be an ongoing dialogue about what quality means and whether the CPE is producing a quality product and a quality service.

The PMO can set the standards, for organizations that use them, and ensure that there is consistency in the definition and metrics of the aspects for quality. If a PMO exists, we recommend that the lead project manager establish them on an ad hoc basis, based on his or her experience. We suggest these items as a minimum list of aspects that are measured and reported on each reporting period:

- *Financial accuracy.* Including the budget at completion, estimate at completion, the size of the contingency reserve, and the variances.
- *Time accuracy.* Include the length of the project, the size of the buffer or reserve, the number of near-critical activities, and the out-of-sequence progress.

- *Scope accuracy*. The number of disputes on scope settled and unsettled.
- *Leadership*. The trust, empathy, transformation, communication (see next item) and cultural intelligence; more on this in Chapter 11.
- *4 Cs—cooperation, commitment, coordination, and communication.* Communication assesses the promptness and completeness of the replies to inquiries, the perceived transparency of communications, and the effectiveness of the communications with key stakeholders.

10.3.3 CPE Quality

As with the other parts of the project plan, once the lead project manager has sorted out his or her view of the project, it is time to get the partners and value chains input. As we noted, if the partners have their own quality plan, a joint program can be negotiated. If the partners do not have one, then the lead project manager's program can be utilized and adjusted if necessary. For example, the partner firm may not have the resources or technology to measure responses as shown in Figure 10.3. In such a case, the lead project manager can offer either to do the measurements for the partner or to take the lead in responding to the customer and user. In practice, most customers prefer to have the lead project manager speak for the partnership.

Once the standards, aspects, and metrics have been established, the CPE quality assurance (QA) program can be developed. The processes to be utilized must be connected to the metrics so that variances can be identified and the sources of those variances chased to their causes and corrected. As we said earlier, organizations that have PMOs or have project management standards will find this easy to do. Organizations that rely on ad hoc project management will need to sketch out how they do their version of project management. The QA portion of the work connects the definition of quality, the voice of the customer, and the metrics to the processes that create it. The QA program must address who is responsible for QA and for quality control (QC). On international projects, QA and QC are often left to each organization. We recommend that the lead project manager accept the overall responsibility for QA and QC. Since the standards are for the CPE, the ultimate quality results will be driven by the weakest link in the chain. The lead project manager must therefore take ownership for the CPE QA and QC. Think of it as a holistic quality management plan.

The trick is to get enough detail to monitor and control, but not so much as to become a slave to the details. Part of the lead project manager's efforts is of course auditing the processes that have been established or having the audit done independently. If the project is large and complex, we prefer independent auditors because they can be totally transparent and unbiased. From experience, we also prefer them because they often uncover issues that we sometimes are too close to see. We also prefer that the audits be random, and not scheduled. Customers resonate well with this approach for it provides an independent assessment of their project.

The next step consists of planning for the QC effort. Using the QA processes and metrics, the testing and accumulation of the data should be straightforward. The largest challenge at this point likely will be implementing the necessary communication channels to ensure the data are reported at the required time. On large international projects, some value chain organizations may need to report the data for the month three weeks early. In fact, planning the acquisition of information for a monthly report can be thought of as a mini project. A monthly report may require a month of preparation, and on some projects there are teams assigned to just this task. In addition, it is likely that the format for the customer will be different from that for the organization's lessons learned. To save time, make sure that the formats are established well in advance and that the data coming from the value chain are in the customer's requested format.

Although it should go without saying, make sure that there is time to look at and evaluate the data before giving them to the customer. This is especially important if the metrics show that performance was not adequate to meet the customer's quality criteria. You must provide the customer with the break in the process and the planned corrective measures along with the data. Too many times, data are presented that have not been reviewed until the customer's technical team brings issues to light. Customers want to trust that the lead project manager has scrutinized the details and then summarized them into a management-level dashboard. A lead project manager can destroy trust by not doing this.

Likewise, the lessons learned need the same level of attention. The data may be rolled up into quite different sets of metrics for the organization itself, so it may be necessary for the lead project manager to do two reviews and to weather internal audits. Audits are a conundrum at times for answers to the questions, "What did we learn?" and, "What adjustments will we make?" are not always obvious, and sometimes are rather disturbing. For example, your QC indicates that the prompt response metric in our example is not being met. You do your Ishikawa diagram and find that the problem seems to center around the communications regarding tourism. As you bore down using the nominal group technique, you find that the problem seems to be a knot in the customer's system. Further details indicate that the tourism people do not have e-mail access. The Indonesian tourism department receives e-mails from the minister's office by post and then responds by post. Questions are asked at an executive level, flowed down to a technical level, and when the interaction occurs between the lead project manager's organization and the customer's organization, the posting problem kicks in, which delays the response at the executive level. You must imagine that the size of the project prohibits the lead project manager from knowing each person involved on the project and even knowing how many people there are.

Whose problem is the lack of prompt communication? Well, of course, it is obviously internal to the customer, but remember, this is a new system to connect the tourism office with the SAP. To do this, everyone would need computer systems and e-mail access. As the lead project manager, shouldn't you have planned the procurement and training for the new computers before executing the project? Time to get out the contract. As the old saying goes, no one reads a contract until there is a dispute. The next section addresses this issue, and the issues of procurement and contracts.

10.4 MONITORING AND CONTROL

Monitoring and controlling quality is a critical aspect of international project management, to assure customer satisfaction in the product and the service. We recommend that quality be a section of every progress report, just as are cost and time. The baseline is a combination of dashboards that address the aspects of quality that the customer believes are critical to quality (CTQ). The prioritization of these aspects, as with the filtering of risks, should enable the lead project manager to focus on no more than a dozen priorities. Think of the aspects of quality being passively and actively managed, just like risk. If a prioritization system of 1 to 10 is used, the actively managed CTQs would be ranked above 7.

The matrix will help in building the list of CTQs, and the dashboards can be generated easily from the data gathered. The real challenge is making sure the data are collected diligently throughout the value chain.

Chapter 11

Procurement Management

Fairness is what justice really is.

—*Potter Stewart*

Ethics and equity and the principles of justice do not change with the calendar.

—*D. H. Lawrence.*

11.1 THE BASICS

In procurement, most all of the techniques that have developed are aimed at the transfer of work, scope, or risk from one organization to another. At the beginning of a project, the customer will know more about the project product desired than does the service provider, unless the project is a collaborative one. The service provider will know more about the project management processes and technology relating to the product. At the end of the project, knowledge has been transferred, and both parties understand exactly what the other party needed and exactly what the project was to have been. To use a well-worn metaphor, think of a project as a marriage between the customer and the service provider, and the contract as the wedding. Imagine a project that you have undertaken, and think about what you knew at the start and what you knew at the finish. In procurement, the perfect world is to know at the beginning what you would know at the end.

In some cultures, marriages are arranged by the parents, where the couple may have met but know very little about each other. In these cases, the couple will spend their initial years building trust and learning about each other's personality, communication techniques, habits, needs, and ethics. In other cultures, the couple chooses their partners, sometimes after many years of living together. Such couples live through the same trust building and learning period, but they know each other very well when they finally choose to formalize the relationship. It is said that one does not just marry another person but marries the other person's family, in this case the organization. In a globalized economy, the family can be a great asset or a liability.

One essential ingredient in all marriages and relationships is trust. The definition of trust we prefer is "the willingness of a party to be vulnerable to the actions of another party based on the expectation that the other will perform a particular action important to the trustor, irrespective of the ability to monitor that other party" (Mayer, Davis et al. 1995). The amount of trust you are willing to give another party then is a function of your ability to monitor and of the consequences if the other party fails to perform. Said another way, it is the amount of risk you are willing to endure. Contracts are a formal way to assign this risk.

If people and organizations trusted one another completely, there would be no need to use contracts to transfer obligations or risks. Societal and organizational cultures deal with risk in different ways. In some societal cultures, such as Asia in general, personal relationships must be constructed before business is conducted. In such cultures, contracts tend to be short, general, and secondary to the relationship. In other cultures, such as the United States and Nordic Europe in general, business dealings are more transactional, one-time business relationships, and the contracts tend to be longer, more detailed, and primary. Some organizational cultures, such as governmental agencies and banks, tend to be risk averse, and some tend to be risk seekers, such as Google.

The level of trust, or willingness to be vulnerable, in organizations is tied in part to the repercussions of failure and who would bear them. Think of trust as a bank account opened by the trustor, where the trustee makes deposits and withdrawals. For example, consider two rock climbers doing a rock face together for the first time. The climbers may have different attitudes toward risk. One is more conservative, safety focused—risk averse—while the other is more aggressive, adrenaline focused—risk seeker. There are external risks, such as animals, spiders, snakes, and birds, that could distract or injure climbers. There are also equipment risks, such as rope failure, "friend" (cam-locking device for crevices) that fails or pulls out, a sit harness that fails, and others. Then there are the risks associated with experience and training, such as ability to set friends and tie knots properly, knowing the limits of one's strength and endurance, ability to become distracted, and the like. The bank account was opened before the climbers met, and the initial deposit was made based on reputation.

Imagine the climbers get ready for the first pitch; let us say a rope length of 50 meters. The conservative climber belays (holds the rope in case the other climber falls) while the aggressive person climbs. Halfway up the pitch there is an overhang that can be skirted safely and easily, but the direct route over the overhang is challenging, and past the limits of the equipment and abilities of both climbers. If the aggressive climber moves toward the overhang, the bank account will realize a withdrawal. If the aggressive climber skirts the overhang, the bank account will realize a deposit. Both bank transactions are in the eyes of the risk-averse climber. The complexity in such interactions is clear, as is the comparison to project management. Trust and risk are two critical components of procurement, but there is also equity in the mix.

In our example, the issue of equity is one of fairness and consideration. Is it fair or considerate for the aggressive climber to place both climbers at risk, when she or he knows that the other climber is conservative? Imagine that the climbers, both project managers, plan the climb in advance and plot their route. In selecting the route, they discuss the option of the overhang and decide that for their first adventure together; they will skirt it but will add on a traverse to keep the route interesting. Here the equity is a less dangerous route on their first experience. There could also have been a discussion to attack a more challenging climb on their next experience. Each climber knows his or her limits, the risks, and the ways to mitigate the impact and probability of an accident, and both know who owns which risks at which times. One of the great similarities between rock climbers and project teams is that the participants are tied to the same rope. If one fails, the others must catch them, or everyone will come off the rock.

In a global economy, we recommend that you get the contract documents in electronic form. Make certain that they are signed by a duly authorized person by each party, that all comments are initialed by all parties, that all pages are initialed, and that they are scanned so that the text, signatures, notes, and execution date are legible. As the lead project manager, make certain that all signatories have a copy of the same version. For the value chain, make certain that the most current version of the documents is provided, either 100% or excerpts. These steps will save large amounts of time and cost.

As the lead project manager, it is your responsibility to make certain that a complete listing of all contract documents, including drawings, reference specifications and documents, technical specifications, general conditions, and special conditions, are created as a document baseline before execution. We suggest in Chapter 12 that this information be posted on a server that is accessible to the collaborative project enterprise (CPE). As the project progresses, changes will occur and the contract documents will need to be revised as necessary. On physical projects, often value chain organizations are required to do design or to detail the conceptual design in the baseline contract. These designs then become contract documents and must be organized, posted, or distributed to all of the organizations that may be affected. Then there are the changes made to the contract through formal contract change orders. These designs and changes, along with the attachments, need to be posted or distributed to those affected.

Last, there is the issue of submittals. These can be processes, technical details, schedules, work breakdown structure (WBS), risk registers, drawings, and more. Submittals are required in most contracts, and while they can be a byzantine maze of detail, they are necessary. Our recommendation is to establish an inventory control system during the initiating and planning phase and include the input required from each value chain in their contracts. Submittals are a contractual obligation and require that level of effort and attention. On many international projects, we see completely different versions of documents, especially submittals, in the possession of the different organizations. This waste time and money, and can divert the CPE's attention if not planned.

Keep the rock-climbing metaphor in mind as we progress through the remainder of this chapter. Now let us begin by discussing the documents that underpin contracts.

11.2 TECHNICAL CONDITIONS

The technical conditions of a contract describe the details of the product to be provided. For an information technology (IT) product, they would provide the specifications for such things as functionality, dependability, ease of use, speed, and compatibility. For a development bank, such as the Asian Development Bank, technical assistance project, they might include such things as strengthened implementation capacity of governmental agencies, strengthening financial management, and support for improved policy analysis within the government and civil society. For a nongovernmental organization (NGO), such as Médecins sans Frontiers, it might be to reduce the deaths due to starvation, treat those suffering from tuberculosis, or provide emergency medical care in a war zone. The technical conditions describe the details of the product, its characteristics, operational capabilities, and the grade, quality, and metrics for determining if the product meets the customer's stated and implied needs. On power projects, the metrics might be heat rate or output; on NGO projects, a reduction in infant mortality; for IT projects, throughput speed; for development bank projects, decrease in government spending; and for highway projects, number of people served.

Often customers hire consultant expertise—subject matter experts (SMEs)—to write technical specifications, because the customer does not have the expertise. Sometimes customers hire consultants in order to shift risk to the consultant. Other times customers engage in public-private partnerships (PPPs) or hire a "turn-key" service provider to provide the specifications and the product. Sometimes, though, customers will write the technical specifications themselves. This is often the case when a government agency maintains adequate expertise on staff. The technical specifications often reference other technical standards, such as the International Organization for Standardization (ISO), American Society of Mechanical Engineers, Institute of Electrical and Electronic Engineers, or World Health Organization, to mention a few. In making such references, the writer is pointing to an international or national standard to provide a more detailed account of a product's attributes. In addition to the written technical specifications, certain types of projects require pictorial illustrations or drawings. This is the case on physical projects, such as buildings and bridges.

In cases where the product incorporates or makes use of proprietary information, such as patents, the technical specification may simply refer to the component by trade name—Microsoft, for example. One of the technical specification techniques that we mentioned in Chapter 5 was an "equal-to" approach. In technical specifications, this relies on an existing product to convey the required technical aspects and performance attributes. For example, rather than specify all of the components and attributes of a notebook computer, you could say "equal-to or better than a Dell® Latitude model 630." It may be necessary to describe options but it would clearly be much easier and cheaper to use an equal-to approach

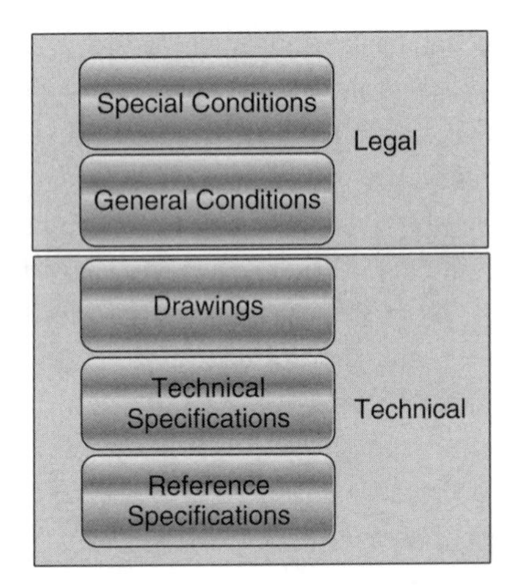

Figure 11.1 Contract Documents

rather than writing a set of technical specification for a notebook from scratch. Figure 11.1 provides a hierarchy of typical contract documents including the legal documents, reference specifications or standards, technical specifications, and drawings. As shown, these documents provide the foundation for a contract.

Globalization has changed the way contract documents are assembled, so let us take a quick look at a few of the components of the contract starting with design. Taking advantage of the Internet, good quality, and low labor rates, it is common on international projects to have design work performed in multiple countries. We have used India, Romania, Brazil, and China to mention a few. This globalization has increased the surveillance and review time required and the need for excellent cross-cultural communication skills. The lead project manager no longer can walk down the hall to check on the work and interact with the designers face to face. In a virtual world, checking relies more heavily on detailed electronic review of the design. We do not find this more difficult or challenging; it simply entails using different skills and setting aside time to do the review. In collaborative environments, the lead project manager must make certain that the design standards and units utilized are consistent and that the format or layout is compatible. On one project, we received a piece of custom equipment designed in four different countries and manufactured in four other countries. What we found was that there were four different types of pipe threads on the same machine, and none of the piping would fit.

For general and special conditions, globalization makes life interesting:

- Different legal systems, with lawyers trained in those systems
- Negotiating partners scattered about the globe, with sales and commercial teams in multiple countries
- Global organizations with different practices in different cities
- Customers made up of consortiums with radically different attitudes toward risk

Quite frankly, it is amazing that contracts ever get done. The local legal conditions must be synchronized, the commercial terms must be agreed on, the risk must be apportioned, and the rules must be agreed on. The next two sections discuss the rules governing work under the contract. Most often the technical conditions are the easy part; now we turn to the tougher part, the legal conditions.

11.3 LEGAL CONDITIONS

Continuing with Figure 11.1, the general conditions form the heart of the legal requirements of the contract. Often they are written by the customer's legal department, but occasionally they are provided by a SME if the customer has no legal resources. Many governmental agencies and NGOs have standard conditions that appear in all of their contracts. If a customer is using an SME to write general conditions, make certain you check its work, and get another SME to do likewise if you are intent on using an adversarial environment. If the customer is using a collaborative environment, the terms will be developed bilaterally.

As we have discussed before, we recommend that the conditions of the primary contract between the customer and the lead project management organization flow down to the partners and value chain to the greatest extent possible. For certain clauses, such as liquidated damages (LDs), it may not be possible, but we recommend that the default be to pass along the same rules as the ones used by the lead organization. It is the responsibility of the lead project manager to do this for his or her organization and to encourage, persistently, partners and value chain organizations to do likewise. Imagine that the lead project manager has a contract that is conducted under the laws of the United Kingdom and a second-tier value chain has a contract conducted under the laws of Malaysia. If a dispute occurs, it will be a mess to sort out.

The legal conditions typically cover these types of issues (note: this list is not all-inclusive).

Contract structure. From adversarial to collaborative. The next section provides a detailed discussion of this item.

Parties. Signatories to the contract. On international projects, it is not unusual to have groups of customers or owners and PPPs. Think back to the example of the highway project in South Africa where the government held a partial share of the project. Signatories have explicit rights and obligations under a contract that will be different from those that are not signatories. This situation must be considered when structuring a contract.

Laws. States the governing laws for the contract—for example, South Africa in our example. This is a key legal issue on international contracts, for it dictates the laws that govern the conduct of the parties. There are numerous legal systems, and it is critically important that legal expertise is sought out and employed in each country before the contract is completed and signed. The lead project manager must ensure that the legal requirements in each country are consistent with those of the primary contract.

The CPE must comply with the laws in *all* countries involved. Such things as insurance, labor laws, benefit packages, work hours, work conditions, and overtime pay are different in most countries. Global organizations also may have the internal policies that may conflict with local laws. In addition, the governing laws are important, as they will govern in the event of a conflict between laws and in the event of a dispute. One of many issues is enforcement; if a judge in Zimbabwe rules on a particular issue, that ruling will not necessarily be enforceable in the United Kingdom, for example.

Title. Defines when title transfers normally using International Commercial (INCO) terms. Table 2.2 provided examples of the standards for international title. One approach for smaller projects is for the seller to provide delivered duty paid (DDP). In this case, the seller pays the insurance on carriage, customs, and local taxes. On large international projects, customers may realize savings from other approaches to insurance, customs, and duty. Furthermore, in countries where corruption is a problem, having the local customer responsible for taxes and customs is sometimes a good option for a service provider.

Insurance. Defines who is responsible for insurance (e.g., builder's risk, shipping, auto, personal indemnity, etc.). Sometimes customers hold umbrella policies, such as all-risk and marine insurance, and they are in the best position to take advantage of scale. In such cases, it makes far more economic sense to have service providers added as additional insured to save project cost. If this is the case, then the lead project manager must obtain copies of the policies so that she or he knows the terms and the deductibles, and can pass them along to the value chain and partner. Lloyds of London, for example, offers a wide range of insurance coverage options for special conditions.

Risks. Weather, force majeure, escalation, currency fluctuations, and the like. As we discussed in Chapter 8, the risk register must be used to assign ownership, which should be formally addressed in the contract and in the value chain agreements. Here are two real-world examples of the types of clauses that cause concern regarding risks:

- Malaysia: "In the event the service provider encounters obstructions, conditions, pollutants, or archaeological items . . . which could not have been foreseen by an experienced service provider . . . "
- Israel: "The service provider shall provide all of the work required to make the facility complete and fully operational whether or not the information is shown on the drawings or called for in the specifications." (This is known as exculpatory language, and you should be mindful of such clauses when reviewing contract terms.)

Contract schedule. Defines the start, finish, and milestones for the contract. Some customers and users have very specific preferences on scheduling. In our experience, some even insist on the software to be utilized and participate actively in the process. On one project, we were engaged to provide expertise for the service provider, as it was unable to get the customer to accept the schedule. The reason was that the customer had more experience in the product and the software than did the service provider. It is also important to look for the presence of important milestones. On a cogeneration boiler project, there was no seasonal milestone for loss of load during extended holidays, so there was nowhere to utilize the steam.

Contract start and acceptance. Obviously it is important to know when a project begins and ends, but it is not always clear. On memorandum of understanding contracts, the start may be vague and work often begins before a formal contract is executed. On cost-plus contracts for speculative investments, the start date can also be somewhat blurry.

We have seen many contracts where the parties want to get started before the final negotiations are concluded. Our suggestion is to agree to and commemorate a start date in writing. Everyone then knows when the project began, because many contract provisions, such as payments, bonuses, penalties, and milestones, are linked to start dates.

Knowing when the work is complete is often very difficult, for it is left to the opinion of the user, customer, lead project manager, and partner. Not surprisingly, their perspectives tend to be quite different. Here are two real examples that define completion and acceptance:

- India: "The date occurring 31 months after the date of issue of the Notice to Proceed or such other date fixed upon the Notice to Proceed pursuant to Clause 5 of the Agreement subject to Clause 24.2, and in each case as adjusted thereafter under Clause 25."
- Germany. "In the case of interruptions to operation... which last for more than 3 months continuously or in the case of more than 6 interruptions each lasting longer than one month, and for which the supplier is responsible, the guarantee period shall be recommenced upon notification that readiness for operation has been restored (in the case of multiple interruptions, after the sixth interruption)."

Big warning! As the two examples demonstrate, knowing when a project is accepted may require a bevy of attorneys. In a collaborative environment, negotiate acceptance definitions that are clear and unambiguous. In an adversarial environment, price the ambiguity in the tender. Also, *always* get acceptance from the customer in writing.

Contract amount. Defines the price of the contract. The contract amount can range from a simple fixed-price guaranteed maximum, to a revenue stream based on operations on a build-operate-transfer or PPP type project. Often lead project managers must have a good understanding of international finance to manage a project effectively. In PPP projects, the financial value for money can be complex and will rely on quite different financing and revenue streams during different phases of the project. One such project in our experience relied on export/import bank financing for part of the initial capital investment. On this type of project, the nationality of the value chain organizations will be determined in large part by the bank, and that may have an impact on the procurement processes normally utilized by a service provider.

On projects that have durations of decades, as do some PPP, NGO, and development bank projects, the contract amounts must take account of escalation, market conditions, stock and bond markets, and, increasingly, sovereign wealth funds. In addition, many international projects require the service provider to provide the project financing and sometimes to issue bonds denominated in multiple currencies. This requires far more complex financial planning and a balance between revenues and expenditures. These are true CPEs for they must be run as you would an international business with shareholders, not just stakeholders. When this situation is juxtaposed with a complex partnership or alliance, a lead project manager with outstanding cross-cultural leadership intelligence (XLQ) and experience.

Payment provisions. Date and what backup is required. One issue associated with payments is the contract amount discussion presented earlier. Clearly, service providers must balance costs with revenues to avoid uncovered short-term borrowing, and customers must know their payments requirements to plan their finances. This is especially important on international projects, as most deal in multiple currencies. For example, imagine a project that utilizes bank funding from Japan for a project in Uzbekistan. The draws on the loan facility by the customer would need to be transferred into SOM (UZS) for the funds needed in that country. Assuming that the bank takes the currency risk, it would hedge its position by purchasing the right to buy SOM futures so that if the currency devalues against the yen, the loss will be mitigated. If the yen depreciates against the SOM or stays unchanged, the futures contract would not be used, and the price of the position would be a sunk cost—that is, of course, assuming that a futures contract market exists for the currency. There are pros and cons regarding currency hedges, and we recommend that lead project managers at least become conversant with the concepts and processes.

One other issue deserves mention: letters of credit (LOCs). A LOC often is the source of payment for an international transaction of significant value for deals between a supplier in one country and a customer in another. LOCs also can be used in the land development process to ensure that approved public facilities will be built. The parties to a LOC are usually a beneficiary, the service provider who is to receive the money, the issuing bank to which the customer is a client, and the advising bank to which the service provider is a client. Almost all letters of credit are irrevocable, meaning they cannot be amended or canceled without prior agreement of the service provider, issuing bank, and confirming bank. If the customer's willingness or ability to pay is uncertain, the service provider can insist on an irrevocable LOC to ensure that it will be paid. We have used the LOC approach in developing countries where the financial systems were immature or when a customer's ability or intentions to pay were questionable.

Penalties and incentives. Defines such things as liquidated damages or incentives. In Figure 3.2, the milestone called "Initial Acceptance" needs to be broken down a bit further for this discussion. Figure 11.2 provides a view of the project life cycle from the perspective of the customer and the service provider on a project that tendered using a competitive public bid. In this case, the customer normally would have done some planning before tendering for the work, and the seller would do the initiation, planning, and tendering concurrently. If the project was a negotiated arrangement, where the service provider will do some or all of the design, then the planning and tendering for the buyer and seller would be concurrent.

At the end of the project, "Preliminary Acceptance" in Figure 11.2, the product becomes an asset for the customer. The service provider's involvement in the project would be complete when the warranty period has expired, unless the project is a PPP. In this case, the involvement would continue through operations, or the useful life of the product. The *Project Management Body of Knowledge* (*PMBOK* 2004) also calls this the product life cycle. Figure 11.2 shows milestones called "Beneficial Use" and "Preliminary Acceptance." "Beneficial use" means the product can be

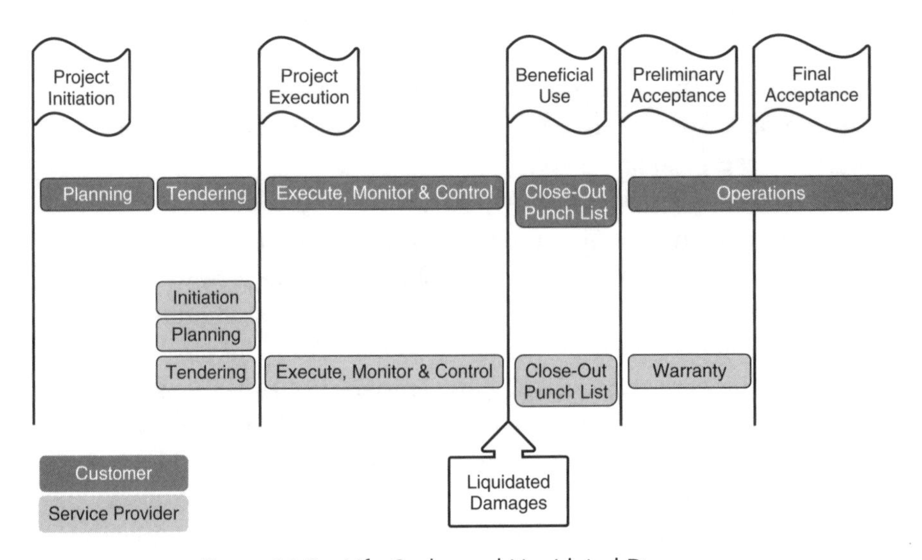

Figure 11.2 Life Cycles and Liquidated Damages

used for its intended purpose—carry traffic, dispense meds, run a new IT platform, and so on—but that some of the detailed components of the product are either not completed or need some adjustments ("Close out punch list" in the figure). A punch list may also be called a deficiency list, open items list, or something similar. At preliminary acceptance, the work on the product is complete, and the service provider's only remaining obligation is the warranty period.

"Liquidated Damages" shown in Figure 11.2 are normally tied to the beneficial use milestone. The reason is that the inability to use the product deprives the customer of the benefits, income, tolls, repayment of bonds, and so on. Liquidated damages must be "fairly" estimated in advance and agreed to by both parties at the time the contract is signed. Liquidated damages are not punitive damages. Punitive damages can be applied in legal systems to teach a lesson; for example, cigarette manufacturers that knew of the health risks of smoking but failed to inform the public, the court could award punitive damages.

For example, if the project were a toll road, failure to open the toll road would deprive the customer of the start of revenues. Thus the costs of increased financing costs and extended overhead might be totaled to arrive at a "fairly" estimated damage of €20,000 per day. In such a case, the service provider would pay the customer this amount for each day the toll road was not opened for use. The largest LDs we have seen on an international project were $3 million per day. As you can imagine, this makes quite an impression on the service provider. We have also seen smaller LDs attached to the preliminary acceptance milestone to encourage the seller to complete the details. Make certain to negotiate such clauses if the environment permits, or to include the necessary funds to cover delays if the due diligence indicates that there is not enough time to perform the work.

Transparency makes a very big difference here. Imagine that a service provider provides a price in a collaborative environment based on a due diligence schedule created with the Program Evaluation and Review Technique. Assume that that schedule shows it will take 30 days longer to complete the work than the customer has requested. If the LDs are €20,000 per day, the total costs of the project would need to be increased by €600,000. If you were the customer, would you not want to know this? It is always possible that the LDs were estimated by an external SME who did not know the cost or time or that the customer may have other options for financing . In collaborative environments, withholding such information will destroy trust. In competitive environments, there is an ethical decision to be made.

Schedule requirements. Start, finish and milestones. Figure 11.2 shows the major project milestones, and these need to be specified in the contract. Often the official commencement of the contract clock begins with a notice to proceed (NTP), or similar document, expressly to pin down the start of the project. The finish dates, beneficial use, preliminary acceptance, and final acceptance are also normally specified in writing—a practice we strongly recommend. Often the start date is used to establish due dates for other project deliverables; such as the CPE baseline plan. Depending on the customer, the schedule requirements can be quite specific, reaching down to the level of scheduling practice, logic, float, and calculation methods. On projects with large value chains, it is critically important to make certain the information coming from the value chain can be incorporated almost automatically into the master CPE plan. Remember, our recommended strategy is to build the CPE plan from the detail of the value chain organizations up to and including the customer.

We strongly recommend that the CPE schedule and the individual organizational schedules be resource loaded. We know firsthand how much work this is and how much difficulty it creates with partners, the value chain, and especially the customer and user. We also know, from both sides, the amount of resistance a lead project manager will face in attempting to resource load schedules. However, it is an essential tool in leading successful projects. As we have said, the most difficult portion of the task is keeping the information up to date. If you do not, though, metrics become entirely subjective, arguments are more prevalent, and lessons learned are poor, at best. In addition, without resource-loaded schedules, earned value assessment is not possible.

Float. As we discussed in Chapter 9, having a float buffer at the end of the critical path for a project is a wise planning tool, just as it is with a contingency fund for costs. We recommended it be utilized in a transparent manner, despite the disadvantages of doing so. The ownership of float must be defined in the contract, or it will likely be the property of the project. We believe that float should belong to the organization that creates it, for it represents a very real and tangible liability if it is used. On large capital projects, it is very easy to have a daily overhead cost exceeding €30,000, so the cost of staying on a project can be large.

Reporting requirements. Frequency and format. On international projects, stakeholders likely will be spread around the globe. The emerging standard is to utilize electronic formats, although some still some prefer paper. Printed copies of the details or compact discs (CDs) containing them still must required to be mailed in circumstances where the local Internet will not permit transmission of large files. In 2009, it was estimated that 22% of the world's population (www.internetworldstats.com/stats.htm) was connected to the Internet. Depending on the country, it may be necessary to deliver paper reports. Some executives still do not use e-mail; they have their secretaries print them out and type the responses. The point here is to explore the needs of the stakeholders, especially the key stakeholders, and adjust accordingly. Our preference is to utilize a project server rather than fuss with paper, CDs, or e-mail.

One way to streamline the reporting and communication process is to use a server as a communication hub for the CPE. Each organization in the CPE would have its own secure area for project records, and the information to be shared is stored in a common area. The customer reports would be stored in the common area, along with the current scheduling information, cost information, risk information, quality information, and other data pertinent to the project. If there is a need for higher levels of security for political or proprietary information, there can also be a high-security common area just for the lead project manager and the customer. The key is deciding who needs what, in what format, and when.

For frequency and format, the value chain is the critical component. With large value chains, it takes time to assemble the information, especially if it is in a format that makes cut-and-paste impossible. Assembling data from a value chain can easily take two or three weeks. In such situations, members of the value chain must start summarizing the next report immediately after filing the current report. The project can become a continual reporting project for the value chain and for the lead project manager.

Another problem that most complex international organizations face is the amount of time it takes to close corporate books. For example, assume the customer needs a monthly report delivered on the twenty-first of each month so that it can take it to the board of directors on the twenty-fifth of each month. Assume that an organization's books close on the first of each month and that the monthly cost reports are finalized by the twenty-sixth of each month. Now imagine that the partner has a closing on the twenty-sixth of each month and final cost reports on the twenty-sixth of the following month. The issue is one of synchronizing the accounting clocks and deciding the cut-off date for data.

This one detail can lead to excessive work, inaccuracy, confusion, and consternation on the part of the customer. Returning to our sailing metaphor, imagine having no sextant or GPS, and just guessing where you are using the stars and sun. Despite best efforts, because of inaccurate data, one month can show that there is variance, as in Figure 10.2. Possibly worse is the amount of time the CPE must spend guessing what the actual numbers are. When the lead project manager knows the different timings, he or she can provide the options, and accuracy, to the customer and key stakeholders. Here again, the lead project manager needs to understand something about the internal processes of the organizations that are part of the CPE.

Generally, the key stakeholders will want an executive summary report that tells them the status of the project in one or two pages, but they will also need the detailed backup information for their internal teams. The key stakeholders will also know when they need the information. For example, on some projects, the key stakeholders may want a preliminary report for an internal executive meeting that occurs before the regular monthly report is issued. Obviously, the lead project manager needs to synchronize this with the value chain and partner, and must establish a way to explain any differences automatically. One point on communication needs to be made here. Never, ever, provide bad news in writing without explaining it verbally beforehand. Monthly reports should serve to formalize discussions with key stakeholders, not replace them. Be attentive to the issue of face, and never surprise a key stakeholder.

It is also quite possible that weekly, monthly, and quarterly reports may be required on the same project for different needs. Our recommendation is to take the time to map out the needs and synchronize the accounting clocks first. Then make sure that the reports build from weekly into quarterly, meaning that the quarterly report is a summary of the weekly and monthly reports. Stay with the same structure, just different levels of detail. Our experience has shown that this will save hundreds of hours of work and will avoid a lot of confusion when, inevitably, the weekly report conflicts with the quarterly report. Also, though it should go without saying, make sure that value chain and partner disputes are not just copied and forwarded. We are not saying they should be ignored; rather, if the lead project manager chooses to pass disputes along to the customer, he or she needs to add value by explaining them and providing an opinion—one that has been discussed with both parties in advance.

Roles and responsibilities. Permits, design, oversight, and the like. Normally, the overall roles and responsibilities are described in the legal conditions. For capital investment projects, the organization responsible for doing the design, acquiring the permits, constructing the project, testing the product, and operating the facility will be set forth. For humanitarian projects, the organizations responsible for public relations, customs, transportation, and so on would be identified.

Title. Refer back to Chapter 2. The risk of loss must be established on international projects, and it is normally included in the legal conditions of a contract.

Training provisions. Who, number, when? Training is always difficult because of the burden it places on the customer and the user. Their employees may be novice or experienced, accessible or inaccessible (no roads, no facilities, no Internet), they may need to be hired or may be on staff, they may be readily available or busy producing revenue. We recommend that the lead project manager understand the processes and conditions inside customer and user organizations before attempting to plan for training. The best time to understand is when the negotiations are ongoing because the financial changes caused by not understanding can be significant.

Disputes. How are disputes resolved? As noted, numerous legal systems exist, and it is critical that the lead project manager understand any laws that must be followed in the various countries. On disputes, it is a bit different. We recommend that disputes be dealt with in steps.

First, if a dispute occurs, the two project managers will attempt to resolve the dispute. If they cannot resolve the dispute within 15 days, then the two project sponsors will attempt to resolve the dispute. If the sponsors cannot resolve the dispute within 15 days then the dispute goes to the lead project manager, who attempts to facilitate an agreement within 15 days. If all of these attempts prove fruitless, the dispute goes to mediation using an internationally recognized organization, such as the International Chamber of Commerce (ICC) or the International Center for Dispute Resolution. Mediation is relatively inexpensive, quick (matter of days), and works about 95% of the time. If mediation does not resolve the dispute, then it would go to binding arbitration in Singapore, London, or New York. We prefer these locations because of the infrastructure available. Other countries can be used, but make certain they are signatories to the International Center for Settlement of Investment Disputes, which is charged with the task of resolving disputes between contracting states and nationals of other contracting states.

Know however, that just because an arbitrator in Singapore rules in favor of one party, that does not mean automatically that the court in, say, Uganda (signatory in June 1966) will enforce the award on a Ugandan business. We prefer to refer to arbitration as "l'arbitration" because it is becoming more expensive and procedural, like litigation. Just going through the process for large, complex disputes can take years and cost millions of Euros. Then there is the cost of the settlement itself. Binding arbitration means the parties agree in advance to accept the decision of the arbitrators. We recommend a panel of three arbitrators, where each party selects one arbitrator from a prescreened list, and the ICC assigns the "neutral" arbitrator, for example. Our recommendation is to resolve the disputes between the project managers or sponsors or through mediation.

Warranties. Duration and if they are "evergreen." Figure 11.2 shows the buyer's use of the product in operations and the seller's transfer of the product to warranty. The *PMOBK* does not appear to view the warranty period as part of the project; rather it considers that managing warranties is a part of the service provider's internal operations. In fact, many firms take this same view; for it is not effective to have project managers tied up for lengthy periods dealing with occasional warranty matters. Most firms have an established warranty group to manage this as an operational issue. Of course, the service provider has responsibility through the warranty period to the customer, but the operations group normally fulfills this obligation. The customer sees it as an integral part of the project.

An evergreen warranty is a warranty that never expires. For example, imagine that the product of the project is a power facility, and the service provider's warranty on the generator is 24 months as a standard offer. Some customers write the warranty clauses so that if the unit were to experience a problem at month 13, it would be repaired, and the warranty would reset itself to 24 months—thus evergreen. We recommend that warranty clauses be read carefully, and a conscious decision be made on their acceptability and pricing.

Proprietary materials. Patents and proprietary rights must be recognized and protected, ethically if not legally. The challenge is that enough information needs to be provided to the partner, customer, user, and value chain to enable them to do their work and to use the product, but while keeping the proprietary portions of the information secret. It is sometimes a difficult balance to strike, and requires a good understanding of the product and the information needs.

Precedence of documents. Most contracts include a clause that says something like "If a conflict is discovered in the contract documents, then the precedence of the documents shall be. . . ." Figure 11.1 is organized from top to bottom to reflect the precedence we have seen most often.

And more. To mention one more item, corporate social responsibility includes how a CPE deals with the public and the environment. The legal conditions need to set forth the responsibilities for these two issues. Who will be the public spokesperson for the CPE, the lead project manager or the customer? It is of course normally something the customer and user prefer to handle, but not always. On sensitive projects, customers sometimes prefer to have the lead project manager in front of the press. We can tell you from painful experience that it is best to be transparent, prepared, and honest when dealing with the media and the public. Many firms have tried to do otherwise, and many have suffered as a result. One of many such examples was the trouble Nike had with child labor.

The legal conditions are the rules for the project, and they may require more attention than do the technical terms. Make certain the conditions are flowed down to the value chain so that everyone is signed up to the same set of rules, as we mentioned earlier.

Having addressed the documentation, now let us look at the different types of contract structures.

11.4 STRUCTURES AND STRATEGIES

To begin, refer back to Figures 3.4, 5.1, and 6.1. Figure 3.4 provides a view of contracts from the perspective of the amount of collaboration between the parties. You will recall that it looked at the structure of contracts relative to the environment of the CPE and the participation level of the customer. The types of structures are adversarial, quasi-adversarial, collaborative, and quasi-collaborative.

Figure 5.1 provides a view of the different levels of intimacy in international agreements for joint ventures, alliances, and the like. These structures are all normally within the collaborative box of Figure 3.4. However, that does not mean they all have the same levels of cooperation or the same goals as we discussed. Figure 6.1 provides a view of bidding strategies, from fixed price to negotiated bids, in relation to customer participation and risk. It also suggests that when the scope is known in detail, a fixed price strategy may be appropriate, whereas a cost-reimbursable strategy would be better suited to a vague scope.

Now let us connect these variables, as shown in Figure 11.3. The vertical dimension is trust and the horizontal dimension commonality of goals, and their value is determined in part by the bidding strategy, whether competitive or negotiated. Their level is also determined by the relationship between the customer and the service providers, whether transactional, joint venture, consortium, partnership, or alliance. Their level is further determined by the certainty, or lack of vagueness, on the scope.

As we have said throughout this book, the level of trust in a CPE is critical. Here are the considerations for building and maintaining trust in a CPE:

- *Leadership.* The key to leading successful international projects is XLQ, and trust is the foundation. The more XLQ the lead project manager has, the better. Using a cost-reimbursable (CR) collaborative contract enables the lead project manager to engage early and often. Fixed price (FP) contracts do not mean the lead project manager cannot build trust; they do mean that it will happen later and

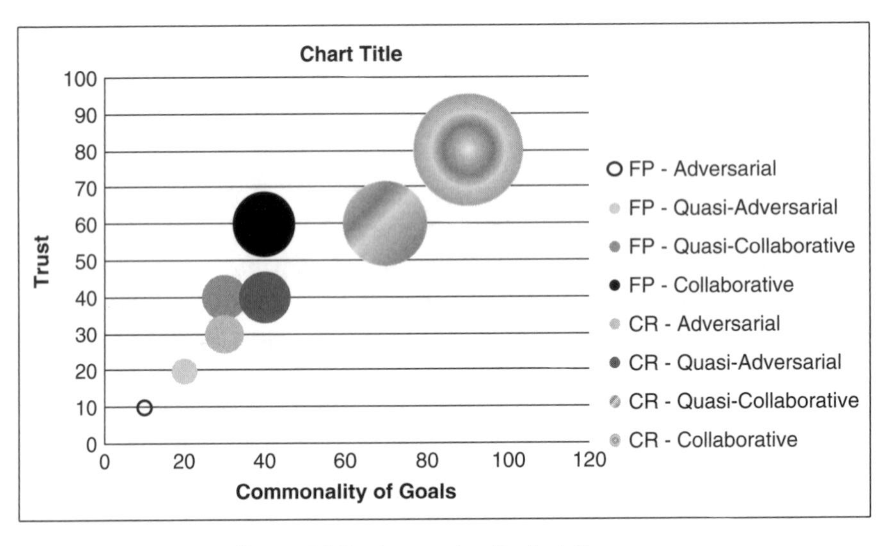

Figure 11.3 Improving Project Success

take longer. Therefore, a lead project manager with high XLQ can possible overcome the disadvantages of a FP contract. Conversely, a lead project manager with lower XLQ could be helped by a CR contract strategy.

- *Change*. The willingness to change depends in large measure on the trust that followers have in the leader. If a contract is adversarial, quasi-adversarial, or quasi-collaborative, the lead project manager will need higher levels of XLQ. Building a collaborative team will greatly enhance the probability of success on a project; anyone who has ever managed a project knows this. Building a collaborative team in an environment that is not collaborative in nature requires much higher levels of XLQ skills. The reason is that the courage to risk change must come first from the lead project manager.
- *Knowledge*. The willingness to share knowledge, not just information, openly and voluntarily demands that the followers trust the leader. This will overcome some of the resistance, but only some. In this case, the environment—adversarial, quasi-adversarial, or quasi-collaborative, collaborative—plays a very important part in the sharing of information and knowledge, as does the view of the organizations about whether the time horizon is short-term transactional or a long-term relationship. An adversarial structure sets a ceiling on the possible range of collaboration, for in many cases the parties are discouraged from communicating with the other participants in the project (Grisham and Walker 2008). Therefore, if 100% is possible, the environment could limit the maximum to 50%, and then XLQ could help keep this number from eroding further. The same rationale applies to the strategy, where similar limits would exist on permissible dialogue on an FP competitively bid contract.

  Sharing knowledge means that people may have to give up what they may perceive to be an advantage, whether personal or business. To do so, at a minimum, requires enough trust in the leader and other participants that the person sharing the knowledge will not be damaged. At the other extreme, it may be a feeling of wanting to damage others by withholding the information or knowledge. XLQ will only go so far in mitigating this issue, for many knowledge-sharing opportunities occur without the participation of the lead project manager.
- *Conflict*. The ability to resolve conflicts quickly and equitably depends in large measure on the level of trust in the leader and

other participants. As with knowledge, the lead project manager's ability to manage conflict will be restricted by the environment, strategy, and relationships of the organizations.

Our experience has shown that the quicker a leader can build trust in him or herself, the quicker trust will infect the CPE, and the stronger and more functional the CPE will become. A lead project manager with low XLQ can get lucky, and a lead project manager with high XLQ can be unlucky, so there are no guarantees. Figure 11.3 provides a relative ranking of the ways to maximize probable project success. As it shows, a CR—collaborative approach coupled with a lead project manager who has high XLQ will lead to project success more often. We have only our experience and our conversations with hundreds of international project managers to prove this—no hard empirical data. There are thousands of combinations of strategy, environments, relationships, XLQ, and complexity, and each one will have its own unique characteristics.

Figure 11.3 will, we hope, help you think about how to maximize success on your specific project. In general, we favor structures and strategies that enable more effective and voluntary communications and knowledge sharing, more equitable and inclusive attitudes toward CPE members, and an atmosphere of mutual respect and understanding. These provide a fertile field in which the seeds of trust may grow. In our experience, CPEs are organic so the metaphor is apt. For PPPs this is especially the case as they tend to span decades, not years, and they require a transparency that is completely different from more conventional project structures (Grisham and Srinivasan 2009). Lessons from the development of PPPs is, however, directly applicable to the growth of a CPE on conventional projects.

With the understanding that CPEs and PPPs require time to grow and develop, the next section addresses a sequence of bidding and negotiation that we suggest as a template for international projects.

11.5 BIDDING AND NEGOTIATIONS

There are many restrictions on the procurement of goods and services, and there is an increasing need for more transparency in both the private and the public sectors. In our experience from competitive bids opened formally and read in public, to handshakes for multimillion-dollar projects, negotiated agreements are a more equitable and efficient way to procure projects. The primary reasons are the following:

- *Risk*. In our experience as a service provider, the more we understood about the project, the better we could estimate the risks, and the more cost-effective bid we could prepare. In Chapter 8, we discussed the importance of assigning risk to the party in the best position to manage it. To do this requires discussion of the risk, the capabilities of the parties, and the least expensive way to deal with the risk.
- *Specifications*. Contract documents are always approximate and often are created by people who are not experts in the implementation of a project. This can result in imposed hindrances to means and methods, which will increase the cost. Specifications also can eliminate the opportunity for value engineering and again increase the cost of a project. It is far better to have an open exchange of ideas to make certain that the customer's goals are clear to the service provider and that the service provider can provide the customer with the benefit of its expertise in order to reduce costs.
- *Culture and relationships*. The time spent discussing the project is invaluable. It is an opportunity to implement the project virtually. The lead project manager can demonstrate her or his leadership by creating the norms and values that will carry through the entire project life cycle. It is a time to build trust, learn about the other personalities, work out communication challenges, get accustomed to working as a team, and discover where potential problems exist. In our experience, all CPEs must go through the stages described by Tuckman and Jensen (1977): forming, storming, norming, and performing. On international projects, the forming, storming, and norming can be accomplished mostly prior to the start of the implementation of the project (performing). In this way, the CPE culture can be established during the negotiations, so when it is time to implement, the CPE can be more efficient from the beginning—more on this in the next chapter.
- *Guidelines*. Control is required for the development and nurturing of alliances (Das and Teng 2001). Change is a given on international projects, and CPEs need some continuity if they are to accept the risk of change. Guidelines—some call it agile project management—can provide this continuity, so long as they are not dogmatic, rigid, and

excessively detailed. CPEs must adopt guidelines if the differences between the business techniques of each organization are to be respected and blended into a functional whole.

In fact, in our experience, one ancillary benefit of developing CPE guidelines is the critique of existing corporate processes. We had been dispatched to a problem project in Thailand to sort out the problems. The initial comments were that the partner did not understand project management. After spending time with the partner, we realized we were the problem. Our recommendation was to learn from that small Singaporean organization, and we integrated its techniques as best practices into our organization. If you think about it, what better way to build rapport than to use the most sincere form of flattery—imitation? To build the project virtually, the CPE must explore the details of how communications and knowledge will flow, how metrics will be established, and how the course will be set and checked. This practical level of interaction is essential to build the team.

- *Emotions*. Negotiations often expose the personal characteristics of the participants and their true intentions. Of course, those practiced in the art can cloak their emotions, but only for a time. Think back on our metaphor of comparing a project to a marriage. Imagine the difference in trust, understanding, and communications between a newlywed couple on their honeymoon and a couple who has been married for many years. Negotiations can be a virtual project crisis if not managed, as they expose how people will react when problems arise and when change occurs. It can be a wonderful opportunity to grow the CPE culture, but it can also be a disaster if not lead properly.

Many government agencies are bound to competitive bidding statutes, but there are ways to meet the requirements of such statutes while still permitting the building of relationships. A number of such processes have been used in the international markets.

We believe that the most important phase of a project is in the bidding and negotiations. On competitively bid projects, this is can be an opportunity lost. On a number of projects, we have been successful in working with the legal authorities in adjusting the procurement requirements synergistically. In doing so, the statutes are followed and some of the advantages of CPE creation can be realized. It requires added attention to transparency but can be accomplished even in very restrictive legal environments. In our experience, such an approach can produce less cost to the public, shorter project life cycles, higher quality, and fewer disputes. In short, it is well worth the effort. In developing countries, such synergy can provide a template for creating a set of statutes and procurement processes that can increase efficiency and transparency.

11.6 MONITORING AND CONTROL

The monitoring and control is part of each area that we have described, such as scope. In this book, the CPE subsumes the projects of each organization into a whole enterprise. There is one exception, however. As the project progresses, organizations may find the need to change their anticipated procurement plans. Scope that they had planned to provide with internal resources may need to be outsourced, and vice versa. Likewise, value chain organizations may need to outsource scope that they had planned to provide. It is very important to keep a complete listing of *all* organizations that are proving services. With a global value chain, organizations can be added that do not respect the same values and norms as do the other CPE members.

This can prove damaging if these organizations employ child labor, have a history of human rights violations, practice open bribery, and such. In the eyes of the customer, the lead project manager is responsible for these value chain organizations. As many others have learned, customers look to the point of contact as the responsible party for the actions of the entire CPE. Like branding, if the logo of a company is on a product, who made it and where are immaterial to a customer. The logo represents a level of quality and care, whoever had a part in create the product.

Chapter 12

CPEs in the Future

> I start with the premise that the function of leadership is to produce more leaders, not more followers.
>
> —*Ralph Nader*

> Great leaders are almost always great simplifiers, who can cut through argument, debate, and doubt to offer a solution everybody can understand.
>
> —*Colin Powell*

> In the course of history, there comes a time when humanity is called to shift to a new level of consciousness, to reach a higher moral ground. A time when we have to shed our fear and give hope to each other. That time is now.
>
> —*Wangari Maathai*

We have described the concept of a collaborative project enterprise (CPE) throughout the book and the idea that it must cross organizational boundaries on international projects. There are no good recognized terms to describe this in the industry except program management. So to connect the CPE idea back to recognized standards, let us have a look at how the Project Management Institute (PMI) and the International Project Management Association (IPMA) see program management.

12.1 PROJECT, PROGRAM, CPE

According to the *Project Management Body of Knowledge* (*PMBOK* 2004), a program is "a group of related projects managed in a coordinated way to obtain benefits and control not available from managing them individually." Also according to the *PMBOK* program management is "the centralized coordinated management of a program to achieve the program's strategic objectives and benefits." The *PMBOK* provides a graphical view of a program that looks something like Figure 3.1 in this book. The difference is that a program is a single organization in the *PMBOK*.

According to the IPMA:

> [A] programme of projects is put together to realize a strategic goal set out by the organization [CPE]. To achieve this, it initiates a group of interrelated projects [the organizations shown in Figure 3.1] to deliver the products/outcomes needed to attain this goal and it defines the organizational changes needed to facilitate the strategic change.

As we indicated in this book, cross-cultural leadership intelligence (XLQ) is required to lead such change, for in a CPE, the largest change is to facilitate people being comfortable and trusting enough to step outside of the boundaries of the organization that employs them. In this case, that is exactly the strategic change described by IPMA. Think of a public-private partnerships (PPPs) with a 30-year life cycle, and the need to build a long-term CPE. This is more easily accomplished because the horizon permits the organizational changes time to develop. On shorter cycle project of less than 4 years, the challenge for the lead project manager is much more acute; on extremely short projects, it is a superhuman challenge. As we have also discussed, the structure of the CPE can seriously help or hinder a lead project manager's ability to utilize all of his or her XLQ.

The IPMA continues by saying:

> [T]he programme defines business benefits management process as well as tracking the business benefits. The programme manager usually directs the projects through project managers, facilitates the interaction with line managers to realize the change and is responsible for benefits management; not for the realization of the benefits, which is again the accountability of line management.

We have discussed the need for transparency in the CPE. The business benefits are that the CPE, and all of its members, should be able to make a fair profit for their work. An effective CPE cannot function in an atmosphere of unbridled competition for profit. The lead project manager must work to establish a set of processes and metrics for the CPE and must control the CPE through individual project managers.

In Figure 3.1, the project manager for service provider B might well be a program manager in the view of his or her organization. For example, if service provider B divides internal functions into projects and utilizes multiple project managers who report to the project manager, this would then be a position of program manager according to the definitions above. IPMA considers benefits in an intra-organizational way; we are looking from an interorganizational point of view. We would broaden that definition to include XLQ—for example, empathy and transformation. Clearly, the lead project manager has no organizational authority over the employees of another organization. Yet we argue that by demonstrating an interest in each individual—"inspire achievement beyond expectations"—the lead project manager can infect the members of a CPE (infect them with the XLQ disease across organizational boundaries).

There is another difficulty confronting the lead project manager: politics. As shown in Figure 3.1, the lead project manager is often part of a service provider's team, so he or she may not be invested with the political power of a customer's representative. In addition, the user is often the strongest partner in a project or PPP, and as the figure shows, there may not be a contractual relationship between the lead project manager's organization and that of the user. In our experience, some users do not want any involvement during the time the project life cycle is ongoing. They have very heavy influence during the conceptualization of the project and particularly heavy participation during the preliminary acceptance of the asset. Look back at Figure 3.2 for a moment. The life cycle phases described are shown in Figure 12.1, but in miniaturized version.

Figure 12.1 displays first the conventional sequence for an overall project or program, depending on the definition you use. From the view of the customer, this is really more like a program since the customer will be involved with different groups, internally and externally, during the different segments of the life cycle for the asset. Imagine the project is the creation of a sewage treatment system for the city of Mumbai. If the city were the customer and user, the first sequence would be a likely series of segments with the planning and procurement group taking the planning and design phase, and the operations group taking the operations segment. Internally it could be seen as a program. If a consultant or joint venture performs the execution, they would see it as a project life cycle for execution. In the first row of Figure 12.1, the length of the arrows is intended to reflect the amount of time that might be involved. Of course, the operations would be very much longer, perhaps 20 years, but the concept should be clear.

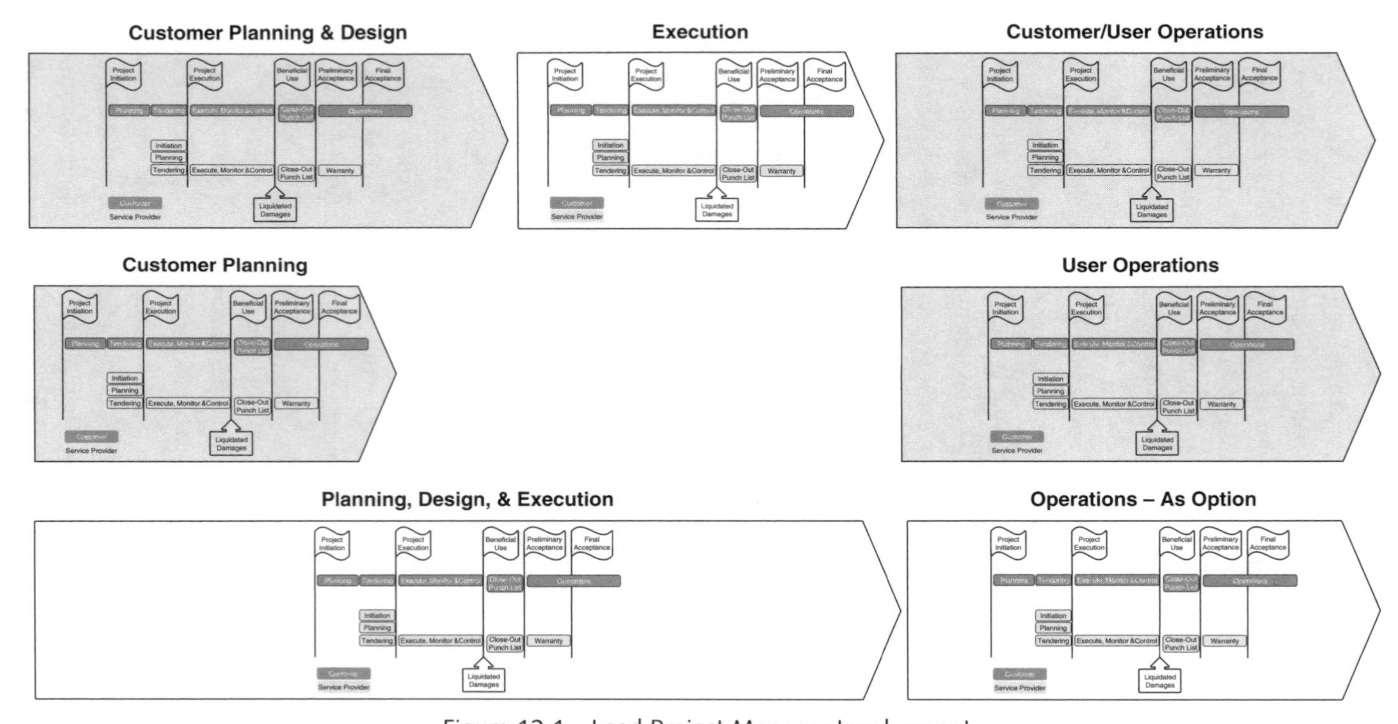

Figure 12.1 Lead Project Manager Involvement
A miniaturized version of the life cycle phases described in Figure 3.2.

The second row in Figure 12.1 shows a different strategy for creating the asset, which is to bring in a service provider for the planning, design, and execution phase. Then there would be a handover to the customer's operations group. Again, the views of the customer and consultant are much the same as the first row, but here the lead project manager for the consultant has a much longer participation in the project and more time to build the CPE.

The last row in Figure 12.1 shows the possibility of having the seller provide cradle-to-grave services. This is what a PPP would look like, where the PPP provides the design, funding, execution, and operation.

Therefore, perhaps the best way for those of you who are credentialed by PMI or IPMA to understand the concept of a CPE, as used in this book, is to think of it as a program. The adjustment needed is that the program manager—lead project manager in this book—is responsible to lead all of the organizations in the CPE. This is where the requirement for XLQ is most important. One of the largest challenges for most people is the fact that the lead project manager has no position power over any of the other organizations. Oddly enough, in many international organizations, project managers will have little or no position power over those who work in different profit and loss centers within their own organization. In this way, there are similarities between being a program manager in a large global firm and being a lead project manager for a CPE—at least in the way the PMI and IPMA describe project and program management.

First then, a lead project manager must have high XLQ. The next challenge will be his or her values. We have discussed the criticality of transparency throughout the book, and the ethical standards apply. Here again is PMI's mandatory standard:

> [W]e do not engage in or condone behavior that is designed to deceive others, including but not limited to making misleading or false statements, stating half-truths, providing information out of context or **withholding information that, if known, would render our statements as misleading or incomplete**. (Emphasis added)

If other organizations see the lead project manager as one who regularly withholds information, there will be no trust, and others will copy the behavior.

Imagine that the lead project manager has high XLQ and intends to divulge information that is needed by others, but in doing so will possibly reduce the profit margin of the firm he or she works for. The organization that employs the lead project manager may insist that such information not be divulged. In such a case, even a person with high XLQ will be unable to build the trust that he or she knows is required. It is not to say that such a person will be a failure but that he or she will be unable to build a culture of trust in the CPE, and that will potentially result in project failure. This issue of ethics is a very difficult one, because often personal ethics and corporate ethics are not the same.

Another common argument against transparency is that if we divulge, who is to say the other organizations will do so as well? If the other organizations withhold information, then the organization that chose to divulge could be at a competitive disadvantage and could realize a reduced margin on the project. We have been fed this logic many times and agree that it is possible. The difficulty with leadership is that you must display the behavior that you wish to imbue. As we indicated in the XLQ model, one aspect of trust is fearlessness, and this situation is one of the tests. Can a project succeed without trust? Possibly, but from our experience, only with luck on your side. You can get far better odds taking the money to Las Vegas or Monaco.

One other element is required alongside XLQ, and that is political astuteness. Imagine being the president of a country that needs to negotiate a trade agreement with 100 other countries, and think of this as a CPE for a moment. The lead negotiator has no position power over others; must convince others of her or his point of view; must accept that not everything desired by his or her country will be achieved; must not destroy the long-term relationship for some short-term gains; and must demonstrate the types of values and norms that the others will be inspired to follow. The analogy is a good one for a lead project manager. As Plato said, "Those who are too smart to engage in politics are punished by being governed by those who are dumber." Political ability is a skill that each lead project manager must cultivate. Politics is how groups make decisions in most cultures but not all.

12.2 CPE STRUCTURE

As we said earlier, there are many ways to structure a CPE, and they must be aligned with the statutes, laws, or policies of the organization(s) that are buying and providing the services. The needs and restrictions of the project must address each of these phases during the concept phase:

- *Concept and feasibility.* The need for a project, such as privatization of the country's water supply, may come from an intercountry evaluation or may be required by an external agency, such as the International Monetary Fund. Such a project will require technical, operational, political, and financial expertise. Imagine that the customer decides, or is forced to, purchase these services externally. There are now external consultants who will define the boundaries for the project, the value for money, and the possibly the structure of the project. Sometimes these participants are hired on a short-term basis, because of costs, and have no further involvement in the project. We think this is a mistake and that they should be retained as advisors at least through project design.
- *Structure.* Once the concept and feasibility have been determined and buy-in has been obtained, the method of procuring the project must be considered. Refer back to Figure 11.3. The project could be conducted as a PPP with the design, procurement, financing, execution, testing, training, and operations by a single contract entity (also called design-build-operate, turnkey, etc.). In this format, the customer would become a shareholder in the ownership of the project and likely would relinquish control to a lead project manager. In such a case, the structure might look much like that in Figure 2.5. In Figure 11.3, this would be a collaborative structure. One entity, the CPE, would be responsible for all activities from cradle to grave and would have a single, perhaps fully empowered, lead project manager.

  At the other extreme is an adversarial structure. In our example, the customer would choose to take the general information from the concept phase and move forward with a design professional to develop drawings, specifications, and tender documents. Subject matter experts (SMEs) may or may not have the expertise to provide the full range of services that are required, so they may subcontract portions of the effort, such as financial requirements, to other professional firms. The SME must also undertake to understand as much of the customer's explicit and implied needs as is possible within the budget. By the way, often SME services are procured on a negotiated basis rather than on a competitively bid basis so the relationship is more like a fixed price collaborative arrangement. The closer the relationship, and greater the trust, the better job a design professional can do to translate the needs of the customer into a design.

- *Design*. As noted, the design can be accomplished in a multitude of ways between the extremes that we described. Regardless of the structure chosen, the design effort will most often require multiple technical disciplines. Let us imagine building a government office building to illustrate the point. There would need to be geotechnical, structural, architectural, electrical, mechanical, heating and cooling (HVAC), fire protection, instrumentation, information technology (IT), hardware, and more. Information must be exchanged among these disciplines for them to accomplish their work, and it could be through a single entity that has overall responsibility for the project so that the technical needs can be checked to the overall implied and stated needs—the PPP-type approach.

  Alternatively, it can be accomplished through a design professional with no knowledge of the other major areas of consideration, such as finance and politics. A design professional would be a short-term relationship and limited in its scope. It is quite different from the long-term PPP approach. The short-term approach would also require the design professional to route information through the customer to the builder, and back again or to hire someone to manage the project (e.g. construction manager). In the adversarial model, this phase would be managed by the design professional.
- *Execution*. If an adversarial relationship was used, there would normally be a hand-off at this point. The customer would contract with a builder, to continue our example of the office building, and would turn over control to the builder. Depending on the contract, the designer's involvement would be limited or nonexistent during the execution, and the customer guarantees the accuracy of the design to the builder. In the collaborative structure, the hand-off is an internal one inside the PPP.

  Another difference between collaborative and adversarial is the interaction between design and execution during the building. It is not feasible for a set of drawings or specifications to be 100% complete in the details, and if so, the customer might forgo possible savings in the market (value engineering). In addition, it is common for design professionals to rely on the builder to complete the details of the design—structural steel detailing, for example. This is one reason why submittals are required during the execution phase. Likewise, any changes made during the execution phase may require the technical approval of the design professional team. In the adversarial mode, this phase normally would be managed by the builder or a construction manager.

- *Start-up or commissioning.* Most often, the commissioning of a new facility or IT platform would be performed as part of the service provider's scope of supply before the product is accepted. During this period, the user becomes more involved and often participates in the commissioning. This is also the time when classroom training for the user has been completed. Often customers do not want user participation in the design and execution process, for they tend to focus on operational details and design issues, such as access to equipment, or functionality, rather than the execution. We encourage user participation in the design phase to minimize surprises at acceptance. A project in Southeast Asia went to completion and user acceptance with a huge number of punch list items. One of them was the need for access to the opacity monitor on the top of the 50-meter stack. The customer had specifically excluded this functionality to save money, but the usuer absolutely needed it for operations.
- *Operation.* If the commissioning is performed as part of a PPP, the user or operations group normally has an earlier and in some cases a continuing role. Building participation into the design and execution helps to ensure that important assets are not missed, that the testing progresses with the execution, and that the durability of the asset is capable of meeting the expected product life cycle and cost profile.

If a governmental agency has strict competitive bidding statutes that require sealed bids and firm fixed price contracts, it is even more important to confront the realities of the phases just described. If the government agency is buying a turnkey project, the risk is that the scope is not adequately described to maximize financial efficiency. Simply, that means that the less you know about what you want to buy, the higher the price will be. Worse, a bidder may decide to low-ball a bid, meaning to submit an artificially low bid with the anticipation of making profit on the

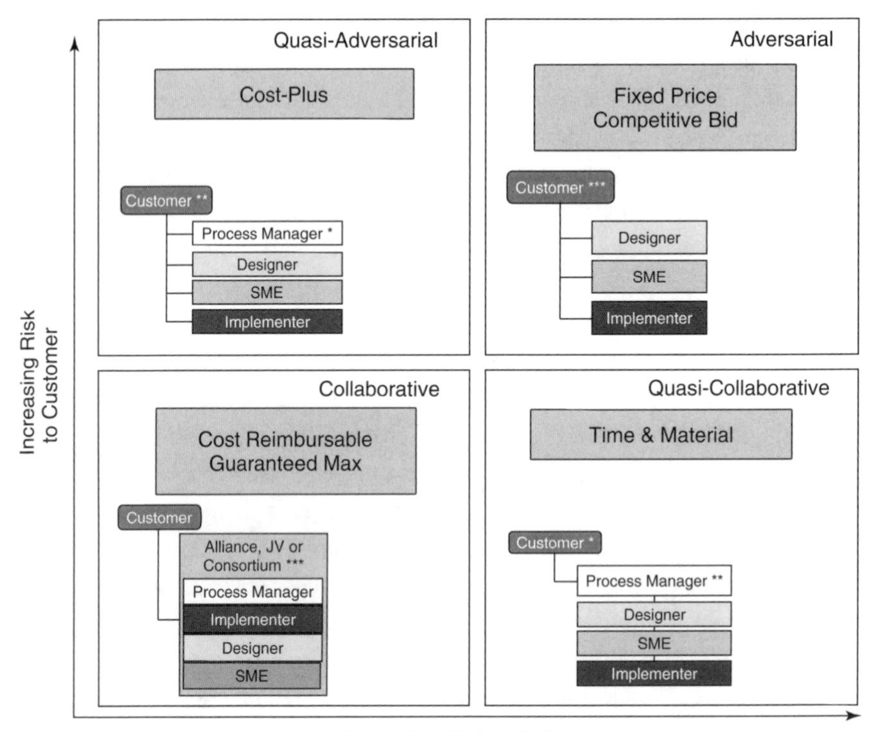

Figure 12.2 Contract Environments

changes. Figure 12.2 provides a view of the categories of envrionments available. The terminology is the same as shown in Figure 11.3, with the chart adjusted to show typical contract structures.

Customers need to be involved in projects, regardless of their structure. The reason is simple: They must help the lead project manager understand their implied needs, and this requires customer interaction throughout the project. In Figure 12.2, which is a copy of Figure 3.4, the SMEs might be financial experts for example, the implementer might be someone like Médecins sans Frontiers and/or an operations organization, and the designers design professionals as described above. The asterisks beside the project managers indicate the level of authority, and the asterisks alongside the customer indicate the level of involvement. The black connector lines are contracts, with the minimal number shown. Therefore, for an adversarial contract structure, there would need to be at least three contracts between the customer and service providers. In a collaborative structure, there would need to be one. In the adversarial structure, the customer functions as a project manager, although not in title, and often not in desire or intent. In an adversarial structure, the customer must ensure that communications occur between the parties. In a collaborative structure, the lead project manager shoulders this task.

Communications are critical, and customers must understand that their responsibilities for communications depend greatly on the structure. We have seen customers attempt to pass this responsibility off onto the design professional as a sort of clerk of the works. In such contracts, the design professional has no authority and just acts as a messenger, not as an agent. In these contracts, it is also typical to require the implementer to submit any questions to the design professional who will then reply. With no contractual agreement between the design professional and the implementer, there is no cause-and-effect reliance, and the design professional has only an advisory role.

This communication issue is shown graphically in Figure 12.3. The dark grey communication valves and filters are established, by contract, by the customer. The customer can open or close these as it sees fit. The organization filters and valves are under the control of each organization. The openness of these valves and the permeability of these filters are dependent on the culture of the organization and the attitude and XLQ of the project manager—assume that means the constructor for the model. It is possible to get lucky and have participants who want open the valves and clear the filters, but that requires taking a large risk by sharing information, voluntarily—the transparency issue.

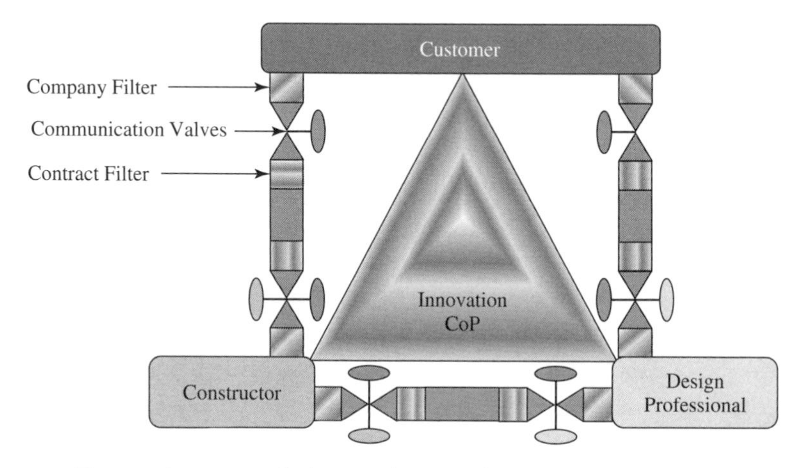

Figure 12.3 Knowledge Pipeline (Grisham and Walker 2008)

One frequent example under an adversarial contract is when a design professional makes a mistake. The constructor relies on the design, and the mistake results in the removal of work already completed. The constructor has no relationship with the designer and may not care for its approach, but the constructor must submit a change request to the designer as required in the constructor-customer contract. A design professional may be reluctant to send the problem only to the customer, because the customer will want to know the price of the fix. So imagine that the design professional must go to his or her value chain SME for structural design for a potential fix. Of course, the design professional wants to minimize the cost, since he or she ultimately may be culpable to the customer. Time is used up, perhaps weeks, and then it goes back to the contractor for pricing. Assume it takes a few days for the cost, then it may need to go back to the SME: another week, then the constructor, then finally to the customer. Imagine this first round of correspondence takes a month, and the customer has a question. The cost of the delay in the response frequently costs far more than the fix itself.

Consider a constructor who has taken a fixed price contract and discovered that he or she has underestimated the cost and one who finds that the personalities of the design professional and the constructor's project manager are incompatible. The constructor might likely be incentivized to withhold knowledge of the error until it is in his or her best interest to inflate the cost with delays—unfortunately, we have seen this more than a few times. It is a contract structure that is designed for potential failure, especially on international projects. Such a structure will likely eliminate the cooperation and innovation shown in Figure 12.3. If the communications are in fact reduced or eliminated, knowledge sharing will be nonexistent, and so will the opportunity for the CPE to possible avoid mistakes and reduce costs.

12.3 PROJECT MANAGEMENT INFORMATION SYSTEM

A lead project manager must devote the majority of her or his time to leading people. Fortunately, there is technology to help with the other tasks. This can be accomplished by enterprise software, project management information system (PMIS) software, or by using a library of software products. These products help to improve productivity on international projects by making information available on demand. Normally they are hosted on a single server with partitioning for individual organizations and a common knowledge area. One fairly recent PMIS comparison that we know was done by at the University of Osnabrück (Ahlemann and Backhaus 2006).

We were not satisfied with the products available, so a few years ago we developed PMIS architecture with a colleague. The concept was based on utilizing the technology that exists and a single server platform as described. The elemental component for the system was a detailed CPE work breakdown structure (WBS), digital design, bar coding, global positioning system (GPS), and of course a project plan. The concept was that all of the project documents would be available via a Palm Pilot equipped with location, communication, scanning, and photographic capabilities. Figure 12.4 provides a summarized view of the IPMS architecture.

The IPMS model is based on a WBS number that meets the criteria set forth in this book: One WBS = One Work Package, One WBS = One Schedule Activity. The concept is simple: Each WBS is located in three spatial dimensions back to the design and has its own unique barcode. Then scope, cost, time, quality, risk, communication, procurement, and human diversity can be related back to a single point. The details are beyond the scope of this work, but it is easy to see the idea of how the tasks would integrate. The idea is that a person could walk a construction project with a Palm Pilot in hand that had a database, communication, photo, barcode, GPS, and data input applications. It the person was, say, inspecting steam piping for an industrial facility, he or she could bring up the drawings on the Palm Pilot, from the GPS coordinates, for the proper routing then:

- Recall the last welding test inspection performed.
- Check the schedule for the planned status.
- Photograph the current installation.
- Scan the barcode on the welding rod to see that it has been properly stored and that it is the proper material.
- Enter a note into the project log (date, location, name, etc.) for the results of the inspection.
- Post the information on the server for open access by the CPE members.
- And more . . .

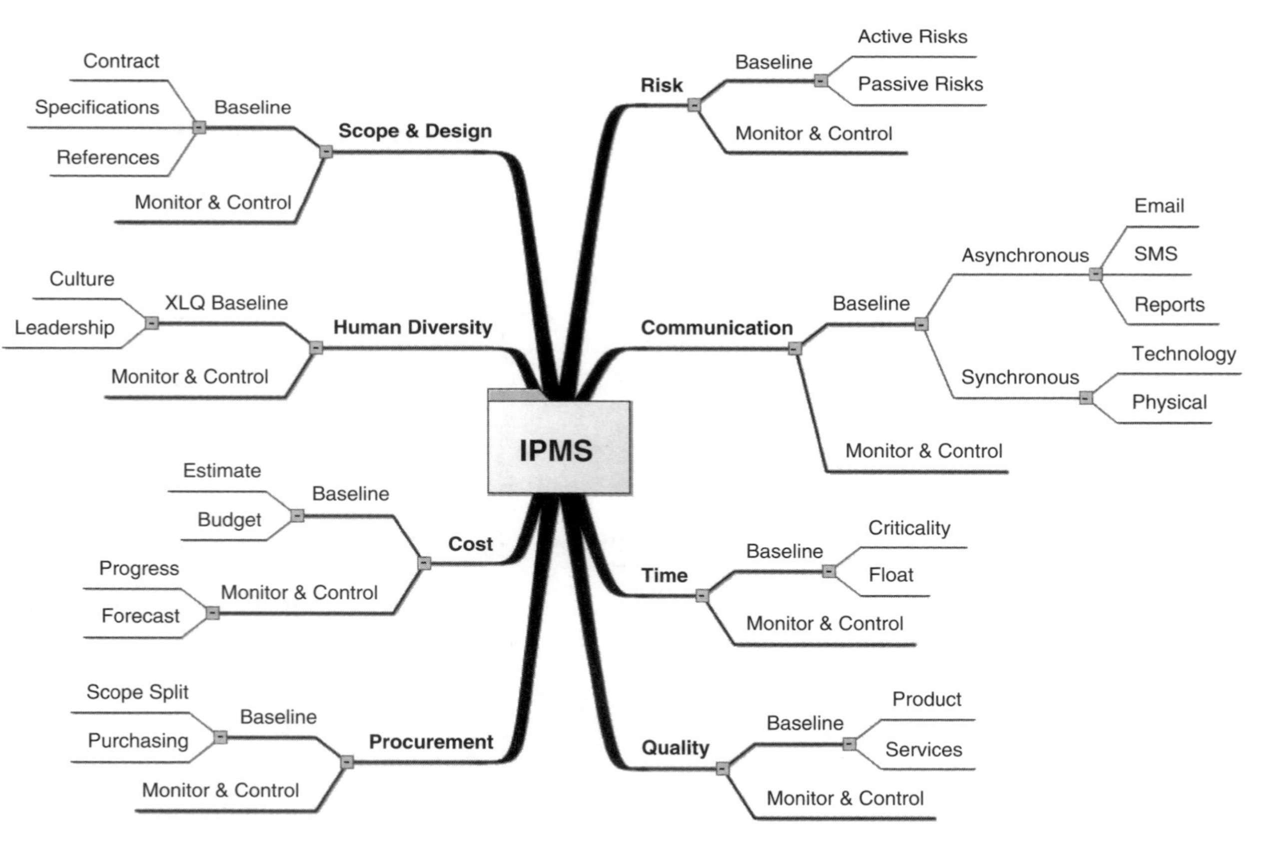

Figure 12.4 IPMS

It was designed to drastically reduce clerical time spent on projects, which can easily add up to three to four hours per day. It makes information available 24/7 and eliminates the need for hundreds of thousands of e-mails. It connects the procurement system to the payment system, to the bills of lading, and to the inventory of consumables on a site. These are a few of the savings on offer but by no means all.

The system was designed for use on construction projects but has obvious applications for most other types of projects. The IPCS model was designed to enable high-XLQ international project managers to leverage their abilities on multiple projects effectively. It would be especially useful on IT projects. Competition on international projects is based on high quality at low cost. To meet this demand, it is necessary to have the flexibility to apply high XLQ on projects that would normally not be able to afford them. It also permits the use of a large, and growing, population of 60+-year-olds who will be retiring in the next few years with huge amounts of experience but little interest in moving.

We are working on developing the model and hope to have it operational within the next few years.

12.4 INTERNATIONAL PROJECT MANAGEMENT AND BUSINESS IN THE TWENTY-FIRST CENTURY

Peter Drucker said that there are five certainties for the twenty-first century: aging populations in the developing world, changes in the distribution of income, the need to redefine performance, global competition, and the divergence of globalization and politics (1999). Certainly all of these certainties are, well, certain. Globalization, and the competition it engenders, is driven by consumers who want high-quality products at low prices, and by local markets desiring to participate. The demands of people who want a better life conflict with governmental policies that are unable, or unwilling, to adapt. The current mediator is business, the entity least capable of bridging the gap and the least willing to do so.

Drucker also said that organizations should be considering the shape of the future corporation, including people policies, external information, and change agents (2002). He was also clear that income disparity above a ratio of 20-1, management to workers, is morally unforgivable. Karl Marx argued that capitalism will produce internal tensions that will lead to its destruction. Marx argued that just as capitalism replaced feudalism, so will socialism replace capitalism and lead to a stateless, classless society, or a society where radical income disparity is unacceptable.

Drucker also wrote often about knowledge workers and said that these people will need to know the tasks and processes (project management), have autonomy (power), produce high quality at a low price (communication and transformation), be considered an asset (trust and empathy), and be engaged in continuous learning (change) (1999 and 2002). The items in parentheses come from the conjunction of the XLQ model, project management structure, and the need for continual change.

What will these knowledge workers want in the twenty-first century? The answer is much of what people have always yearned for, with education playing a central role: health, sustenance, shelter, safety, clean environment, laws, education, income. What will business want in the twenty-first century? Business will have amazing opportunities and face huge challenges: globalization, politics, social constraints, knowledge, profitability. The question is: Are these needs in conflict, are there opportunities for fulfilling both in a symbiotic future? Government currently plays a role as well, but as the century wears on, that role will change.

12.4.1 Visions of Twenty-First-Century Society

Nations exist to distinguish themselves from others and to define the differences in culture, values, norms, and laws. Nations also exist to define physical boundaries and thereby natural resources. According to Michael Porter (1998), the competitive advantage of a nation was based on factor and demand conditions, related and supporting industries, and the strategy, structure, and rivalry of the business. The factor conditions includes the natural resources, both physical and societal—countries that have copper can use this physical resource to compete; countries that have educated people can use this resource to compete.

Nations must attend to the needs of their people, for laws, security, health, the environment, infrastructure, education, and other services. Nations are looking more to business to provide most of these services, except laws. The International Monetary Fund has long insisted on privatization as a means of improving economic efficiency in developing countries. Now developed countries are increasingly using businesses to outsource a variety of services from healthcare administration to prisons.

The reasons are economic, but the execution needs to be humane. This is one factor driving the increase in corporate social responsibility.

As nations devolve responsibilities, businesses will be challenged to understand the national cultures and to offer services that are respectful and responsive to citizens' changing needs. This is what transnational firms see as a local face to a global service. Said another way, it is a local customization of a global product, such as Coca-Cola. Also, as the government of Peru discovered, when it privatized its water services, the cost must be within the reach of the citizens. Citizens who were paying US$0.45 with the government subsidies were paying US$3.22 after privatization, and this proved unsustainable. Conflict will occur when perfect efficiency meets subsidies. Businesses must understand the societal, environmental, and economic conditions of a nation. They will need the power of a local heart.

Governments are struggling to keep pace with emigration and immigration. Drucker (1999 and 2002) pointed to the changes caused by demographic trends, and the pendulum of isolationism will swing back toward more permeability of borders. The people emigrating will need to understand the culture of the society if they are to be successful. Governments will continue to put their metaphorical fingers into the dike of globalization, to maintain tax revenues to fuel their economies, and to maintain their services: healthcare, sustenance, shelter, safety, the environment, laws, and of course education. Transnational firms have business in multiple countries, and their profitability derives in part from tax advantages globally.

Businesses will be the conduit for societal change, scary thought that may be. Business is global and leverages its competitive knowledge advantage to increase market share in a larger global market. Businesses must understand cultures, markets, governments, and resources to survive. In doing so, businesses must adapt to laws, policies, and norms that they cannot change and lobby to adjust those that they can. This interaction changes both a business and the nation, sometimes symbiotically, sometimes not. For the twenty-first century, this dance will yield more sustainable results; it must. Businesses listen, question, think, and then act, transferring information into knowledge. Businesses will need knowledge workers who are equally nimble and accepting of change.

Businesses will be a surrogate for government. Those people who work with a global firm will see the benefits that are available in the developed work and the more developed world. The *Economist* produced a Quality of Life Index ("Intelligence Unit's quality-of-life index", 2005) that measured the yearnings of knowledge workers and more. Drucker said income disparity will cause social changes, and we fully concur. Knowledge workers see the benefits and disadvantages in other societies through the lens of business, which will change the view already gained through global social networking. Incomes need not be equal—that is not what Drucker was saying—but rather equitable.

Businesses will also provide the opportunity for knowledge workers to live in their desired nation and have an income adequate to mitigate government shortcomings. Knowledge workers will have connectivity, comparatively high income, and an understanding of the world. The Internet and income will build a platform for the necessary lifelong learning for knowledge workers and for their children.

Then there is the environment. The twenty-first century will see the change from conspicuous consumption, to measured conservation and protection. But this will take until the end of the century. In the interim, citizens and business will bear the burden. Businesses will be viewed as having deep pockets and a direct responsibility for pollution of the planet. After all, no business, no pollution. There will be national pressure to deal with local pollution, and these pressures will be focused on business. Pressure to deal with global warming, eradication of species, and extreme weather will focus more on government and international institutions. When sea rise displaces millions, business will be asked to mitigate the problems.

Then there are the knowledge workers: persons under the age of 30, who have been playing with technology since age 6, master's degree in a discipline, speak two or more languages, have a social network in a dozen countries, want an income that will support their view of a quality life, and are willing to work the hours necessary. They are likely born in one country and raised in multiple countries. They are curious, in need of acknowledgment, and on-demand connectivity to family and friends. They are people who are not afraid of change but will accept it only if they have a safe and stable platform from which to venture out.

Knowledge workers will live where they choose, and many will remain with or in proximity to their family. Some will choose to explore the planet, living in many locations. They will expect that businesses and governments either support them in acquiring their needs or at least not be an impediment. On this score, they will become more involved politically and socially than their parents and will take to the streets if necessary. Look to the elections in Iran for an example.

Knowledge workers will be a diverse group that will prefer to work virtually. They will be able to lead or follow, comfortably, and will be capable of performing more than one function for the organization. Businesses will adjust to these preferences out of necessity and will be challenged to guide flatter organizations with few core employees and many fluid associates. These organizations will utilize a blend of the young and old knowledge workers, energy and wisdom. As a result, businesses will be more fragmented, more nimble, and, without leadership, more chaotic. But one profession offers an option for the transition in the twenty-first century: project management.

12.4.2 Business and International Project Management

Nimble businesses must have a culture of change, built on a foundation of leadership and trust. Nimble businesses will be forced to see into the future, anticipate market changes, and redefine their strategy. Core team leaders must first fully understand the vision and communicate it effectively to the fluid and core knowledge workers. Businesses will adjust the number of their employees to follow the markets by utilizing the fluid workforce. The result will be rapid swings in the number of resources available in any given month. Such changes will require short-term goals and priorities to be planned and then adjusted frequently.

Businesses will have virtual knowledge workers in different time zones. These workers must understand the new plan and priorities, what role they are required to fulfill, and who they will be leading or following. Businesses will also have knowledge management systems that enable them to find the expertise they require.

The discipline of international project management holds the offer of a platform that can guide change with a globally dispersed virtual workforce. At its best, project management reflects a codification of business thinking and concepts into practice. When businesses adopt a project management approach, they employ guidelines that can be utilized in multiple countries and cultures, with a culturally diverse and dispersed workforce. As Drucker and many others have said, to have a culture of change requires a rock of security. People need stability and the ability to see the ground.

Flexibility and customization will be the keys to success. The rock is in part represented by guidelines that are the same in Seoul, Bogota, Kinshasa, New York, Mumbai, Moscow, Beijing, and Tokyo. Imagine that last month a team from these countries was assembled to work on a project in Milan. The product of their endeavors will be customized to the local customers, so there will be change from the previous project in San Francisco. But the process guidelines will be the same. Some stability; not too much. Also, in this way, as the team size fluctuates, the new members who enter the project already know the process, so they can focus on learning the team members and the customer. It saves time—costs down, quality up.

Knowledge workers require a continuous flow of information and the time to convert it into knowledge. If communications, planning, estimating, risk, procurement, and quality have guideline processes, the team members can again save time by not creating ad hoc operating conditions each time a project is undertaken. In virtual teams, information and structure are simply critical if efficiency is to be achieved. Imagine having a virtual team member come through a project for a week to perform a certain task. If she is, say, 10 time zones away, it could take a week just to connect her to the information highway, introduce her to the team, describe the project, and find the virtual facilities. Four days piddling and one day working will not build an efficient, nimble global organization. International project management can provide this rock. We see it as a tool utilized by leaders with high XLQ in geographically dispersed, fluid teams of knowledge workers who interact in a virtual work environment.

Processes enable knowledge, which often springs from mistakes, to be harvested from lessons learned, and metrics. Metrics enable a kaizen process, which enables a reduction in cost and an increase in quality. Project management holds the hope for a quality product or service, without the cost of a full total quality management or six sigma undertaking.

Drucker saw the trajectory of business in the twenty-first century. The enhancements we have suggested are a more detailed look at these trends. We have trained many people internationally who work in global and transnational businesses. Most of them are enhancing their project management skills at the demand or request of their employers. Many tell us that their businesses are striving to implement project management processes globally. All are seeking ways to reduce the time that they spend working so that they can spend more time enjoying family, friends, and avocations. A majority work in virtual teams, and they come from dozens of different countries.

12.5 CONCLUSION AND GRISHAM'S LAWS

As we said earlier in the book, transparency is essential: to build trust, to improve communications, to leverage knowledge, and so much more. It is like recognizing that float exists in a schedule; few are willing to do so for it takes a leap of faith on some projects. Transparency also can put an organization at risk if the other participants in the project are predatory in nature. However, without transparency, success is not probable. We also made the point earlier that risks must be discussed openly, taken by the organization in the best position to manage the risk, and managed as a WBS item.

We recommend that lead project managers keep the next overview in mind when building a successful international CPE. It is from the viewpoint of optimum effectiveness and the ability to turn aspirations into reality—the perfect project, if you would. The prescription is:

- *Lead project manager*. Select a person with high XLQ, a minimum of 10 years of experience in the type of work being undertaken (power, water, pharmaceutical, medical, capacity building, etc.). Select a person who is compassionate, enthusiastically curious, adaptable, and has the patience of Job. Select someone who also is effective in political environments. Make certain that the person on the proposal is the one who shows up to be the lead.
- *Structure*. Use a collaborative structure and make certain that plenty of time is spent selecting the organization that will lead the CPE. Preapprove a short list of possible organizations based on their abilities (financial, experience, technical, etc.), depth of staff, methodology, and references. Tendering is expensive; by discussing the approach to the project first, the potential sellers can save large amounts of costs. Once a short list has been prequalified, select a number of firms to submit value for money proposals for PPPs or financials for other projects. Then negotiate in good faith and with enough transparency to assure the other bidder(s) that it is being dealt with fairly.

- *Contracts*. Fight the urge to tilt the contract too far in your favor if the customer. The goal should be fairness and equity. Make the discussions about risk, disputes, costs, time, quality, communications, insurance, and politics open, detailed, and transparent. Attorneys are trained to be champions and advocates, and customer may need to reign in their aggressive pursuit of winning. In our experience, it is best to have attorneys provide an outline of topics that need to be addressed and then commemorate what has been agreed to in understandable language. It is also wise to bar attorneys from meetings until their expertise in writing a contract can be productive. We are by no means anti-attorney, but like all professions, attorneys should add value when it the timing is right and when they do not disrupt the process.
- *Do as I do.* Treat other organizations and people the way you would want them to treat you. The golden rule is a norm recognized in many religious traditions, so most people will recognize it regardless of their parent culture (Harris, Pritchard et al. 2000):

  Buddhist version. Hurt not others with that which pains yourself *(Udanavarga,* v. 18).

  Christian version. Treat others as you would like them to treat you (Luke 6:31, New English Bible).

  Confucian version. Do not do to others what you would not want them to do to you (Analects, Book xii, #2).

  Hindu version. Let not any man do unto another any act that he wisheth not done to himself by others, knowing it to be painful to himself (*Mahabharata*, Shanti Parva, cclx.21).

  Jewish version. What is hateful to yourself do not do to your fellow man. That is the whole of the Torah (Babylonian Talmud, Shabbath 31a).

  Muslim version. No man is a true believer unless he desires for his brother that which he desires for himself (Hadith, Muslim, imam 71–72).

In conclusion, we offer in Figure 12.5 Grisham's laws of international project management. We have discussed most of them specifically, and we hope you find them useful. They come from experience and have served us well throughout our career. The selection of the photograph was purposeful as it shows water in its three states, solid, liquid, and

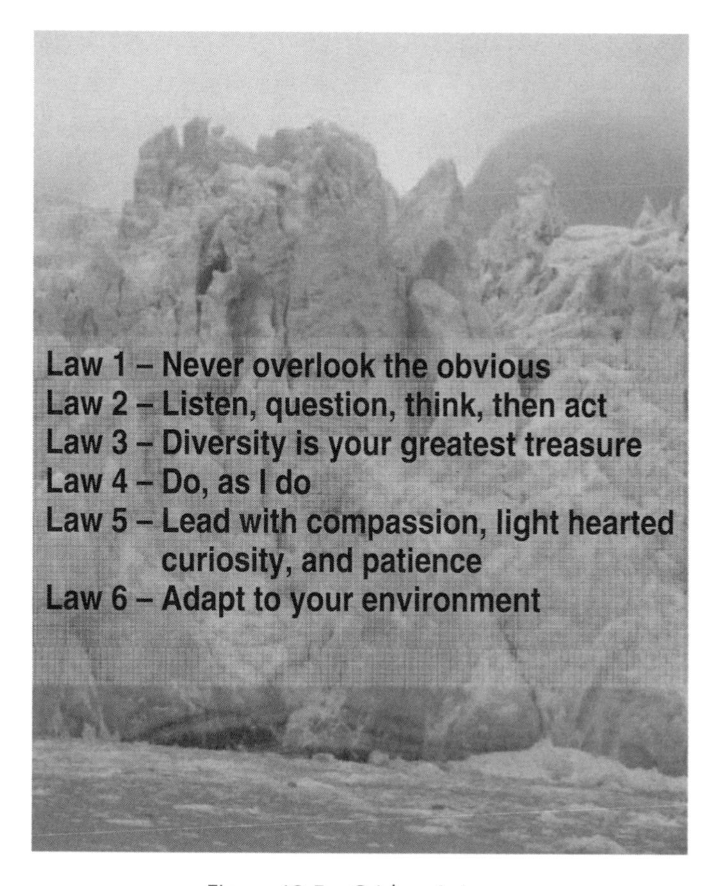

Figure 12.5 Grisham's Laws

gas. One more quotation from Lao Tsu that puts my laws into an ancient context (Hendricks 1989) page 49:

In the whole world nothing is softer and weaker than water.
And yet for attacking the hard and strong, nothing can beat it,
Because there is nothing you can use to replace it.
That water can defeat the unyielding –
That the weak can defeat the strong–
There is no one in the world who doesn't know it

References

Ahlemann, F., and K. Backhaus. "Project Management Software Systems Requirements, Selection Process and Products," Research Center for Information Systems in Project and Innovation Networks, 2006.

APM. *Association for Project Management Body of Knowledge*, 5th ed. Frome, Somerset, UK, Butler and Tanner, 2006.

Bartlett, C. A., and S. Ghoshal. *Managing across Borders: The Transnational Solution*. Boston: Harvard Business School Press, 1998.

Bass, B. M. *Leadership and Performance Beyond Expectations*. New York: Free Press, 1985.

Bass, B. M., and R. M. Stogdill. *Bass & Stogdill's Handbook of Leadership: Theory, Research, and Managerial Applications*. New York: Free Press, 1990.

Bigley, G. A., and K. H. Roberts. "The Incident Command System: High-Reliability Organizing for Complex and Volatile Task Environments," *Academy of Management Journal* 44, no. 6 (2001): 1281–1299.

Blaken, A., and G. Dewulf. "Financial Crisis Show Stopper for DBMFO Projects in the Netherlands." Presented at Revamping PPP's Symposium, University of Hong Kong, 2009.

Bower, D. C. "New Directions in Project Performance and Progress Evaluation." M.A. Thesis, 2007, http://adt.lib.rmit.edu.au/adt/public/adt-VIT20080130.135140.

Brown, J. S., and P. Duguid. "Organizational Learning and Communities of Practice: Towards a Unified View of Working, Learning, and Innovation," in *Organizational Learning*, ed. M. D. Cohen and L. S. Sproull. Thousand Oaks, CA: Sage, 1996.

Buzan, T., and B. Barry. *The Mind Map Book*. New York: Plume, 1993.

Carroll, A., and M. Meeks. "Models of Management Morality: European Applications and Implications," *Business Ethics: A European Review* 8, no. 2 (1999): 108–116.

Cavaleri, S., and S. Seivert. *Knowledge Leadership: The Art and Science of the Knowledge-Based Organization*. Jordan Hill, Oxford, Elsevier Butterworth Heinemann, 2005.

Chomsky, N. *Language and Problems of Knowledge*. Cambridge, MA: MIT Press, 1988.

Crossan, M. M., H. W. Lane, et al. "An Organizational Learning Framework: From Intuition to Institution," *Academy of Management Review* 24, no. 3 (1999): 522–537.

Das, T. K., and T. Bing-Sheng. "Building Confidence within a Strategic Alliance," *Chemtech* 29, no. 3 (1999): 13.

Das, T. K., and B.-S. Teng. "Trust, Control, and Risk in Strategic Alliances: An Integrated Framework," *Organization Studies* 22, no. 2 (2001): 251.

Davenport, T. H., and L. Prusak. *Working Knowledge: How Organizations Manage What They Know*. Boston: Harvard Business School Press, 1998.

DeFillipi, R. J., and M. B. Arthur. "Paradox in Project-Based Enterprise: The Case of Film Making," *California Management Review* 40, no. 2 (1998): 125–139.

Denning, S. *The Leader's Guide to Storytelling: Mastering the Art and Discipline of Business Narrative*. San Francisco: Jossey-Bass, 2005.

Drucker, P. F. *Management Challenges for the 21st Century*. New York, HarperCollins, 1999.

Drucker, P. F. *The Age of Discontinuity: Guidelines to our Changing Society*. New Brunswick, NJ: Transaction Publishing, 2000.

Drucker, P. F. *Managing in the Next Society*. New York: Truman Talley/St. Martin Press, 2002.

Earley, C. P., and S. Ang. *Cultural Intelligence: Individual Interactions across Cultures*. Stanford, CA: Stanford University Press, 2003.

Ekman, P. *Emotions Revealed: Recognizing Faces and Feelings to Improve Communication and Emotional Life*. New York: Times Books, 2003.

Elkington, J. *Cannibals with Forks: The Triple Bottom Line of 21st Century Business*. Oxford: Oxford University Press, 1998.

Emhjellen, M., K. Emhjellen, et al. "Cost Estimation Overruns in the North Sea," *Project Management Journal* 34, no. 1 (2003): 23.

Ferrell, O. C., J. Fraedrich, et al. *Business Ethics: Ethical Decision Making and Cases*, 5th ed. Boston: Houghton Mifflin, 2002.

Fitzgerald, J. P., and C. F. Duffield. "Value for Money in a Changing World Economy," Presented at Revamping PPP's Symposium, University of Hong Kong, 2009.

Fleming, Q. W., and J. M. Koppelman. *Earned Value Project Management*. Newton Square, PA: Project Management Institute, 2000.

Fromm, E. *To Have or to Be?* New York: Harper & Row, 1976.

Gannon, M. J. *Understanding Global Cultures—Metaphorical Journeys through 23 Nations*. Thousand Oaks, CA: Sage, 2001.

Ghoshal, S. "Bad Management Theories Are Destroying Good Management Practices," *Academy of Management Learning & Education* 4, no. 1 (2005): 75–91.

Gibson, C. B. "Do You Hear What I Hear? A Framework for Reconciling Intercultural Communication Difficulties Arising from Cognitive Styles and Cultural Values," in *New Perspectives on International Industrial / Organizational Psychology*, ed. C. P. Earley and M. Erez. San Francisco: New Lexington Press, 1997.

Gladwell, M. *Tipping Point—How Little Things Can Make a Big Difference*. New York: Little, Brown, 2000.

Global Reporting Initiative. *Sustainability Resporting Guideline*. Boston: Author, 2002.

Goldratt, E. M. *Critical Chain*. Great Barrington, MA: North River Press, 1997.

Grabher, G. "Cool Projects, Boring Institutions: Temporary Collaboration in Social Context," *Regional Studies* 36, no. 3 (2004): 205.

Greenleaf, R. *Servant Leadership*. New York: Paulist Press, 1997.

Grisham, T. Cross Cultural Leadership. Melbourne, Australia, RMIT University: 323. 2006.

Grisham, T. *Cross-cultural Leadership (XLQ)*. Germany, VDM Verlag, 2009.

Grisham, T. "Do Like I Do: XLQ and Behavior," Presented during Project Management Days, PMTAGE, Vienna, Projekt Management Group, 2007.

Grisham, T. "Global Project Management Communication Challenges & Guidelines," in *PMI '99 Seminars & Symposium*. Philadelphia: PMI USA, 1999.

Grisham, T. "Metaphor, Poetry, Storytelling, & Cross-Cultural Leadership," *Management Decision* 44, no. 4 (2006): 486–503.

Grisham, T., and P. Srinivasan. *Designing Communications on International Projects*. Cape Town, South Africa: CIB World Building Congress, 2007.

Grisham, T., and P. Srinivasan. "PPP's as Temporary Project Organizations." Presented at Revamping PPP's Symposium, University of Hong Kong, 2009.

Grisham, T., and D. H. T. Walker. "Communities of Practice: Techniques for the International Construction Industry." Presented at the first International Conference on Information and Knowledge Management, Lisbon, Portugal, CIB, 2005.

Grisham, T., and D. H. T. Walker. "Developing Communities of Practice for International Construction Organizations." In *Enhancing Knowledge Management in Construction Organizations: A Practitioner Guide*, 2008.

Grisham, T., and D. H. T. Walker. "Nurturing a Knowledge Environment for International Construction Organizations through Communities of Practice," *Construction Innovation Journal* (2005).

Grun, B. *The Timetables of History*. New York: Simon & Schuster, 1982.

Gyatso, T. "Compassion and the Individual." Accessed April 6, 2009, from www.dalailama.com/page.166.htm.

Harkins, P. J. *Powerful Conversations: How High Impact Leaders Communicate*. New York: McGraw-Hill Professional, 1999.

Harris, C., M. Pritchard, et al. *Engineering Ethics: Concepts and Cases*. Wadsworth, 2000.

Hastings, C. "Building the Culture of Organizational Networking: Managing Projects in the New Organization," *International Journal of Project Management* 13, no. 4 (1995): 259–263.

Hauser, M. D. *Moral Minds—How Nature Designed Our Universal Sense of Right and Wrong*. New York: HarperCollins, 2006.

Hendricks, R. G. *Lao-Tzu Te-Tao Ching*. Toronto: Random House, 1989.

Herzberg, F., B. Mausner, et al. *The Motivation of Work*. New York: John Wiley & Sons, 1959.

Hofstede, G. *Culture's Consequences: International Differences in Work-Related Values*. Beverly Hills, CA: Sage, 1984.

Honoré , C. *In Praise of Slowness—How a Worldwide Movement Is Challenging the Cult of Speed*. San Francisco: HarperSan Francisco, 2004.

House, R. J., P. J. Hanges, et al., eds. *Culture, Leadership, and Organizations—The GLOBE Study of 62 Societies*. Thousand Oaks, CA: Sage, 2004.

Hoyt, J. Panelists' Remarks, Environmental Ethics Congresses, Washington, DC, World Bank. 1998.

Iacoboni, M., I. Molnar-Szakacs, et al. "Grasping the Intentions of Others with One's own Mirror neuron System," *Plos Biology* 3 (2005): 79 (electronic).

ICDR. *ICDR Guide to Drafting International Dispute Resolution Clauses*. New York: American Arbitration Association, 2007.

"Intelligence Unit's quality-of-life index". *The Economist*, 2005.

IPMA. *ICB-IPMA Competence Baseline, Version 3.0*. Nijkerk, Netherlands, International Project Management Association, 2006

Ismail, K., R. Takim, et al. "The Malaysian Private Finance Initiative and Value for Money." Presented at Revamping PPP's Symposium, University of Hong Kong, 2009.

Jarvenpaa, S. L., and D. E. Leidner. "Communication and Trust in Global Virtual Teams," *Journal of Computer-Mediated Communication* 3, no. 4 (1998).

Jensen, C., S. Johansson, et al. "Project Relationships—A Model for Analyzing Interactional Uncertainty," *International Journal of Project Management* 24, no. 1 (2006): 4–12.

Johnson, S. C. "Detecting Agents." *Philosophical Transactions of the Royal Society of London* B, no. 358 (2003): 549–559.

Kerzner, H. *Project Management: A Systems Approach to Planning, Scheduling, and Controlling*. Hoboken, NJ: John Wiley & Sons, 2006.

Kholberg, L. "Stage and Sequence: The Cognitive Development Approach to Socialization," in *Handbook of Socialization Theory and Research*, ed. D. A. Goslin, pp. 347–480. Chicago: Rand McNally, 1969.

Kovecses, Z. *Metaphor in Culture—Universality and Variation*. Cambridge: Cambridge University Press, 2005.

Krajewski, L. J., and L. P. Ritzman. *Operations Management—Processes and Value Chains*. Upper Saddle River, NJ: Pearson Prentice-Hall, 2005.

Kübler-Ross, E. *On Death and Dying*. New York: Scribner Classics, 1969.

Larson, E. "Project Partnering: Results of Study of 280 Construction Projects," *Journal of Management in Engineering—American Society of Civil Engineers / Engineering Management Division* 11, no. 2 (1995): 30–35.

Lave, J., and E. C. Wenger. *Situated Learning—Legitimate Peripheral Participation*. Cambridge: Cambridge University Press, 1991.

Lawrence, T. B., M. K. Mauws, et al. "The Politics of Organizational Learning: Integrating Power into the 4I Framework," *Academy of Management Review* 30, no. 1 (2005): 180–191.

Lee, C. H., and K. J. Templer. "Cultural Intelligence Assessment and Measurement," in *Cultural Intelligence: Individual Interactions across*

Cultures, ed. P. C. Earley and S. Ang. Stanford, CA: Stanford Business Books, 2003.

Lester, A. *Project Management, Planning and Control*, 5th ed. Oxford: Butterworth-Heinemann, 2007.

Managing Change and Transition. Boston: Harvard Business School Press, 2003.

Maslow, A. H. "A Theory of Human Motivation," *Psychology Review* 50 (1943): 370–396.

Mayer, R., J. H. Davis, et al. "An Integrative Model of Organizational Trust." *Academy of Management Review* 20 (1995).

Mead, M., ed. *Cultural Patterns and Technical Change*. New York: UNESCO, 1955.

Meyerson, D., K. E. Weick, et al. "Swift Trust in Temporary Groups," in *Trust in Organizations—Frontiers of Theory and Research*, ed. R. M. Kramer and T. R. Tyler, pp. 166–195. Thousand Oaks, CA: Sage, 1996.

Milgram, S. *Obedience to Authority: An Experimental View*. New York: Harper Perennial, 1969.

Millstein, I. M. *International Comparison of Corporate Governance Guidelines and Codes of Best Practices, Developed Markets*. New York: Weil, Gotshal & Manges, LLP, 2001.

Mintzberg, H. *Structure in Fives: Designing Effective Organizations*. Englewood Cliffs, NJ: Prentice-Hall, 1983.

Morrison, T., W. A. Conaway, et al. *Kiss, Bow, or Shake Hands—How to Do Business in Sixty Countries*. Holbrook: Bob Adams, 1994.

Mortenson, Greg, and Relin, David Oliver. *Three cups of Tea: one man's mission to fight terrorism and build nations . . . one school at a time*. New York: Viking Penguin, 2006.

Mullavey-O'Brien, C. "Empathy in Cross-Cultural Communications," in *Improving Intercultural Interactions*, ed. K. Cusher and R. Brislin. Thousand Oaks, CA: Sage, 1997.

"The New Titans: A Survey of the World Economy," *The Economist* 380 (2006): 3–30.

Nonaka, I. "The Knowledge Creating Company," *Harvard Business Review* 69, no. 6 (1991): 96–104.

Nonaka, I., and N. Konno. "The Concept of Ba': Building a Foundation for Knowledge Creation," *California Management Review* 40, no. 3 (1998): 40.

Nonaka, I., and H. Takeuchi. *The Knowledge-Creating Company*. Oxford: Oxford University Press, 1995.

Nonaka, I., Toyama, R., and N. Konno. "SECI, Ba and Leadership: A Unified Model of Dynamic Creation," in *Managing Industrial Knowledge*, ed. I. Nonaka. Thousand Oaks, CA: Sage, 2001.

Oleska, F. M. *Communicating across Cultures*. Juneau, AK: KTOO, 2007.

Pease, A., and B. Pease. *The Definitive Book of Body Language*. New York: Bantam, 2004.

Pike, K. L. *Language in Relation to a Unified Theory of the Structure of Human Behavior*. The Hague: Mouton, 1967.

PMBOK. *A Guide to the Project Management Body of Knowledge*, 3rd Edition. Newtown Square, PA: Project Management Institute, 2004.

Podsakoff, P. M., and C. A. Schriesheim. "Field Studies of French and Raven's Bases of Power. Critique, Reanalysis, and Suggestions for Future Research," *Psychological Bulletin* 97 (1985): 387–411.

Porter, M. *Competitive Advantage—Creating and Sustaining Superior Performance*. New York: Free Press, 1998.

Ramlogan, R., and N. Persadie. *Commonwealth Caribbean Business Law*. Coogee, Australia, Cavendish Publishing, 2004.

Redfern, A. "Having Confidence in International Arbitration," *Dispute Resolution Journal* (November 2002–January 2003): 60–61.

Rosen, R., P. Digh, et al. *Global Literacies—Lessons on Business Leadership and National Cultures*. New York: Simon & Schuster, 2000.

Roumboutsos, A., and N. Chiara. "Public Private Partnerships: A Strategic Partnering Approach," Presented at Revamping PPP's Symposium, University of Hong Kong, 2009.

SAHDR. *Sustainable Development*. Johannesburg, South Africa: United Nations, 2004.

SAI. *Social Accountability 8000*. New York: Author, 2001.

Santayana, G. *The Life of Reason; or, the Phases of Human Progress*. New York: Scribner, 1953.

Soliño, A. S., and J. M. Vassallo. "Using Public-Private Partnerships to Expand Subways: Madrid-Barajas International Airport Case Study," *Journal of Management in Engineering* 21 (January 2009).

Speth, J. G. *Time for a Revolution: The UN, the US, and Development Cooperation*. New York: Foreign Policy Association, 1977.

Spicer, A. "Cultural and Knowledge Transfer: Conflict in Russian Multi-National Settings." 194. 1997.

Stiglitz, J. E. *Globalization and Its Dicontents*. New York: W. W. Norton, 2003.

Sveiby, K. E. "Measuring Intangibles and Intellectual Capital—An Emerging First Standard." Electronic. 1998.

Sveiby, K. E. "A Knowledge-Based Theory of the Firm to Guide in Strategy Formulation," *Journal of Intellectual Capital* 2, no. 4 (2001): 344–358.

Szulanski, G., and R. J. Jensen. "Overcoming Stickiness: An Empirical Investigation of the Role of the Template in the Replication of Organizational Routines," *Managerial & Decision Economics* 25, no. 6/7 (2004): 347–363.

Tagore, R. *Stray Birds*. Montana, Kessinger Publishing, 2004.

Toffler, A. *Future Shock*. New York: Bantam Books, 1997.

Tuckman, B. W., and M. A. Jensen. "Stages of Small Group Development Revisited," *Group and Organizational Studies* 2 (1977): 419–427.

Turner, J. R. *Handbook of Project-Based Management*. London: McGraw-Hill, 1999.

Turner, J. R., and R. Müeller. "On the Nature of the Project as a Temporary Organization," *International Journal of Project Management* 21, no. 1 (2003): 1.

Tzu, S. *The Art of War*. London: Oxford University Press, 1963.

Wagner, R. K., and R. J. Sternberg. "Practical Intelligence in Real-World Pursuits: The Role of Tacit Knowledge," *Journal of Personality and Social Psychology* 49 (1985): 436–458.

Walker, D. H. T. *Having a Knowledge Competitive Advantage (K-Adv): A Social Capital Perspective*. Information and Knowledge Management in a Global Economy CIB W102. Lisbon: DECivil, 2005.

Walker, D. H. T., and K. D. Hampson. *Procurement Strategies: A Relationship Based Approach*. Oxford: Blackwell, 2003.

Walther, J. B. "Computer-Mediated Communication: Impersonal, Interpersonal, and Hyperpersonal Interaction," *Communication Research* 23, no. 1 (1996): 3.

Wenger, E. C., R. McDermott, et al. *Cultivating Communities of Practice*. Boston: Harvard Business School Press, 2002.

Winter, M., C. Smith, et al. "Directions for Future Research in Project Management: The Main Findings of a UK Government-Funded Research Network," *International Journal of Project Management* 24 (2006): 638–649.

Wysocki, R. K. *Effective Project Management—Traditional, Adaptive, Extreme*, 4th ed. Indianapolis, IN: Wiley Publishing, 2007.

Yeung, J. F. Y., A. P. C. Chan, et al. "The Definition of Alliancing in Construction as a Wittgenstein Family-Resemblance Concept," *International Journal of Project Management* 25, no. 3 (2007): 219–231.
Yukl, G. *Leadership in Organisations*. Sydney: Prentice-Hall, 1998.
Zack, M. H. "Developing a Knowledge Strategy," *California Management Review* 41, no. 3 (1999): 125–145.

Glossary

actively managed risks: An actively managed risk must come from the filtering process, must have a probability times impact (P × I) greater than the corporate risk attitude, must have a mini–project plan constructed for it, and must be preapproved by the key stakeholders. It is then managed the same way a work breakdown structure activity would be managed.

ADB: Asian Development Bank.

adversarial environment: A competitively bid, lowest-price-wins contract in which communications between the parties is limited or nonexistent prior to award of the contract.

AfDB: African Development Bank.

agile project management: Project management performed under general guidelines rather than strict processes.

AIPM: Australian Institute of Project Management.

APM: Association of Project Managers.

ASME: American Society of Mechanical Engineers.

BAC: Budget at completion. The budget, or total cost of the project, accepted by the customer.

baseline: A time-phased plan. Baselines should be created for scope, cost, time, quality, procurement, diversity, risk, and communications. Using a sailing metaphor, the baseline is the course. Baselines should be constructed from period information (e.g., monthly) and summarized into a cumulative format. Also called a need.

BATNA: Best alternative to a negotiated agreement. An offer at which one party is better served by removing itself from the negotiations.

burn rate: A relationship between the resources used to generate the progress achieved. Similar to earned value.

buyer: An organization that purchases a service and/or product from another organization.

CMC: Computer-mediated communication.

coach: A person who demonstrates to another person how a task is to be performed.

collaborative environment: A negotiated contract where all of the parties participate in the construction of the contract scope, time, cost, and quality.

conceptual phase: When the original need for a project is conceptualized; for example, "We need more electricity."

confirmed estimate: A participant organization has met with the collaborative project enterprise and the lead project manager, reviewed their estimate in detail, and has established an agreed-on contingency reserve for their work. Accuracy +/-20% for the contingency has been established to account for estimating errors.

consequential damages: Damages that result from action, or inaction, by a party to a contract. They include such things as business interruption costs and opportunity costs.

contingency: Money, time, or resources set aside to establish the boundaries of accuracy.

CoP: Community of practice. A group of people who share a common interest, hobby, or avocation (e.g., people who enjoy sailing, golf, literature, cooking, etc.). It also includes people who are of a similar technical discipline, political leaning, or the like.

core resources: People on the payroll of an organization who have a long-term relationship and receive wages, healthcare, vacation, retirement, and other such benefits.

CPE: Collaborative project enterprise.

CPE WBS: Collaborative project enterprise work breakdown structure. A detailed list of activities required for the complete project, regardless of performing organization.

CPI: Corruption perception index; consumer price index.

CQ: Cultural intelligence. Understanding the differences between the values and norms of different cultures.

crashing: Adding resources, hours, or days in some combination to shorten the duration of an activity.

critical chain: The critical path of a project that is task dependent and resource constrained.

CSR: Corporate social responsibility. CSR is, at this point, a self-regulated business policy that commits the organization to follow laws and international standards of ethics and norms. Normally it is concerned with stakeholders (e.g., profitability), society (e.g., social consciousness), employees (e.g., equal opportunity), and the environment.

CTQ: Critical to quality. The stated and implied needs of a customer, prioritized and defined.

currency hedge: An insurance policy against a devaluation in a currency that will need to be traded at some time(s) in the future.

customer: A person or organization that actively contracts for, pays for, and accepts the product of the project.

customer centric: According to the Wharton Business School at the University of Pennsylvania, customer centricity means that companies consider themselves a portfolio of customers rather than a portfolio of products. Their business is organized around organizational customer segments, and they know exactly what revenue is associated with customers. They are also firms that focus on sustainability.

customer satisfaction: Achieved by fulfilling the implied and stated needs of the customer in both the services and the products provided. It is the aggregate of all the touch points that customers have with the collaborative project enterprise.

DDP: Delivered duty paid.

deep packet: An information technology system that enables filtering of data.

detailed estimate: A participant organization has completed its detailed work breakdown structure, has had its scope questions answered, and has thoroughly reviewed its estimate. The participant organization can define its scope boundaries. Accuracy +/-25%, excluding contingency.

diligence: The attention and care legally and professionally expected or required of a person.

DRB: Dispute resolution board. A panel of stakeholders or subject matter experts that hear and rule, binding or otherwise, periodically on disputes on large projects.

EAC: Estimate at completion. The total cost of the project estimated by the lead project manager without bias or predisposition. Some call this a fair cost estimate.

earned value: The budgeted cost of the work actually performed (BCWP).

EQ: Emotional intelligence. Understanding oneself and being able to control one's emotions.

equal-to: A technical specification that refers to an existing product to describe the performance and physical characteristics of a product.

EU: European Union.

evergreen warranty: A warranty that never expires.

fast-tracking: Overlapping activities in a schedule. Also called multitasking or concurrency. For example, starting to test code before the writing is complete; or building foundations for a building before the building design is complete.

flow down: The process of passing the contract requirement of the primary contract between the customer and service providers down through the value chain.

fluid resources: People hired on a short-term basis and paid on an hourly or monthly basis, who receive no benefits. Sometimes called contract employees.

force majeure: Hurricanes, tornadoes, cyclones, earthquakes, floods, landslides, tsunamis, and the like.

foreign direct investment: The investment of capital in a country from external source.

GDP: Gross domestic product.

GPS: Global positioning system.

Hand-off: Transfer of responsibility from one group to another. For example, the transfer of project responsibility from the lead project manager to a lead warranty manager or a lead operations manager.

ICB: International competency baseline.

ICC: International Chamber of Commerce.

ICDR: International Center for Dispute Resolution.

ICIDS: International Center for Settlement of Investment Disputes.

IDB: Inter-American Development Bank.

IFC: International Finance Corporation.

IMF: International Monetary Fund.

INCO: International Commercial Terms.

IEEE: Institute of Electrical and Electronic Engineers.

international lead project manager: The person responsible to the customer for the successful completion of the project for the entire collaborative project enterprise.

international project: A unique transient endeavor undertaken to create a unique product or service that utilizes resources from, or provides product or services in, more than one country.

international project manager: The person responsible for the successful completion of the project within one organization.

IPMA: International Project Management Association.

Ishikawa diagram: A diagramming technique that explores the causes of problems, or variance.

IT: Information technology.

joint project planning meetings: A series of meetings conducted to prepare the project plan, first by each individual organization, then by the collaborative project enterprise. These meetings culminate in a meeting with the customer and the user.

key stakeholder: A person who can significantly help or hinder the project and must be identified and managed actively.

lead international project manager: The project manager of the lead organization for a project. Ideally, the project manager for the organization that is responsible for the success of the project although not directly in control of each aspect of the project.

lead project manager: Project manager.

Level A: Level A, B, C, and D are certifications of project management competencies issued by the International Project Management Association.

LIBOR: London Interbank Offered Rate. It is a daily reference rate based on the interest rates at which banks borrow unsecured funds from other banks in the London wholesale money market (or interbank market).

liquidated damages (LDs): Penalties imposed by the customer for completing the project late, for failure to meet contract performance requirements, or for failure to meet a preestablished project milestone.

LOC: Letter of credit.

make or buy: A decision taken by an organization to produce a service or product with its own resources or to procure them from an external organization.

means and methods: The techniques, sequences, methods, and resources to perform the project.

mentor: A person who develops an interpersonal relationship with another person and provides them with guidance, not instruction.

MOU: Memo of understanding. An informal contract.

NGO: Nongovernmental organization.

nominal group technique: A brainstroming session that enables the participants to prioritize ideas that they think most important. It is a way to rank ideas.

NTP: Notice to proceed. A formal document issued by the customer that commemorates the official start date of the contract.

organization: A publicly traded firm, a privately held firm, a nonprofit organization, a governmental agency, a public opinion group, a tribe, etc.

partnering: A consortium, joint venture, or alliance.

passively managed risk: Risks identified during the risk filtering process and have a P × I less than the corporate risk attitude. A mini–project plan does not have to be created, nor is key stakeholder approval required.

PDM: Precedence diagramming method. A technique that uses rectangles to represent activities and arrows to represent relationships.

PERT: Program Evaluation and Review Technique. A technique to estimate time or cost. The formula is (Best Case + 4 (Most Likely Case) + Worst Cast / 6.

PEST: Political, economic, social, and technological analysis.

plan-do-check-act: Dr. Demming and Shewart's process for quality. Plan the operation, do the operation, check the results, make necessary corrections.

PMBOK**:** *Project Management Body of Knowledge.*

PMI: Project Management Institute.

PMMM: Project management maturity model. A model that describes the steps that an organization would achieve in realizing a fully mature project management organization.

PMO: Project management office. Any organization that has a PMO or similar function to systematize project management processes across the organizational boundaries.

PMP: Project management professional. A credential offered by the Project Management Institute.

PPP: Public-private partnership.

PR firm: An organization that provides public relations expertise.

preliminary estimate: The stage at which a participant organization needs a detailed scope explanation, more time to prepare a detailed work breakdown structure, but has acquired the expertise necessary to do the work. Accuracy +/-40%, excluding contingency.

prime time: The best time for synchronous communications on international projects recognizing the cycle of life for the different participants.

PRINCE2: Projects in Controlled Environments, second major revision. Information technology projects conducted under processes, or controlled environments, to facilitate the kaizen or continuous improvement process.

program, programme: *PMBOK* definition: A group of related projects managed in a coordinated way to obtain benefits and control not available from managing them individually.

IPMA definition: A programme of projects is put together to realize a strategic goal set out by the organization. To achieve this, it initiates a group of interrelated projects to deliver the products/outcomes needed to attain this goal and defines the organizational changes needed to facilitate the strategic change. The programme defines business benefits management process as well as tracking the business benefits. The programme manager usually directs the projects through project managers, facilitates the interaction with line managers to realize the change and is responsible for benefits management, not for the realization of the benefits, which is the responsibility of line management.

program management: *PMBOK* definition: The centralized coordinated management of a program to achieve the program's strategic objectives and benefits.

project: *PMBOK* definition: A temporary endeavor undertaken to create a unique product, service, or result.

project charter: A document prepared by the lead project manager and signed by the sponsors that formally authorizes the start of the project. It provides the lead project manager with the authority to lead the project on behalf of the collaborative project enterprise, defines the goals and objectives, project scope, deliverables, budget, and time frame, and establishes the ethical standards for the collaborative project enterprise.

project directory: A listing of all stakeholders on a project. The directory should include the value chain, governmental officials, and public interest groups. It should include this information for each person (within reason): Name, title, and firm name; CV; Location and time zone; Phone and fax numbers, and cell phone number; Skype address if used; e-mail address. We also suggest that each person be encouraged to share a bit about him- or herself, such as culture, religion (so that holidays may be celebrated by the collaborative project enterprise), and interests; photographs.

project management plan: Includes all of the subsidiary management plans needed to manage the project, including scope plan, communication plan, risk plan, time plan, cost plan, quality plan, procurement plan, and diversity plan.

project team: The people from all firms and organizations involved in the project.

projectized: *PMBOK* definition: Any organizational structure in which the project manager has full authority to assign priorities, apply resources, and direct the work of persons assigned to the project. We prefer the definition of any organization that has a project management office or similar function to systematize project management processes across organizational boundaries.

punch list: A list of scope work that has not been completed or that has been completed and is deficient. Also called a deficiency list.

RegPM: Registered project manager credential for the Australian Institute of Project Management.

retention: Money withheld by a customer until the punch list or deficiency list work items have been completed.

risk attitude: An organization's attitude toward risk. The three attitudes used in this book are risk seeker, risk neutral, and risk averse.

risk register: A listing of risk including the probability and impact and the P × I. For active risks, it also includes the trigger event, response strategy, time horizon, costs, and any assumptions.

ROI: Return on investment. The profit made on an investment over a period of time.

ROM: Rough order of magnitude. A participant organization does not have the details of the scope, has not had enough time to prepare a detailed work breakdown structure, does not have the expertise and will need to find a partner, etc. For example, this is an estimate prepared in a hurry for a tender. Accuracy +/-50%, excluding contingency.

scope split: A document that lists the work breakdown structure activities for the entire project and assigns responsibility to one or more organizations.

seller: An organization that provides a service and/or product to another organization.

service provider: An organization or groups of organizations that provide services to a customer. In the *PMBOK,* the seller in a transaction.

SME: Subject matter expert. A person or organization that is a recognized expert in a particular discipline, process, business, or other knowledge area.

sponsor: An upper-level management person responsible for the project, who has the ability to commit company resources, alter profit and loss metrics, and remove barricades.

stakeholder: A person or organization that has an interest in the project, or that could be impacted by it.

subsidiary management plans: Plans created for each *PMBOK* knowledge area including: scope management plan, schedule management plan & baseline, cost management plan and baseline, quality management plan and baseline, process improvement plan, staffing management plan, communication management plan, risk management plan, and procurement management plan.

supply chain: Represents a hierarchy of buyers and sellers involved in transactions and self-interest. We use the term "value chain" in this book.

SWOT: Strength, opportunity, weakness, and threat model.

synchronous and asynchronous: "Synchronous" means communicating with others at the same moment, though in different time zones through phones, meeting software, Skype, chat rooms, text messaging, and so on. "Asynchronous" means communicating with others at different moments using such things as e-mail or discussion boards.

TBL, TBL+1: Triple bottom line plus one, or TBL, like corporate social responsibility, commits the organization to focus on stakeholders, the environment, and society. Organizations that strive to meet a TBL or CSR goal recognize a responsibility to make money, to protect the environment, and to be good citizens. At RMIT University in Melbourne Australia the need for corporate governance—the care and nurturing of the workforce and transparency—was added to produce TBL+1.

TQM: Total quality management. The concept of providing quality in all aspects of a business or undertaking, not just in the product itself.

triple constraint: One way to look at project success: on time, under budget, to the scope specified.

user: A person or organization that will use the product of the project.

value chain: A hierarchy of buyers and sellers involved in relationships that result in a quality project.

variance: A deviation outside of established control limits. Not a lack of precision.

VFM: Value for money. Value is how much a desired object or condition is worth relative to other objects or conditions in economic terms. Value for money is the relative value in corporate social responsibility terms.

WBS, WBS Dictionary: Work breakdown structure. In the *PMBOK,* the WBS is a list of tasks only. The remaining information on such things as resources is what the *PMBOK* calls a WBS dictionary. In practice, the term "WBS" is often used to describe both.

WHO: World Health Organization.

win-win: In a dispute, where both parties feel that they have been fairly treated and won enough to satisfy their needs.

work package: The lowest level of detail in a work breakdown structure. Also the recommended lowest level of detail for activities.

XLQ: Cross-cultural leadership intelligence.

Index

Q

R

S

T

U

V

W

X

Y

Z